Nine Mathematical Challenges

An Elucidation

Proceedings of Symposia in
PURE MATHEMATICS

Volume 104

Nine Mathematical Challenges

An Elucidation

Linde Hall Inaugural Math Symposium
February 22–24, 2019
California Institute of Technology,
Pasadena, California

A. Kechris
N. Makarov
D. Ramakrishnan
X. Zhu
Editors

AMS AMERICAN MATHEMATICAL SOCIETY
Providence, Rhode Island

2020 *Mathematics Subject Classification.* Primary 03Exx, 14Fxx, 11G05, 20D05, 30A99, 35P15, 35Q35, 37A17, 57R22.

Library of Congress Cataloging-in-Publication Data

Names: Linde Hall Inaugural Math Symposium (2019: California Instiute of Technology, Pasadena, Calif.) | Kechris, A. S., 1946– editor. | Makarov, Nikolai G., editor. | Ramakrishnan, Dinakar, editor. | Zhu, X. (Xinwen), 1982– editor.

Title: Nine mathematical challenges: An elucidation: Linde Hall Inaugural Math Symposium, February 22–24, 2019, California Institute of Technology, Pasadena, California / A. Kechris, N. Makarov, D. Ramakrishnan, X. Zhu, editors.

Description: Providence, Rhode Island: American Mathematical Society, [2021] | Series: Proceedings of symposia in pure mathematics, 0082-0717; volume 104 | Includes bibliographical references.

Identifiers: LCCN 2021023568 | ISBN 9781470454906 (paperback) | ISBN 9781470467463 (ebook)

Subjects: LCSH: Mathematics–Congresses. | AMS: Mathematical logic and foundations – Set theory. | Algebraic geometry – (Co)homology theory in algebraic geometry. | Number theory – Arithmetic algebraic geometry (Diophantine geometry) – Elliptic curves over global fields. | Group theory and generalizations – Abstract finite groups – Finite simple groups and their classification. | Functions of a complex variable – General properties of functions of one complex variable. | Partial differential equations – Spectral theory and eigenvalue problems for partial differential equations – Estimates of eigenvalues in context of PDEs. | Partial differential equations – Equations of mathematical physics and other areas of application – PDEs in connection with fluid mechanics. | Dynamical systems and ergodic theory – Ergodic theory – Homogeneous flows. | Manifolds and cell complexes – Differential topology – Topology of vector bundles and fiber bundles.

Classification: LCC QA1 .L436 2021 | DDC 510.5–dc23

LC record available at https://lccn.loc.gov/2021023568

DOI: https://doi.org/10.1090/ramakrish/104

∞ The paper used in this book is acid-free and falls within the guidelines
established to ensure permanence and durability.
Visit the AMS home page at `https://www.ams.org/`

10 9 8 7 6 5 4 3 2 1   26 25 24 23 22 21

Contents

Preface

Problems are the lifeblood of mathematics. These are the immortal words of Hilbert, uttered during the 1900 International Congress of Mathematicians, which are well known and oft repeated. This volume is a similar venture, albeit more modest, but still with the same purpose. What we have here are nine beautiful articles by distinguished mathematicians, each discussing a specific open problem, or a set of such, in nine different fields within mathematics, each very important in its own right.

The idea for this volume began with the plans for a special event at Caltech celebrating the opening of the newly remodeled Ron and Maxine Linde Hall of Mathematics and Physics, which mainly houses the math department. There was an Inaugural Math Symposium, which was held at Caltech during February 22–24, 2019. Nine distinguished mathematicians representing different areas of the subject spoke, each giving a one-hour lecture during the symposium. There were two lectures on the afternoon of Friday, February 22, five lectures during Saturday, with three in the morning and two in the afternoon, and two lectures on the morning of Sunday, February 24. During lunch on Saturday there was a poster session for students and postdocs. The speakers were, in alphabetical order, M. Aschbacher, H. Esnault, B. Gross, U. Hamenstädt, P. Michel, A. Poltoratski, B. Simon, T. Tao, and W.H. Woodin. The program of the Symposium can be found right after this preface.

The Lectures were such a resounding success, giving rare and exciting glimpses into a slew of mathematical challenges, that it was decided by the organizing committee–composed of A. Kechris, N. Makarov, D. Ramakrishnan (chair), and X. Zhu–to try to bring out a special volume presenting the contents of the Lectures with suitable expansion. It was enthusiastically supported by Elena Mantovan, then the Executive Officer, a.k.a. chair, of the math department. We asked the lecturers if they would contribute expanded write-ups of their lectures, but unfortunately B. Gross and B. Simon were unable to provide articles for the volume. It was decided that the other speakers could write about anything they wanted, preferably related to the subject of their talks.

We approached the American Mathematical Society about the possibility of bringing out the volume in the Proceedings of the Symposia in Pure Mathematics series. After due evaluation, they agreed and we began putting together the volume with invitations to the authors. Hélène Esnault decided to write about something completely different and submitted a joint article with V. Srinivas on ramification in covers and curves. Meanwhile, the organizing committee decided to ask two distinguished mathematicians to supply articles in the general areas of B. Gross and B. Simon. Fortunately for us, Chris Skinner agreed to write about the same subject matter as Gross's talk (on the Birch and Swinnerton-Dyer Conjecture) in collaboration with A. Burungale and Ye Tian. And R. Frank agreed to write a survey on a different topic from the lecture of B. Simon, but still in the general area of Mathematical Physics.

Now we will briefly touch upon the contents. In the first article, M. Aschbacher explains the problem of classification of finite groups and explains where it stands. In the second, A. Burungale, C. Skinner, and Ye Tian explain the recent progress on the Birch and Swinnerton-Dyer Conjecture for rational elliptic curves E, which relates the order of vanishing of its L-function at $s = 1$ to the rank r of $E(\mathbb{Q})$ as an abelian group, focusing on $r = 0, 1$. In the third article, H. Esnault and V. Srinivas analyze ℓ-adic local systems of bounded rank and ramification on a smooth variety in characteristic p (different from ℓ), then introduce and study ramification outside codimension 2 by a finite separable extension of bounded degree. In the fourth, R. Frank gives a survey of problems and results concerning the Lieb-Thirring inequalities. In the fifth article, U. Hamenstädt studies topological properties of surface bundles over surfaces which result from the Milnor-Wood inequality for the Euler class of flat S^1-bundles over surfaces. In the sixth contribution, P. Michel explains substantial recent extensions of the fundamental equidistribution theorem of W. Duke. The seventh article, by A. Poltoratski, describes the Gap and Type problems in Fourier analysis with new results. In the following (eighth) article, T. Tao discusses quantitative bounds for higher regularity norms of the classical solutions to three-dimensional Navier-Stokes equations. In the final (ninth) article, Hugh Woodin gives a survey of Cantor's Continuum Hypothesis starting from the work of Gödel and Cohen, and leading to the current developments and approach of the author on this problem.

The editors would like to thank Michelle Vine and Stephanie Cha-Ramos for their help with the Linde Inaugural Math Symposium at Caltech, and also acknowledge invaluable help from Christine Thivierge of the AMS in the putting together of this volume.

The Linde Hall Inaugural Math Symposium at Caltech

The symposium will be held on February 22–24, 2019, to celebrate the opening of the newly remodeled Ron and Maxine Linde Hall of Mathematics and Physics. Nine distinguished speakers representing different areas of Mathematics will each give a one-hour lecture during the symposium. There will be two lectures on the afternoon of Friday, February 22, five lectures on Saturday, with three in the morning and two in the afternoon, and two lectures on the morning of Sunday, February 24. During lunch on Saturday there will be a poster session for students and postdocs.

Lectures

(with authors in alphabetical order)

(1) Michael Aschbacher, Caltech
The finite simple groups and their classification

(2) Hélène Esnault, Free University of Berlin
Vanishing Theorems for étale sheaves

(3) Benedict Gross, UCSD
On the conjecture of Birch and Swinnerton-Dyer

(4) Ursula Hamenstädt, University of Bonn
Amenable actions and rigidity

(5) Philippe Michel, École Polytechnique Fédérale de Lausanne
L-functions, moments and subconvexity

(6) Alexei Poltoratski, Texas A & M
Gap and Type problems in Fourier analysis

(7) Barry Simon, Caltech
Fifty Years of the Spectral Theory of Schrödinger Operators

(8) Terence Tao, UCLA
The global regularity problem for Navier-Stokes

(9) W. Hugh Woodin, Harvard
Cantor's Continuum Hypothesis

Organizing Committee:
Dinakar Ramakrishnan (chair), Alexander Kechris, Nikolai Makarov, and Xinwen Zhu

Proceedings of Symposia in Pure Mathematics
Volume **104**, 2021
https://doi.org/10.1090/pspum/104/01871

The finite simple groups and their classification

Michael Aschbacher

This is a slightly expanded version of a talk given at the Linde Hall Inaugural Math Symposium at Caltech in February of 2019. The topic is the finite simple groups and their classification. The paper is expository and hopefully not too technical. I'll briefly describe the simple groups, supply some history, and introduce some of the important notions that underly the classification of the finite simple groups (which I'll abbreviate CFSG). I'll touch upon approaches currently in progress to improve the proof of the CFSG, and I'll point out the sort of information about the simple groups needed to apply CFGS to solve problems in finite group theory — problems that often arise in other areas of mathematics.

For a much more detailed discussion of CFSG see [**ALSS**].

Motivation

I'll start slowly with some motivation. A group G is *simple* if G and 1 are the only normal subgroups of G. In the general group G we can consider *subnormal series* $\mathfrak{S} = (1 = G_0 \lhd G_1 \lhd \cdots \lhd G_n = G)$, and if G is finite there exist *maximal* series: those that can't be extended by inserting an extra term. Observe $\mathfrak{S}$ is maximal if and only if each of the *factors* G_{i+1}/G_i of $\mathfrak{S}$ is simple. The maximal series are called the *composition series* of G and the family $(G_{i+1}/G_i : 0 \leq i < n)$ of factors of $\mathfrak{S}$ are called the *composition factors* of G. By the Jordan-Holder Theorem, the composition factors are independent of the choice of composition series. Thus the finite simple groups are the building blocks of finite group theory, in that each finite group is "built" from its composition factors.

Indeed from the nineteenth century we have the two-step *Holder Program* for studying finite groups:

(1) Determine the finite simple groups.

(2) Solve the *Extension Problem:* Retrieve G from its composition factors; ie classify all groups with a given family of composition factors.

Step (1) – the CFSG – has been achieved, although given the current state of the art, the proof is barely within the boundary separating the possible from the impossible. However the Extension Problem is almost certainly too complex to admit a solution in general. To see why this might be so, let us consider the example of p-groups.

2020 *Mathematics Subject Classification.* Primary 20D05.
This work was partially supported by DMS NSF-1601063.

Let p be a prime; recall that a (finite) *p-group* is a group of order p^n for some nonnegative integer n. It was probably implicit in your first algebra course that $|G| = p^n$ if and only if G has n composition factors, each of order p. As the groups of prime order are the most well behaved of simple groups, one would expect that if the Extension Problem is approachable for any class of groups, then it would be approachable for p-groups. And indeed, viewed from a distance, p-groups are well behaved with many nice properties. But if one attempts to look more closely and classify p-groups, difficulties occur.

For example in my office is a copy of a book by Marshall Hall and the chemist James Senior [**HS**] that I inherited from Marshall. The book determines and describes all 2-groups of order at most 2^6. In particular we find there are exactly 267 groups of order 2^6. Further I have some vague memory that the Hall-Senior project was extended by some people at JPL to determine the groups of order 2^n for somewhat larger values of n. One of the referees did a google search, and determined that the number of groups of order 2^{10} is known, and exceeds 10^{11}. In other words as n gets large the number of groups of order p^n gets very large.

Thus it doesn't seem to be in the cards to classify all finite groups. But what does seem to be the case is that many problems in finite group theory can be reduced to the case where the group is "nearly simple", with the exact notion of "near simplicity" depending on the problem. Then to complete the solution, one normally needs some information about the nearly simple group, such as a description of its irreducible linear representations or its maximal subgroups. Later in the paper we encounter *quasisimple groups* and *almost simple groups*, that are examples of "nearly simple" groups in the context of representation theory and subgroup structure, respectively.

Eugene Wigner wrote about the "unreasonable effectiveness" of mathematics in solving real world problems [**W**]. One could say the same about how often problems in finite group theory can be reduced to suitable facts about the simple groups.

That is my attempt at motivation. From time to time I'll hint at what is known about the subgroup structure and linear representations of the simple groups, and some of what remains to be done.

Now its time for the statement of the Classification Theorem:

THEOREM 1. *Each finite simple group is one of the following:*
(1) *A group of prime order.*
(2) *An alternating group.*
(3) *A group of Lie type.*
(4) *One of 26 sporadic groups.*

Of course this statement has little content unless we can provide a good description of each of its four classes of groups. See Robert Wilson's book [**Wi**] for a more detailed discussion of the finite simple groups than the brief treatment supplied below.

Groups of prime order

The groups of prime order are the abelian simple groups. For each prime p there exists a unique group of order p, which can be regarded as the group of integers modulo p under addition. Its subgroup structure is trivial and it is easy to describe its irreducible linear representations.

Alternating groups

Let X be a set of finite order n. The group of all permutations on X is the *symmetric group* of degree n; this group is the automorphism group of X. From your first algebra course, the symmetric group S_n has a normal subgroup A_n of index 2: the *alternating group* of degree n, and A_n is simple if $n \geq 5$. The object X is highly homogeneous, so that the action of S_n on X can (with some effort) be used to give a satisfactory description of the maximal subgroups of S_n and A_n [**LPS**]: maximal subgroups are either stabilizers of natural structures on X (eg. partitions or product structures), or almost simple subgroups acting primitively on X. Here G is *almost simple* if it has a unique minimal normal subgroup L and L is nonabelian simple. Thus for permutation groups, one good notion of "near simplicity" is almost simplicity.

There is a classical linear representation theory of S_n over the complex numbers, which extends to some extent to the alternating group and representations over other fields.

Groups of Lie type

The groups G of Lie type are linear groups on finite dimensional vector spaces V over suitable fields F. When G is finite, F is finite. Indeed G is essentially the group of automorphisms of a suitable collection of forms on V. The best example is the general linear group $GL(V)$ (where the set of forms is empty), which is usually not simple, but close enough for purposes of this paper.

Alternatively one can consider a corresponding simple algebraic group $\bar{G}$ over an algebraic closure $\bar{F}$ of F, and regard G as the fixed points of a suitable endomorphism of $\bar{G}$. For example $G = GL(V)$ is the fixed points of a "field automorphism" σ of $\bar{G} = GL(\bar{F} \otimes_F V)$, where for a matrix $(a_{i,j}) \in \bar{G}$, we define $\sigma(a_{i,j}) = (\sigma(a_{i,j}))$, induced by some $\sigma \in Gal(\bar{F}/F)$.

The study of groups of Lie type is often called *algebraic Lie theory*. Algebraic Lie theory has its early roots in the work of Galois and Jordan on linear groups, and later in the classification of simple Lie groups by Killing and Cartan. Modern algebraic Lie theory might be said to begin with Chevalley's paper on the "Chevalley groups" [**C**] in the mid fifties. Major contributors also include Steinberg and Tits.

The finite groups of Lie type consist of the *ordinary Chevalley groups*, obtained as fixed points of field automorphisms of simple algebraic groups, together with the *twisted Chevalley groups* obtained from more exotic endomorphisms. The twisted groups are of two types: Steinberg variations and Ree groups. The unitary group $GU(V)$ is an example of a Steinberg variation, obtained from the product of a field automorphism with the transpose-inverse map.

Finite groups of Lie type can also be regarded as automorphism groups of finite buildings. The notion of a *building* is due to Tits [**Ti**]. Buildings can be defined in various ways. For example buildings can be regarded as a certain class of simplicial complexes, or as a certain class of incidence geometries. One can characterize the groups of Lie type of Lie rank at least 2 as automorphism groups of buildings. Also the representation of a group on its building can be used to study the subgroup structure of the group.

The maximal subgroups of G can, to a large extent, be described from its action on the object X consisting of V and the defining forms; for example if G is classical (ie. special linear, symplectic, orthogonal, or unitary) then maximal subgroups of

G are either stabilizers of natural structures on X, or G is primitive and tensor indecomposable, and its image in the projective group $PSL(V)$ is almost simple [**A**]. However much work remains to be done to complete this theory.

Irreducible linear representations of groups of Lie type over the complex numbers are described using Deligne-Lusztig theory. Representation over fields of characteristic char(F) can be obtained from the rational representations of $\bar{G}$; while this theory still has major open problems, it is strong enough to determine the irreducibles for G of small degree. The theory of representations in prime characteristic $r \neq$ char(F) with r dividing the order of G is a work in progress.

Sporadic groups

Finally we come to the sporadic groups. Unlike the groups in the first three classes, sporadic groups are not members of any naturally defined infinite class of simple groups known to finite group theorists.

Historically the initial treatment of a sporadic group G fell into one of three major categories: discovery, existence, and uniqueness. Roughly speaking, G was regarded as discovered when enough self-consistent information about a group satisfying the hypotheses of G was established; such information might include the order of G or the character table of G. Existence and uniqueness require a proof of the existence or uniqueness of a group satisfying the hypothesis. Typically the group was named after its discoverer, although some people use both the name of the discoverer and the name of the person proving existence.

The first five sporadics were discovered by Mathieu in the mid nineteenth century as multiply transitive permutation groups. The first of the modern sporadics was discovered around 1965 by Janko via local group theory. After that sporadics were discovered at the rate of roughly two a year for about ten years. Janko discovered the last sporadic in 1976.

Here is an example involving Caltech. The second and third of the modern sporadics were also discovered by Janko, using local group theory; they are often denoted by J_2 and J_3. But J_2 was also discovered at roughly the same time by Marshall Hall as an automorphism group of a strongly regular graph. Since Marshall was a senior faculty member at Caltech, group theorists at Caltech (eg. David Wales and I) called the group the *Hall-Janko group* and denoted it by HJ; indeed at least half of the finite group theorists of the time did the same. Hall and Wales also used the graph to establish existence and uniqueness of the group.

Roughly speaking a graph Γ is *strongly regular* if $Aut(\Gamma)$ is transitive on edges and non-edges. After Marshall, another handful of sporadics were discovered as groups of a strongly regular graph. Indeed the story is that Higman and Sims discovered their own group over dinner after hearing Marshall give a talk on HJ.

The proof of CFSG

Let us now move on to the proof of CFSG. Typically to classify a class of objects, we associate certain "invariants" to each object in the class and show that the invariants determine the object up to isomorphism. From what I've said so far, its not clear what the invariants should be for a finite simple group; after all some of our groups are permutation groups, some are linear groups, and the sporadic groups were discovered from many points of view. Moreover, at least initially we are working with abstract groups rather than groups equipped with an action on some useful object. In the end the invariants used in CFSG are certain "local subgroups" of the group. Thus I need to spend some time discussing:

The local theory of finite groups

Let p be a prime and G a finite group. A *p-local subgroup* of G is a normalizer $N_G(P)$ of some nontrivial p-subgroup P of G. The *local theory of finite groups* studies finite groups from the point of view of their local subgroups.

EXAMPLE. For $X \subseteq G$ the *centralizer in G of X* is the subgroup $C_G(X)$ of all elements of G commuting with each element of X. An *involution* in G is an element of G of order 2. Observe if $t \in G$ is an involution then $C_G(t)$ is a 2-local subgroup of G.

In 1954 at an International Congress of Mathematicians [**B**], Richard Brauer proposed studying finite simple groups from the point of view of their involution centralizers. In support of this suggestion, Brauer and his student Fowler proved: Let H be a finite group. Then there are at most a finite number of finite simple groups G possessing an involution t with $C_G(t) \cong H$. (In fact from CFSG, at most three simple groups can share a common involution centralizer.)

In otherwords Brauer was suggesting that involution centralizers might be among the invariants used in the CFSG.

REMARK. G possesses an involution if and only if G has even order. Around the turn of the century, Burnside conjectured that each nonabelian finite simple group is of even order; if Burnside's conjecture were true then Brauer's approach might be viable. In 1963, Feit and Thompson [**FT**] proved Burnside's conjecture, using a mixture of character theory and local group theory.

If we adopt Brauer's approach and focus on involution centralizers, or more generally on local subgroups, then we must address at least three questions/problems:

PROBLEM. What do involution centralizers (or local subgroups) look like in simple groups, and how do they differ from those in more general finite groups?

PROBLEM. *Prove* that each simple group contains an involution centralizer (or local subgroup) resembling one in a known simple group.

INVOLUTION CENTRALIZER PROBLEM. Given a potential involution centralizer (or local subgroup), determine all the simple groups with such an involution centralizer.

Each of the four sporadic groups discovered by Janko was discovered via the solution of an involution centralizer problem.

At this point, to simplify exposition, I'm going to state a theorem that combines a number of difficult results. But first we need some more notions. Write $O_2(G)$ for the largest normal 2-subgroup of G. Define G to be of *characteristic 2-type* if for each 2-local H of G we have $C_H(O_2(H)) \leq O_2(H)$.

REMARK. If G is of Lie type over a field of even order then G is of characteristic 2-type. Some of the sporadic groups and few groups of Lie type over the field of order 3 are also of characteristic 2-type.

G is *quasisimple* if $G = [G, G]$ is perfect and $G/Z(G)$ is simple. Quasisimple groups are an example of "nearly simple" groups; they are particularly important as minimal objects in the class of linear groups.

Let p be a prime. The *p-rank* $m_p(G)$ of G is the maximum n such that G has a subgroup that is the direct product of n copies of the group of order p.

The next result is not quite correct, but will serve for purposes of this paper.

THEOREM 2. *Let G be a finite simple group with $m_2(G) > 2$. Then either*

(1) G is of characteristic 2-type, or

(2) there exists a quasisimple subgroup L of G such that a Sylow 2-subgroup Q of $C_G(L)$ is nontrivial, $L \trianglelefteq N_G(X)$ for each $1 \neq X \leq Q$, and Q is cyclic, quaternion, or a 4-group.

There are no nonabelian simple groups of 2-rank 1. Because the simple groups of 2-rank 2 are "small"; their treatment requires special methods.

The groups of Lie type over fields of even characteristic appear in case (1), while almost all the groups over fields of odd characteristic appear in case (2). Thus the Theorem provides criteria for assigning a "characteristic" to abstract simple groups (in terms of their local structure and lumping all odd primes together into a single box labled *odd*) that, in the case of the groups of Lie type, corresponds to the characteristic of their field of definition as a linear group.

In case (2), for each involution $t \in Q$, $L \trianglelefteq C_G(t)$ and Q is Sylow in $C_{C_G(t)}(L)$, so $C_G(t)$ is dominated by L, and it remains to solve the corresponding involution centralizer problem for each known quasisimple group L. In that effort there is a

COMPLICATION Let L be a known simple group and W the wreath product of L by the group of order 2. That is W has a normal subgroup $L_1 \times L_2$ with $L_i \cong L$, and W is L_1L_2 extended by an involution t with $L_1^t = L_2$. Hence $C_W(t) \cong C_2 \times L$. This shows that case (2) is difficult when $|Q| = 2$. For while W is not simple, it is hard to see this until deep into the analysis.

I believe the proof of the CFSG should be restructured so as to avoid this difficulty and others. I'm in the midst of a program that, if successful, would bypass these and other problems. A novel feature of the approach is that much of the analysis takes place in the category of fusion systems rather than the category of groups.

Let p be a prime and S a finite p-group. A *fusion system* on S is a category $\mathcal{F}$ whose objects are the subgroups of S, and for $P, Q \leq S$ the set $\hom_{\mathcal{F}}(P,Q)$ of morphisms consists of injective group homomorphisms from P to Q, with the morphism sets satisfying two weak axioms. Let G be a finite group and $S \in Syl_p(G)$. The motivating example is the fusion system $\mathcal{F}_S(G)$ on S whose morphisms are those induced by conjugation in G. The system $\mathcal{F}_S(G)$ is *saturated*: that is it satisfies two more axioms which are easily checked using Sylow's Theorem. The notion of a fusion system is due to Lluis Puig [**P**], although he uses different terminology.

There are saturated fusion systems realized by no finite group; such systems are said to be *exotic*. For example there is an infinite family of exotic simple 2-fusion systems: the Benson-Solomon systems.

The map $(G,S) \to \mathcal{F}_S(G)$ extends to a functor from the category of pairs (G,S) with $S \in Syl_p(G)$, to the category of saturated fusion systems. This functor can be exploited to study fusion systems using finite group theory, and vice versa. In particular difficulties in CFSG arising from normal subgroups of odd order in 2-locals, can be avoided in the category of fusion systems. Such difficulties caused serious problems in the proof of the Theorem above, where it was necessary to prove an important property of finite simple groups known as the *B-conjecture*.

The simple groups of "odd characteristic" arise in case (2) of the Theorem as solutions of involution centralizer problems, while the simple groups of "even characteristic" arise in case (1); how are these groups of even characteristic treated?

The "small" groups of odd characteristic were those of 2-rank at most 2; a different measure of size is more appropriate when dealing with groups of even characteristic. This measure was introduced by Thompson in his ground breaking classification of the N-groups [**Th**]: the simple groups all of whose local subgroups are solvable.

Define

$$e(G) = \max\{m_p(H) : H \text{ is 2-local in } G \text{ and } p \text{ is an odd prime}\}.$$

If G is of Lie type in characteristic 2 then $e(G)$ is a good approximation of the Lie rank of G, an important measure of the "size" of a group of Lie type.

Roughly speaking, Thompson organized his treatment of N-groups into three cases: the small cases $e(G) = 1$ and $e(G) = 2$, and the generic case $e(G) \geq 3$. The treatment of groups of even characteristic in the original proof of CFSG was organized into three slightly different cases: $e(G) \leq 2$, $e(G) = 3$, and the generic case $e(G) \geq 4$.

Janko and Gorenstein coined the name *quasithin* groups for the groups G of characteristic 2-type with $e(G) \leq 2$. The classification of the quasithin groups was the last step in CFSG to be completed, roughly in 2004 by Aschbacher and Smith in [**AS, AS2**]. The proof involved specialized 2-local analysis introduced by Thompson in the N-group paper and developed further by various mathematicians.

The generic case where $e(G) \geq 4$ was treated in part by switching to an analysis of local subgroups for odd primes. The case $e(G) = 3$ was treated using a mixture of the approaches from the quasithin and generic cases. There is a program in progress (with principals Meierfrankenfeld, Stellmacher, and Stroth [**MSS**]) to rewrite that part of CFSG dealing with the groups of even characteristic using only 2-local analysis. Like the treatment of the quasithin groups, MSS use extensions of Thompson's ideas from the N-group paper, including weakly closed subgroups and factorizations.

Here is an example. Let H be a 2-local in G and $S \in Syl_2(H)$. A subgroup W of S is *weakly closed* in S with respect to H if W is the unique H-conjugate of W in S. For example the *Thompson subgroup* $J(S)$ of S is the subgroup of S generated by all abelian subgroups of S of exponent 2 and 2-rank $m_2(S)$. Observe $J(S)$ is weakly closed. Let Z be the subgroup generated by all involutions in $Z(S)$ and V the subgroup of H generated by all H-conjugates of Z. As $C_H(O_2(H)) \leq O_2(H)$, $V \leq Z(O_2(H))$ and $V \trianglelefteq H$. Thus we can regard V as a faithful module for $H/C_H(V)$ over the field of order 2. If $J(S)$ centralizes V then we have the *Thompson factorization* $H = C_H(V)N_H(J(S)) = C_H(Z)N_H(J(S))$. If $J(S)$ does not centralize V then V is a so-called *failure of factorization module* for $H/C_H(V)$, which imposes strong restrictions on the factor group and its action on V.

The proof of CFSG is very long and complicated, consisting of thousands of pages, in hundreds of journal articles by hundreds of authors. Given the utility of CFSG, it is important that the proof be simplified and placed on as solid a footing as possible. There is a program in progress (with principals R. Lyons and R. Solomon [**GLS1**]) to write down the proof of the CFSG carefully and completely in one place, while at the same time making changes in the proof to simplify it. Eight of a projected twelve volumes in the project have appeared to date.

I close with two commonly asked questions about CFSG:

QUESTION 1. Is it possible to prove a theorem classifying all but a finite number of the simple groups?

QUESTION 2. Why are there only 26 sporadic groups?

The two questions are related. The answer to Question 1 is: not using the tools available today. For CFSG is proved by induction on the order of the simple groups. We consider a counter example G to CFSG of minimal order; thus each simple section of G is known, and the analysis of G depends heavily on properties of the known simple groups appearing on the list in CFSG. If some simple section of G were not known, the presence of this section might vitiate the proof, and as a result a new simple group might be overlooked. Similarly if some important property of a known simple group is miscalculated, the existence of a corresponding extension as a section of G could have the same effect. For example each simple group X has a universal covering group: a largest quasisimple group $\tilde{X}$ with $\tilde{X}/Z(\tilde{X}) \cong X$. Miscalculation of $\tilde{X}$ could lead to missing a simple group G with a subgroup L as in the Theorem with $L \cong \tilde{X}$.

Most of the sporadics have a local section which is a smaller sporadic group, or a subnormal quasisimple subgroup of a local that is an unexpected covering of a group of Lie type. The existence of such accidents could have propagated upward in an infinite tower, or at least a very long tower; if that were the case, CFSG might have been impossible or at least much more difficult. Thus one answer to Question 2 is that there are only 26 sporadics because the number of accidents in small.

We close with an example illustrating this discussion. The largest sporadic is the Monster F_1, and the second largest is the Baby Monster F_2. Let G be F_1; then G has two classes of involutions with representatives z and t. Here $C_G(z)$ is the extension of the extraspecial group $E = O_2(C_G(z))$ of order 2^{25} (a finite version of a Heisenberg group) by the Conway group Co_1, with $E/\langle z\rangle$ the Leech lattice modulo 2 as a module for $C_G(z)/E$. On the other hand $C_G(t)$ is the universal covering group of F_2.

References

[A] M. Aschbacher, *On the maximal subgroups of the finite classical groups*, Invent. Math. **76** (1984), no. 3, 469–514, DOI 10.1007/BF01388470. MR746539

[ALSS] M. Aschbacher, R. Lyons, S. D. Smith, and R. Solomon, *The classification of finite simple groups*, Mathematical Surveys and Monographs, vol. 172, American Mathematical Society, Providence, RI, 2011. Groups of characteristic 2 type, DOI 10.1090/surv/172. MR2778190

[AS] M. Aschbacher and S. D. Smith, *The classification of quasithin groups. I*, Mathematical Surveys and Monographs, vol. 111, American Mathematical Society, Providence, RI, 2004. Structure of strongly quasithin K-groups, DOI 10.1090/surv/111. MR2097623

[AS2] M. Aschbacher and S. D. Smith, *The classification of quasithin groups. II*, Mathematical Surveys and Monographs, vol. 112, American Mathematical Society, Providence, RI, 2004. Main theorems: the classification of simple QTKE-groups, DOI 10.1086/428989. MR2097624

[B] R. Brauer, *On the structure of groups of finite order*, Proceedings of the International Congress of Mathematicians, Amsterdam, 1954, Vol. 1, Erven P. Noordhoff N.V., Groningen; North-Holland Publishing Co., Amsterdam, 1957, pp. 209–217. MR0095203

[C] C. Chevalley, *Sur certains groupes simples* (French), Tohoku Math. J. (2) **7** (1955), 14–66, DOI 10.2748/tmj/1178245104. MR73602

[FT] W. Feit and J. G. Thompson, *Solvability of groups of odd order*, Pacific J. Math. **13** (1963), 775–1029. MR166261

[GLS1] D. Gorenstein, R. Lyons, and R. Solomon, *The classification of the finite simple groups*, Mathematical Surveys and Monographs, vol. 40, American Mathematical Society, Providence, RI, 1994, DOI 10.1090/surv/040.1. MR1303592

[HS] M. Hall Jr. and J. K. Senior, *The groups of order* 2^n $(n \leq 6)$, The Macmillan Co., New York; Collier-Macmillan, Ltd., London, 1964. MR0168631

[LPS] M. W. Liebeck, C. E. Praeger, and J. Saxl, *A classification of the maximal subgroups of the finite alternating and symmetric groups*, J. Algebra **111** (1987), no. 2, 365–383, DOI 10.1016/0021-8693(87)90223-7. MR916173

[MSS] U. Meierfrankenfeld, B. Stellmacher, and G. Stroth, *The local structure for finite groups with a large p-subgroup*, Mem. Amer. Math. Soc. **242** (2016), no. 1147, vii+342, DOI 10.1090/memo/1147. MR3517155

[P] L. Puig, *Frobenius categories versus Brauer blocks*, Progress in Mathematics, vol. 274, Birkhäuser Verlag, Basel, 2009. The Grothendieck group of the Frobenius category of a Brauer block, DOI 10.1007/978-3-7643-9998-6. MR2502803

[Th] J. G. Thompson, *Nonsolvable finite groups all of whose local subgroups are solvable*, Bull. Amer. Math. Soc. **74** (1968), 383–437, DOI 10.1090/S0002-9904-1968-11953-6. MR230809

[Ti] J. Tits, *Buildings of spherical type and finite BN-pairs*, Lecture Notes in Mathematics, Vol. 386, Springer-Verlag, Berlin-New York, 1974. MR0470099

[W] E. Wigner, *The unreasonable effectivenesss of mathematics in the natural sciences*, Comm. Pure Appl. Math. **13** (1960), 1–14.

[Wi] R. A. Wilson, *The finite simple groups*, Graduate Texts in Mathematics, vol. 251, Springer-Verlag London, Ltd., London, 2009, DOI 10.1007/978-1-84800-988-2. MR2562037

California Institute of Technology, Pasadena, California 91125

Proceedings of Symposia in Pure Mathematics
Volume **104**, 2021
https://doi.org/10.1090/pspum/104/01876

The Birch and Swinnerton-Dyer Conjecture: A brief survey

Ashay A. Burungale, Christopher Skinner, and Ye Tian

Abstract. The celebrated Birch and Swinnerton-Dyer (BSD) conjecture connects the structure of the rational points on an elliptic curve defined over the rationals to the analytic properties of its associated Hasse–Weil L-function. In this paper we recall the BSD conjecture (and its various parts) and survey some of the known results towards it, especially recent work.

Contents

1. Introduction

Elliptic curves, in one guise or another, have attracted the attention of mathematicians and especially number theorists for centuries. In the last century, the Birch and Swinnerton-Dyer Conjecture emerged as the most fundamental unsolved problem about the arithmetic of elliptic curves. As Andrew Wiles expressed it [**73**]:

Needless to say for those who have devoted some time to this subject, [elliptic curves] is so full of fascinating problems that it is hard to turn from this to anything else. The conjecture of Birch and Swinnerton-Dyer... made the old subject irresistible.

This note is meant to be a brief introduction to the Birch and Swinnerton-Dyer Conjecture for elliptic curves over the rationals and a similarly brief survey of recent results about this problem, which has both compelled and frustrated number theorists for over sixty years.

2020 *Mathematics Subject Classification.* Primary 11G05, 11G40, 11R23.
The work of the second author was partially supported by the Simons Investigator Grant #376203 from the Simons Foundation and the National Science Foundation Grant DMS-1901985.

The Birch and Swinnerton-Dyer conjecture is primarily about the rational points on elliptic curves. This can be prosaically expressed as being about rational solutions to a class of cubic equations in two variables. As such it is firmly rooted in one of the oldest branches of mathematics: the study of the integer or rational solutions to equations. It has connections to open problems with origins buried in antiquity (such as the congruent number problem) while clearly being a part of the rich landscape of modern number theory (with connections to L-functions, automorphic forms, and the Langlands Program).

After recalling the Birch and Swinnerton-Dyer (or BSD) conjecture for elliptic curves defined over the rationals in Section 2, we very briefly survey a few of the most representative results toward the conjecture in Section 3. These break down roughly by decade. A recent result is then viewed a little more closely in Section 4.

The papers of Cassels [**25**] and Tate [**67**] and the book of Silverman and Tate [**58**] are excellent introductions to elliptic curves, while [**66**], [**72**], [**38**], and [**78**] are instructive surveys of the BSD conjecture.

2. The Birch and Swinnerton-Dyer conjecture

We describe the BSD conjecture for elliptic curves defined over the rational numbers.

2.1. Elliptic curves. An elliptic curve over the rationals is given by a polynomial relation

$$E : y^2 = x^3 + Ax + B \tag{2.1}$$

for $A, B \in \mathbb{Z}$ with $\Delta := 4A^3 + 27B^2 \neq 0$. The associated projective curve

$$\{[X : Y : Z] \ : ZY^2 = X^3 + AXZ^2 + BZ^3\} \subset \mathbb{P}^2_{/\mathbb{Q}}. \tag{2.2}$$

is a smooth algebraic curve of genus one. It is endowed with a distinguished point $[0 : 1 : 0]$ at ∞. By an elliptic curve E defined over the rationals we really mean such a smooth projective curve of genus one defined over the rationals with a distinguished rational point.

The set of points on E over a field has a natural structure as an abelian group with the distinguished point at ∞ being the identity. This group law is usually written as $+$, and for this reason the distinguished point is often denoted O. The group law gives E the structure of a commutative algebraic group, that is, a variety with a group law defined by rational maps.

The set $E(\mathbb{C})$ is a compact Riemann surface of genus one. It has a uniformization $\mathbb{C}/L \xrightarrow{\sim} E(\mathbb{C})$ for some lattice $L \subset \mathbb{C}$ under which $0 \in \mathbb{C}$ is mapped to $O \in E(\mathbb{C})$. Such a uniformization can be described explicitly in terms of Weierstrass $\wp$-functions. The group law on the torus $\mathbb{C}/L$ agrees with the algebraic group law on $E(\mathbb{C})$.

Loosely speaking, by the *arithmetic of* E we mean the problem of

understanding the abelian groups $E(\mathbb{Q})$, $E(\mathbb{F}_p)$, $E(\mathbb{Q}_p)$,

where p is a prime number.

2.1.1. $E(\mathbb{F}_p)$. The equation (2.2) can be reduced modulo p. If moreover $p \nmid 2\Delta$ then this reduction defines a smooth (projective) curve over $\mathbb{F}_p$. Let

$$N_p = \#E(\mathbb{F}_p) = 1 + p - a_p. \tag{2.3}$$

The Riemann hypothesis for $E/\mathbb{F}_p$ (a theorem of Hasse) tells us that the roots of $x^2 - a_p x + p$ all have absolute value $\sqrt{p}$. In particular,

$$|a_p| \leq 2\sqrt{p}. \tag{2.4}$$

The prime p is an *ordinary* prime for E if $p \nmid a_p$, and it is a *supersingular* prime for E if $p \mid a_p$. Note that it follows from (2.4) that if $p \geq 5$ is a supersingular prime then $a_p = 0$.

2.1.2. $E(\mathbb{Q})$. The starting point for questions about the rational points on E is the celebrated theorem of Mordell from 1922 [**51**]:

THEOREM 2.1. *Let E be an elliptic curve over the rationals. The set of rational points $E(\mathbb{Q})$ is a finitely generated abelian group.*

In particular, there exists an integer r_E, the (Mordell–Weil[1]) rank of E, such that

$$E(\mathbb{Q}) \simeq \mathbb{Z}^{r_E} \oplus E(\mathbb{Q})_{\mathrm{tor}}, \tag{2.5}$$

where $E(\mathbb{Q})_{\mathrm{tor}}$ is the (finite) torsion subgroup.

The torsion subgroup is well understood following work of Mazur in the mid-1970's [**49**]: $E(\mathbb{Q})_{\mathrm{tor}}$ is isomorphic to one of $\mathbb{Z}/n\mathbb{Z}$ for $n = 1, 2, \ldots, 10, 12$ or $\mathbb{Z}/2\mathbb{Z} \times \mathbb{Z}/n\mathbb{Z}$ for $n = 2, 4, 8$, and each possibility occurs for an infinite family of elliptic curves.

The rank r_E is more mysterious. It can be zero: the curve $y^2 = x^3 - x$ has only finitely many rational points. It can be non-zero: the curve $y^2 = x^3 - 36x$ has infinitely many rational points. So it is a natural question to ask

What can r_E be?

or even

How large can r_E be? and *How to determine r_E?*

Following on influential conjectures of Goldfeld [**35**], Katz and Sarnak [**44**] proposed that r_E is typically 0 or 1 (see Section 2.2.3).

It used be thought that r_E is bounded as E varies, then it was expected to be unbounded, and now some again propose it to be bounded (see [**52**]).

As for determing the rank: at present there is no algorithm that provably computes r_E.

2.1.3. $Ш(E)$. A group possibly even more mysterious than the Mordell–Weil group $E(\mathbb{Q})$ is the Tate–Shafarevich group:

$$Ш(E) = \ker\left\{H^1(\mathbb{Q}, E) \overset{\mathrm{loc}}{\to} H^1(\mathbb{R}, E) \times \prod_{p \text{ a prime}} H^1(\mathbb{Q}_p, E)\right\}, \tag{2.6}$$

where $H^1(K, E) = H^1(G_K, E(\overline{K}))$ is Galois cohomology with $G_K = \mathrm{Gal}(\overline{K}/K)$ for $\overline{K}$ an algebraic closure of K.

[1]The group $E(\mathbb{Q})$ is generally referred to as the Mordell–Weil group of $E(\mathbb{Q})$. Though Mordell alone proved that $E(\mathbb{Q})$ is finitely generated, Weil generalized this to all abelian varieties over any number field.

The Tate–Shafarevich group can be identified with the isomorphism classes of homogenous spaces of E that have an $\mathbb{R}$-point and a $\mathbb{Q}_p$-point for all primes p. In some ways, it is an analogue of the class group of a number field.

The group $Ш(E)$ is conjectured to be finite. This is known to be true for some curves, but is not known in general. If $Ш(E)$ is finite then $Ш(E) \cong N \oplus N$ for some (finite) torsion group N, and so its order is a square.

2.1.4. $L(s, E)$. Another important player in the arithmetic of E is its (Hasse–Weil) L-function:

$$L(s,E) = \prod_{p \text{ a prime}} L_p(p^{-s})^{-1} \tag{2.7}$$

for $s \in \mathbb{C}$ and

$$L_p(X) = 1 - a_p X + pX^2$$

for $p \nmid 2\Delta$ and $L_p(X)$ as in [**66**] for $p|2\Delta$. In light of (2.4) the L-function is convergent for $\mathrm{Re}(s) > \frac{3}{2}$.

The L-functions of elliptic curves are expected to have analytic continuations to all of $\mathbb{C}$. For elliptic curves over $\mathbb{Q}$, this was fully established by Wiles *et al.* as a consequence of the proof of the modularity of such curves [**74**] [**14**].

2.1.5. *Examples.* The arithmetic of elliptic curves has connections to open problems that go back to ancient Greek geometry.

- *The congruent number problem.* A positive integer is a congruent number if it is the area of a right-angled triangle all of whose sides have rational lengths. For a square-free positive integer n, let
$$E^{(n)} : y^2 = x^3 - n^2x.$$
Then
$$n \text{ is a congruent number} \iff r_{E^{(n)}} > 0.$$
- *The cube sum problem.* An integer is said to be a cube sum if it is the sum of two cubes of rational numbers. For a cube-free positive integer $n \geq 3$, let
$$E_n : y^2 = x^3 - 432n^2.$$
Then
$$n \text{ is a cube sum} \iff r_{E_n} > 0.$$

2.2. The BSD conjecture. We now explain the Birch and Swinnerton-Dyer Conjecture for elliptic curves defined over $\mathbb{Q}$.

The idea behind the conjecture is the following. Let $[x : y : z] \in E(\mathbb{Q})$ be a point. The mod p reduction[2] gives a point in $E(\mathbb{F}_p)$. It seems reasonable to expect that if $E(\mathbb{Q})$ has lots of points, then $N_p = \#E(\mathbb{F}_p)$ is big. We can measure the 'bigness' of N_p by comparing it with its expected value of p. So in the 1960's Birch and Swinnerton-Dyer [**12, 13**] computed the products

$$\prod_{p \leq X} \frac{N_p}{p}$$

[2]The projective coordinates x, y, z of the point can be taken to be integers with no common factor. The reduction of the point mod p is just the projective point with coordinates the mod p reductions of x, y, and z.

for many elliptic curves E. This led them to the guess [**13**, (A)] that

$$\prod_{p\leq X} \frac{N_p}{p} \sim c_E \cdot (\log X)^{r_E} \tag{2.8}$$

for some constant $c_E > 0$.

The asymptotic (2.8) can be dressed up in terms of the associated Hasse–Weil L-function $L(s,E)$ of E.

2.2.1. *The conjecture.* The BSD conjecture, as recorded by Birch [**11**] and Tate [**66**], is the following.

CONJECTURE 2.2 (The Birch and Swinnerton-Dyer Conjecture). *Let E be an elliptic curve over the rationals.*

(a) *The Hasse–Weil L-function $L(s,E)$ has an analytic continuation to the entire complex plane*[3] *and*

$$\mathrm{ord}_{s=1} L(s,E) = \mathrm{rank}_{\mathbb{Z}} E(\mathbb{Q}). \tag{BSD}$$

(b) *The Tate–Shafarevich group $Ш(E)$ is finite and*

$$\frac{L^{(r)}(1,E)}{r! \cdot \Omega_E \cdot \mathrm{reg}(E)} = \frac{\#Ш(E) \cdot \prod_p c_p(E)}{\#E(\mathbb{Q})^2_{\mathrm{tor}}} \tag{BSD-f}$$

for

- $r = \mathrm{ord}_{s=1} L(s,E)$,
- $c_p(E)$ *the Tamagawa number at p: the cardinality of the component group of the special fiber of the Néron model of E over $\mathbb{Z}_p$*
- $\Omega_E \in \mathbb{C}^\times$ *the Néron period: $\Omega_E = \int_{E(\mathbb{R})} |\omega|$, where $\omega \in \Omega^1(\mathcal{E}_{/\mathbb{Z}})$ is a $\mathbb{Z}$-basis of the differentials of the Néron model $\mathcal{E}_{/\mathbb{Z}}$ of E,*
- $\mathrm{reg}(E)$ *the regulator of the Néron–Tate height pairing on $E(\mathbb{Q})$*[4].

The order of vanishing $\mathrm{ord}_{s=1} L(s,E)$ of $L(s,E)$ at $s = 1$ is referred to as the analytic rank of E. So the conjecture (BSD) is that the analytic rank of E equals the Mordell–Weil rank of E. The equality (BSD-f) is often referred to as the BSD formula.

Goldfeld [**34**] showed[5] that if an asymptotic of the form

$$\prod_{p\leq X} \frac{N_p}{p} \sim c_E \cdot (\log X)^{r}$$

exists for some r, then the Generalized Riemann Hypothesis holds for $L(s,E)$, $r = \mathrm{ord}_{s=1} L(s,E)$, and c_E is an explicit multiple of $1/L^{(r)}(1,E)$. In particular,

$$(2.8) \implies (\mathrm{BSD}).$$

In the case $\mathrm{ord}_{s=1} L(s,E) \leq 1$ the left hand side of (BSD-f) is known to be a rational number. It therefore makes sense to ask whether the same power of a prime p appears in both sides of (BSD-f). This 'p-part of the BSD formula' is the focus of some of the recent results described in §3.

---

[3] This is a conjecture due to Hasse, Taniyama–Shimura and Weil.

[4] As explained by Néron and Tate (cf. [**66**]) there is a canonical positive definite bi-linear pairing on $E(\mathbb{Q}) \otimes \mathbb{R}$, which is generally called the canonical height or canonical height pairing.

[5] Assuming that $L(s,E)$ has an analytic continuation.

2.2.2. *Examples.* The BSD conjecture has important consequences for congruent numbers and cube sums. It leads to the following predictions.

- *The congruent number problem.* All square-free integers $n \equiv 5, 6, 7 \bmod 8$ are congruent.
- *The cube sum problem.* All primes $p \equiv 4, 7, 8 \bmod 9$ are cube sums.

2.2.3. *The conjectures of Goldfeld and Katz–Sarnak.* In [**35**] Goldfeld made the following influential conjecture.

CONJECTURE 2.3 (Goldfeld's Conjecture). *For an elliptic curve E over the rationals and $r = 0, 1$, 50% of the quadratic twists of E - ordered by discriminant - have analytic rank r.*

In combination with the BSD conjecture, this implies that 50% of the quadratic twists have a finite Mordell–Weil group and 50% have a Mordell–Weil group of rank one.

A similar conjecture was later proposed by Katz and Sarnak [**44**].

CONJECTURE 2.4 (Katz–Sarnak Conjecture). *Let $r = 0, 1$. When ordered by conductor, 50% of the elliptic curves over the rationals have analytic rank r.*

In combination with the BSD conjecture, this implies that 50% of elliptic curves over the rationals, when ordered by conductor, have a finite Mordell–Weil group and 50% have a Mordell–Weil group of rank one.

2.3. Selmer groups and the BSD conjecture. More accessible than each of $E(\mathbb{Q})$ or $Ш(E)$ separately is an amalgamation: the Selmer group of E.

2.3.1. $\mathrm{Sel}_{p^\infty}(E)$. Let p be a prime and $E[p^n]$ the group of p^n-torsion points of E. As a group $E[p^n] \simeq \mathbb{Z}/p^n\mathbb{Z} \times \mathbb{Z}/p^n\mathbb{Z}$. Let $T = T_pE = \varprojlim_n E[p^n]$ be the Tate module of E for the prime p. As a $\mathbb{Z}_p$-module, $T \simeq \mathbb{Z}_p^2$. Let $A = T \otimes_{\mathbb{Z}_p} \mathbb{Q}_p/\mathbb{Z}_p = E[p^\infty]$.

The p^∞-Selmer group $\mathrm{Sel}_{p^\infty}(E)$ is a subgroup of the Galois cohomology group $H^1(\mathbb{Q}, A)$. It appears as the middle term of the fundamental short exact sequence

$$0 \to E(\mathbb{Q}) \otimes_{\mathbb{Z}} \mathbb{Q}_p/\mathbb{Z}_p \to \mathrm{Sel}_{p^\infty}(E) \to Ш(E)[p^\infty] \to 0, \tag{2.9}$$

where $Ш(E)[p^\infty]$ is the p-primary part of $Ш(E)$.

2.3.2. *The conjecture, again.* In light of the exact sequence (2.9), Conjecture 2.2 suggests the following.

CONJECTURE 2.5. *Let E be an elliptic curve over the rationals. The following are equivalent:*

(a) $\mathrm{rank}_{\mathbb{Z}} E(\mathbb{Q}) = r$ *and* $Ш(E)$ *is finite.*
(b) $\mathrm{corank}_{\mathbb{Z}_p} \mathrm{Sel}_{p^\infty}(E) = r$ *for p a prime.*
(c) $\mathrm{ord}_{s=1} L(s, E) = r$.

Moreover, the BSD formula (BSD-f) *holds under any of* (a), (b), and (c).

Part (b) follows from part (a) by (2.9). We refer to '$(b) \implies (c)$' as a p-converse: a p-adic criterion to have analytic rank r. In §3 we survey results towards this and other implications, including (BSD-f).

REMARK 2.6. There is also a mod p Selmer group $\mathrm{Sel}_p(E) \subset H^1(\mathbb{Q}, E[p])$. This is the p-torsion of the p-adic Selmer groups: $\mathrm{Sel}_p(E) = \mathrm{Sel}_{p^\infty}(E)[p]$. From (2.9) it follows that if $E[p](\mathbb{Q}) = 0$ and $r_p := \dim_{\mathbb{F}_p} \mathrm{Sel}_p(E) \leq 1$, then $\mathrm{corank}_{\mathbb{Z}_p} \mathrm{Sel}_{p^\infty}(E) =$

r_p. This uses the fact that if $Ш(E)[p^\infty]$ is finite, then $Ш(E)[p^\infty] = N_p \oplus N_p$ for some finite group N_p. So by Conjecture 2.5 we should expect

$$(2.10) \quad E[p](\mathbb{Q}) = 0 \text{ and } r_p := \dim_{\mathbb{F}_p} \mathrm{Sel}_p(E) \leq 1 \implies \mathrm{ord}_{s=1} L(s,E) = r_E = r_p.$$

This can be viewed as a mod p criterion for rank 0 or 1.

3. Results

We describe some of the progress towards the BSD conjecture.

3.1. A theorem of Coates–Wiles and Rubin and a p-converse. The first general results towards the BSD conjecture were proved for elliptic curves with complex multiplication (CM elliptic curves).

THEOREM 3.1. *Let E be a CM elliptic curve over the rationals. Then*

$$L(1,E) \neq 0 \implies \#E(\mathbb{Q}),\ \#Ш(E) < \infty.$$

The finiteness of $E(\mathbb{Q})$ was proved by Coates and Wiles in 1976 [**31**]. The finiteness of Ш was proved by Rubin in 1986 [**55**]. This gave the first examples of elliptic curves for which Ш was proved to be finite.

The methods employed by Coates and Wiles made a surprising connection between Iwasawa theory and the BSD conjecture. Since then, Iwasawa-theoretic methods have been one of the main tools in the still-growing toolkit for studying the arithmetic of elliptic curves.

In very rough terms, the Coates–Wiles argument went as follows: Let K be the imaginary quadratic field by which E has complex multiplication. Let ψ_E be a Hecke character of K such that $L(s,E) = L(s,\psi_E)$. There is a CM period $\Omega \in \mathbb{C}^\times$ depending only on K such that $L(1,\psi_E)/\Omega \in K$. On the other hand, if $P \in E(\mathbb{Q})$ has infinite order, then for good primes $\mathfrak{p}$ of K (of which there are infinitely many), by using the complex multiplication of E it is possible to construct from the point P a non-trivial abelian p-extension $L/K(E[\mathfrak{p}^n])$ (p is the rational prime below $\mathfrak{p}$) that is ramified at the primes above $\mathfrak{p}$ and unramified at all other primes. By class field theory this implies the non-triviality of the index of a certain group of global units of $K(E[\mathfrak{p}^n])$ inside a corresponding group of local units. Using elliptic units – explicit global units of $K(E[\mathfrak{p}^n])$ – and methods of Iwasawa theory, Coates and Wiles were able to show that the non-triviality of this extension implies $\mathfrak{p} \mid L(1,\psi_E)/\Omega$. Since this holds for infinitely many primes $\mathfrak{p}$, it follows that if $E(\mathbb{Q})$ is infinite then $L(1,E) = 0$. Later work of Rubin, refining and extending the methods of Coates and Wiles, established the finiteness of $Ш(E)$ as well.

Rubin eventually succeeded in proving the Iwasawa main conjecture for imaginary quadratic fields in the early 1990s. This yielded a p-converse to Theorem 3.1.

THEOREM 3.2. *Let E be a CM elliptic curve over the rationals and p a prime. Then*

$$\mathrm{corank}_{\mathbb{Z}_p} \mathrm{Sel}_{p^\infty}(E) = 0 \implies L(1,E) \neq 0.$$

This p-converse is largely due to Rubin [**56**], at least with the hypothesis $p \nmid \#\mathcal{O}_K^\times$ for K the CM field. The unconditional p-converse was recently proved in [**16, 24**].

REMARK 3.3. In combination with [**65**], the 2-converse leads to the first example of a quadratic twist family of elliptic curves for which the even parity case of Goldfeld's conjecture (Conjecture 2.3) holds. In [**24**] the even parity case of the conjecture is proved for the congruent number elliptic curve:

$$L(1, E^{(n)}) \neq 0 \text{ for a density one set of integers } n \equiv 1, 2, 3 \bmod 8.$$

3.1.1. *The CM/non-CM dichotomy.* The theorems of Coates–Wiles and Rubin are examples of a dichotomy that appears over and over again in the arithmetic of elliptic curves over the rationals. A proof of arithmetic results for CM curves often precedes that for non-CM curves, and the methods employed for one case may not even apply to the other. For example, the analyticity of the Hasse–Weil L-function for CM curves is due to Hecke from the early 1940's whereas for non-CM curves it is a celebrated theorem of Wiles, Taylor–Wiles, Breuil–Conrad–Diamond–Taylor from the mid to late 1990's. Often one studies the arithmetic of CM elliptic curves (or even CM modular forms) via the arithmetic of the CM field, about which much is known.

3.2. The theorem of Gross–Zagier and Kolyvagin. After the work of Coates and Wiles in the 1970's, the next spectacular result towards the BSD conjecture came in the 1980's and was due to Gross and Zagier [**37**] and Kolyvagin [**47**].

THEOREM 3.4. *Let E be an elliptic over the rationals. Then*

$$\mathrm{ord}_{s=1} L(s, E) \leq 1 \implies \mathrm{rank}_{\mathbb{Z}} E(\mathbb{Q}) = \mathrm{ord}_{s=1} L(s, E),\ \#Ш(E) < \infty.$$

In the case $\mathrm{ord}_{s=1} L(s, E) = 1$ the method of proof yields a systematic construction of a non-torsion point in $E(\mathbb{Q})$. In the case $L(1, E) \neq 0$ the result was independently proved by Kato in the early 1990's [**42**],[**43**]. In fact, Kato proved the upper bound for $\mathrm{Sel}_{p^\infty}(E)$ predicted by (BSD-f) for all but finitely many explicit p.

A construction of Heegner, elucidated by Birch, gives a systematic construction of points on E over ring class fields of imaginary quadratic fields. Loosely, the elliptic curve admits a modular parameterization[6] $\phi_E : X_0(N) \twoheadrightarrow E$, where N is the conductor of E, and the theory of complex multiplication produces many 'special points' on $X_0(N)$ that are defined over ring class fields of suitable imaginary quadratic fields. The Heegner points on E are the (sums of Galois conjugates) of images of these points under the map ϕ_E. In particular, if K is an imaginary quadratic field such that every prime $\ell \mid N$ splits in K, then this construction yields a point $P_K \in E(K)$, often referred to as *the* Heegner point. This then yields a point $P_{\mathbb{Q}} = P_K + P_K^{\tau} \in E(\mathbb{Q})$, for $\tau \in \mathrm{Gal}(K/\mathbb{Q})$ the non-trivial automorphism.

Following up on an idea of Birch, Gross and Zagier [**37**] proved a remarkable formula: For an imaginary quadratic field K such that every prime $\ell \mid N$ splits in K,

$$L'(1, E/K) = (*)\langle P_K, P_K \rangle_{NT},$$

where $L(s, E/K)$ is the Hasse–Weil L-function of E over K, $(*)$ is an explicit non-zero constant and $\langle -, - \rangle_{NT}$ is the canonical height pairing of Néron and Tate.

[6]This was an hypothesis in the work of Birch, Gross and Zagier, and Kolyvagin. Following the work of Wiles *et al.*, this is now superfluous.

Since a point $P \in E(K)$ satisfies $\langle P, P\rangle_{NT} \neq 0$ if and only if P is non-torsion, this formula can be interpreted as:

$$\text{(3.1)} \qquad \mathrm{ord}_{s=1} L(s, E/K) = 1 \iff P_K \text{ has infinite order.}$$

If $K = \mathbb{Q}(\sqrt{-d})$ and E' is the $-d$-twist[7] of E, then $L(s, E/K) = L(s, E)L(s, E')$. If the root number w_E of E is -1, then it is possible to choose K so that $L(1, E') \neq 0$. Similarly, if the root number of E is $+1$, then is possible to choose K so that $\mathrm{ord}_{s=1} L(s, E') = 1$. In particular, if $\mathrm{ord}_{s=1} L(s, E) \leq 1$, then K can be chosen so that $\mathrm{ord}_{s=1} L(s, E/K) = 1$.

Around the same time, Kolyvagin [**47**] surprised the number theory world with his proof that

$$P_K \text{ has infinite order} \implies \mathrm{rank}_{\mathbb{Z}} E(K) = 1, \#\text{Ш}(E/K) < \infty.$$

In fact, Kolyvagin's proved that if P_K has infinite order, then

$$\text{(3.2)} \qquad r_E = \frac{1 - w_E}{2}, \; r_{E'} = \frac{1 + w_E}{2}, \quad \text{and } \#\text{Ш}(E), \#\text{Ш}(E') < \infty.$$

Kolyvagin's proof – his method of *Euler systems* – exploits not just the Heegner point P_K but the fact that P_K is the bottom layer of a tower (the Euler system) of points on E defined over ring class fields and related in a manner that reflects the Euler factors of the Hasse–Weil L-function of E.

Theorem 3.4 is obtained by combining (3.1) with (3.2) after making an apt choice of K as outlined above.

The works of Gross–Zagier and Kolyvagin did not make use of methods from Iwasawa theory, though Heegner points and Kolyvagin's Euler system were later combined with Iwasawa-theoretic methods, for example in the works of Perrin-Riou [**53**] and Howard [**40**]. But the methods of Iwasawa theory were once more front and center in the work of Kato [**43**] in the 1990's.

Motivated by a construction of Beilinson, Kato defined special classes in the Galois cohomology groups $H^1(\mathbb{Z}[\mu_n, \frac{1}{p}], T_pE(1))$ for varying integers n and showed that these form an Euler system. Analogously with the work of Coates and Wiles, Kato also proved an explicit reciprocity law that linked these classes with values of the Hasse–Weil L-function of E. By passing to $H^1(\mathbb{Z}[\mu_{np^\infty}, \frac{1}{p}], T_pE(1))$ and then descending, Kato obtained a class $z_E \in H^1(\mathbb{Z}[\frac{1}{p}], T_pE)$ with the property that the localization $\mathrm{loc}_p(z_E) \in H^1(\mathbb{Q}_p, T_pE)$ belongs to the Kummer image of $E(\mathbb{Q}_p)\hat{\otimes}\mathbb{Z}_p$ if and only if $L(1, E) = 0$. The $r = 0$ case of Theorem 3.4 is readily deduced from this, and a bound on $\#\text{Ш}(E)[p^\infty]$ in terms of $L(1, E)/\Omega_E$ comes from the method of Euler systems.

REMARK 3.5. A cyclotomic approach to the the $r = 1$ case of Theorem 3.4 (as opposed to the anticyclotomic Heegner point Euler system) is given in [**20**], inspired by [**43**]. This approach, however, still relies on the existence of the Heegner point P_K and the Gross–Zagier formula.

---

[7]If E is defined by the equation $y^2 = x^3 + Ax + B$, then E' is defined by the equation $-dy^2 = x^3 + Ax + B$. Though generally non-isomorphic over $\mathbb{Q}$, E and E' are isomorphic over $K = \mathbb{Q}(\sqrt{-d})$.

3.2.1. *p-converses.* Recent advances on the Iwasawa theory of elliptic curves has led to p-converses to Theorem 3.4. We describe three of these.

THEOREM 3.6. *Let E be an elliptic curve over the rationals with conductor N. Let $p \geq 3$ be a prime at which E has ordinary reduction. Suppose:*

(irr$_\mathbb{Q}$) *The mod p Galois representation $E[p]$ is absolutely irreducible.*
(ram) *There exists a prime $\ell || N$, $\ell \neq p$, such that $E[p]$ is ramified at ℓ.*

Then

$$\mathrm{corank}_{\mathbb{Z}_p} \mathrm{Sel}_{p^\infty}(E) = 0 \implies L(1, E) \neq 0.$$

This p-converse is largely due to Skinner–Urban around the mid 2000's [**60**],[**61**]. A p-converse with p a prime of supersingular reduction for E has been established by Wan [**71**].

Theorem 3.6 is a consequence of one divisibility in the Iwasawa main conjecture for the elliptic curve E. This divisibility takes the shape:

$$(p\text{-adic } L\text{-function of } E) \mid (\text{characteristic idea of } \mathrm{Sel}_{p^\infty}(E/\mathbb{Q}_\infty)^\vee), \tag{3.3}$$

where $\mathbb{Q}_\infty$ is the cyclotomic $\mathbb{Z}_p$-extension of $\mathbb{Q}$ and the superscript $\vee$ denotes the Pontrjagin dual. The divisibility is in the Iwasawa algebra

$$\mathbb{Z}_p[[\mathrm{Gal}(\mathbb{Q}_\infty/\mathbb{Q})]] \otimes_{\mathbb{Z}_p} \mathbb{Q}_p \cong \mathbb{Z}_p[[T]] \otimes_{\mathbb{Z}_p} \mathbb{Q}_p.$$

If $\mathrm{corank}_{\mathbb{Z}_p} \mathrm{Sel}_{p^\infty}(E) = 0$, then the right-hand side of (3.3) is non-zero under specialization at the trivial character of $\mathrm{Gal}(\mathbb{Q}_\infty/\mathbb{Q})$. The divisibility then implies that the left-hand side also has non-zero specialization. But the specialization of the p-adic L-function of E at the trivial character is a non-zero multiple of $L(1, E)$.

The proof of (3.3) in [**60**] is given by an extensive generalization of the methods employed by Wiles in his proof of the Iwasawa main conjecture for totally real fields: Eisenstein congruences and Galois representations. For elliptic curves this involved Eisenstein conguences on the unitary groups $U(2, 2)$ and the Galois representations associated with cuspidal automorphic representations of these groups (see [**59**] for a longer description of this proof).

The methods of [**60**] do not directly extend to the case where p is a prime of supersingular reduction. Wan succeeded in deducing the analog of the divisibility (3.3) for this case by first proving a divisibility for a different main conjecture: the left-hand side in (3.3) is replaced by a p-adic L-function that interpolates special values of the L-function of E twisted by certain Hecke characters of an imaginary quadratic field that have infinite order, and the right-hand side is replaced by the characteristic ideal of a different Selmer group. This auxiliary divisibility also turns out to be a key ingredient in one approach to the following p-converse (this is explained in the introduction to [**62**]).

THEOREM 3.7. *Let E be an elliptic curve over the rationals with conductor N and $p \nmid 6N$ a prime at which E has ordinary reduction. Suppose:*

(irr$_\mathbb{Q}$) *The mod p Galois representation $E[p]$ is absolutely irreducible.*
(ram) *There exists a prime $\ell || N$ such that $E[p]$ is ramified at ℓ.*

Then,

$$\mathrm{corank}_{\mathbb{Z}_p} \mathrm{Sel}_{p^\infty}(E) = 1 \implies \mathrm{ord}_{s=1} L(s, E) = 1.$$

The first results towards such a rank one p-converse were due to Heegner [**39**] (see also [**68**]). The first general results towards this p-converse were independently due to Skinner [**62**] and Zhang [**77**] a few years back. Other results in the same vein can be found in [**3**], [**26**], [**70**], [**75**], [**30**], [**20**]. The version in this Theorem will appear in [**21**].

There has been some progress in removing the hypothesis ($\mathrm{irr}_{\mathbb{Q}}$) of Theorems 3.6 and 3.7, most notably as a consequence of the work of Greenberg and Vatsal [**36**], but see also [**30**].

If $p \nmid N$ and E is semistable, then the hypothesis (ram) of Theorems 3.6 and 3.7 is always satisfied. However, the hypothesis (ram) is never satisfied by CM curves. So CM curves are excluded by both theorems.

Of course, Theorem 3.2 is the rank zero p-converse for CM curves. A rank one p-converse is given by the following.

THEOREM 3.8. *Let E be a CM elliptic curve over the rationals with conductor N and $p \nmid 6N$ a prime. Then,*

$$\mathrm{corank}_{\mathbb{Z}_p} \mathrm{Sel}_{p^\infty}(E) = 1 \implies \mathrm{ord}_{s=1} L(s,E) = 1.$$

For p also a prime of ordinary reduction, this p-converse was proved in [**23**]. The above p-converse will appear in [**22**]. Another approach, which generalizes to CM elliptic curves over totally real fields, is given in [**8, 19**]. An extended sketch of this last approach is given below in Section 4.

3.3. The BSD formula. Methods employed in making progress towards the rank part (BSD) of the BSD conjecture have also led to progress toward the BSD formula (BSD-f).

For CM elliptic curves we have the following.

THEOREM 3.9. *Let E be an elliptic curve over the rationals with CM by an imaginary quadratic field K and conductor N. If $r := \mathrm{ord}_{s=1} L(s,E) \leq 1$ then the BSD formula* (BSD-f) *holds up to multiplication by an element in $\mathbb{Z}[\frac{1}{\#\mathcal{O}_K^\times}]^\times$ in the case $r = 0$ and up to multiplication by an element in $\mathbb{Z}[\frac{1}{\#\mathcal{O}_K^\times \cdot N}]^\times$ in the case $r = 1$.*

The $r = 0$ case of this theorem was proved by Rubin in the early 1990's [**56**]. For the $r = 1$ case, the p-part of the formula for p a prime of ordinary reduction is also due to Rubin (*op. cit.*), while the p-part of the formula for p a prime of supersingular reduction was proved by Kobayashi in the early 2010's [**46**].

For non-CM curves:

THEOREM 3.10. *Let E be an elliptic curve over the rationals with conductor N. Let $p > 3$ be a prime at which E has ordinary reduction. Suppose:*

($\mathrm{irr}_{\mathbb{Q}}$) *The mod p Galois representation $E[p]$ is absolutely irreducible.*
(ram) *There exists a prime $\ell || N$, $\ell \neq p$, such that $E[p]$ is ramified at ℓ.*

If $\mathrm{ord}_{s=1} L(s,E) \leq 1$, then the p-part of the BSD formula (BSD-f) *holds.*

In the case $L(1,E) \neq 0$ and $p \nmid N$, this is largely due to Kato and Skinner–Urban around the mid 2000's [**43**], [**60**]. The hypothesis $p \nmid N$ was removed in [**61**].

In the case $\mathrm{ord}_{s=1} L(s,E) = 1$, the first general results towards the p-part – but with additional conditions on p – were independently due to Jetchev–Skinner–Wan [**41**] and Zhang [**77**] in the mid 2010's. Other results in the same vein were established in [**3**], [**27**], and [**63**]. The result stated in Theorem 3.10 is proved in [**20**].

Similar results for p a prime of supersingular reduction can be found in [**41**], [**29**] and in the combination of results of Kobayashi [**46**] and Wan [**71**].

3.4. Arithmetic statistics. An elliptic curve E over the rationals is isomorphic to a unique curve (2.1) such that for all primes p:

$$p^6 \nmid B \text{ whenever } p^4 | A. \tag{3.4}$$

The (naive) height $H(E)$ of E is defined to be

$$H(E) = \max\{4|A|^3, 27B^2\}. \tag{3.5}$$

The number of non-isomorphic elliptic curves (defined by equations satisfying (3.4)) with height $\leq x$ is asymptotic to $cx^{\frac{5}{6}}$ for a constant c. So it seems reasonable to ask what proportion of elliptic curves, ordered by height, have some property (eg. rank zero, rank one, finite Ш, etc.).

THEOREM 3.11. *When all non-isomorphic elliptic curves over the rationals are ordered by height,*

- *at least* 16% *have both Mordell–Weil rank and analytic rank equal to* 0,
- *at least* 20% *have both Mordell–Weil rank and analytic rank equal to* 1,
- *at least* 66% *have Mordell–Weil rank and analytic rank both equal to* 0 *or both equal to* 1.

Similarly:

THEOREM 3.12. *When all elliptic curves over the rationals are ordered by height, at least* 66% *have finite Tate–Shafarevich group.*

Both theorems are due to Bhargava–Skinner–Zhang [**10**]. Earlier work of Bhargava–Skinner [**6**] established similar results for the rank 1 case but for an undetermined positive proportion of elliptic curves.

The method of proof uses results of Bhargava–Shankar [**7**],[**9**] to estimate the proportion of curves with $\mathrm{Sel}_5(E) = 0$ or $\mathbb{Z}/5\mathbb{Z}$ and then concludes the corresponding result for the analytic ranks of the $L(s, E)$ from 5-converse theorems such as Theorem 3.7 (cf. Remark 2.6).

Ordering curves by naive height is conjecturally equivalent to ordering them by conductor. So Theorem 3.11 can be seen as evidence and progress toward Conjecture 2.4.

3.4.1. *Other results.* The literature on the BSD conjecture is vast, reflecting the hold the problem has gained on the imaginations of number theorists. The results we have described represent a few of the main theoretical advances toward the conjecture. Brevity has necessitated omitting a discussion of many other important works, such as: the ever-growing body of computational evidence (e.g. [**32**]), results over fields other than $\mathbb{Q}$ and especially totally real fields (e.g. [**76**]), and equivariant results (e.g. [**5**], [**33**], [**45**]).

4. Methods: an instructive example

To give a sense of what goes into the proofs of some of the more recent results, such as Theorems 3.7 and 3.8, we describe in more detail the proof of a p-converse for CM curves from [**19**]. Our discussion of this assumes much more familiarity with methods of Iwasawa theory than we have so far in this paper. A reader can safely skip directly to the brief discussion of some open problems in Section 5.

4.1. The set-up. Let E be an elliptic curve over the rationals and p a prime such that $\mathrm{Sel}_{p^\infty}(E)$ has corank one. It is expected, say from the conjectured finiteness of $\text{Ш}(E)[p^\infty]$, that the corank is explained by $E(\mathbb{Q})$ having rank one. Then the BSD conjecture would predict that $\mathrm{ord}_{s=1}L(s,E)=1$. If the latter held then, by the results of Gross–Zagier and Kolyvagin, $E(\mathbb{Q})$ would essentially be generated by a Heegner point for a suitable auxiliary imaginary quadratic field L. This suggests that to prove a p-converse theorem one might begin by choosing a suitable field L and then try to directly explain the non-torsion-ness of the Heegner point as a consequence of the Selmer group having corank one.

Now suppose that E has CM by an imaginary quadratic field K. To exploit the arithmetic of the field K, it is natural to look to a Heegner point over K or a ring class field of K. However, the root number $w_{E/K}$ of E over K is $+1$, while Heegner points typically arise from auxiliary imaginary quadratic fields L such that $w_{E/L}=-1$. To get around this issue, we introduce an auxiliary Rankin–Selberg convolution over K which leads to a (generalized) Heegner point whose non-triviality (non-torsion-ness) amounts to the p-converse.

Let λ be the Hecke character of K associated to E (in particular, $L(s,\lambda)=L(s,E)$). Fix an algebraic closure $\overline{\mathbb{Q}}$ and embeddings $\iota_\infty:\overline{\mathbb{Q}}\hookrightarrow\mathbb{C}$, $\iota_p:\overline{\mathbb{Q}}\hookrightarrow\mathbb{C}_p$. Let $\tau\in\mathrm{Gal}(\mathbb{C}/\mathbb{R})$ denote the complex conjugation, which induces via ι_∞ the non-trivial element in $\mathrm{Gal}(K/\mathbb{Q})$.

4.2. The auxiliary Rankin–Selberg convolution and a Heegner point. Let χ be a finite order Hecke character of K such that

$$L(1,\lambda\cdot\frac{\chi^*}{\chi})\neq 0, \tag{4.1}$$

where for a Hecke character ψ of K, $\psi^*:=\psi\circ\tau$. Then

$$\mathrm{corank}_{\mathcal{O}_\wp}\mathrm{Sel}_{\wp^\infty}(\lambda\cdot\frac{\chi^*}{\chi})=0. \tag{4.2}$$

Here $\mathcal{O}$ is the ring of integers of the number field associated to $\lambda\cdot\frac{\chi^*}{\chi}$, $\wp|p$ is the prime of $\mathcal{O}$ determined via ι_p, and $\mathrm{Sel}_{\wp^\infty}(\lambda\cdot\frac{\chi^*}{\chi})$ is the associated Bloch–Kato Selmer group. As $\frac{\chi^*}{\chi}$ is an anticyclotomic Hecke character of K, the existence of the desired character satisfying (4.1) is a consequence of [**54**]. That (4.2) follows from (4.1) is just a variant of Theorem 3.1.

Let g be the CM modular form associated to the Hecke character $\lambda\chi^{-1}$. There is a factorization of complex L-functions

$$L(s,g\times\chi^*)=L(s,\lambda^*)\cdot L(s,\lambda\cdot\frac{\chi^*}{\chi}), \tag{4.3}$$

where $L(s,g\times\chi^*)$ is the Rankin–Selberg convolution associated with the (self-dual) pair (g,χ^*). This is just the Rankin–Selberg convolution of g with the CM form associated with χ^*. It follows from this factorization and (4.1) that

$$\mathrm{ord}_{s=1}L(s,g\times\chi^*)=1 \iff \mathrm{ord}_{s=1}L(s,E)=1.$$

The self-dual pair (g,χ^*) satisfies a *generalized* Heegner hypothesis. This relies on the automorphic induction of λ to $\mathrm{GL}_{2/\mathbb{Q}}$ being self-dual with root number equal to -1 and (4.1) (that the root number is -1 is a consequence of the corank one assumption by a proven case of the parity conjecture). A special case of [**76**] then yields a generalized Heegner point

$$P_{g,\chi} \in B(K) \otimes_{\mathbb{Z}} \mathbb{Q},$$

where $B_{/K}$ is a certain abelian variety associated with the pair (g, χ^*). This abelian variety is a quotient of the Jacobian of a Shimura curve, and the Heegner point arises from CM points on the Shimura curve. The L-function $L(s, B_{/K})$ is a product of the Rankin–Selberg L-functions for Galois conjugates of the pair (g, χ^*). In particular, it follows from the Gross–Zagier formula in [**76**] that

$$\text{(4.4)} \qquad \operatorname{ord}_{s=1} L(s, g \times \chi^*) = 1 \iff P_{g,\chi} \neq 0.$$

The abelian variety B is such that $\mathcal{O} \hookrightarrow \operatorname{End}(B)$, and there is a factorization of Selmer groups

$$\text{(4.5)} \qquad \operatorname{Sel}_{\wp^\infty}(B_{/K}) = \operatorname{Sel}_{\wp^\infty}(\lambda \cdot \frac{\chi^*}{\chi}) \oplus \operatorname{Sel}_{\wp^\infty}(\lambda^*)$$

paralleling the factorization of L-functions (4.3). In light of this factorization and (4.1), (4.2), and (4.4), the desired p-converse for E is then equivalent to

$$\text{(4.6)} \qquad \operatorname{corank}_{\mathcal{O}_\wp} \operatorname{Sel}_{\wp^\infty}(B_{/K}) = 1 \implies P_{g,\chi} \neq 0.$$

REMARK 4.1. The generality of the Gross–Zagier formula in [**76**] is crucial to this approach, as it allows access to self-dual pairs (g, χ^*) with g having non-trivial central character.

4.3. A Heegner point main conjecture. The proof[8] of the implication (4.6) goes via Iwasawa theory.

Suppose now that p is a prime of ordinary reduction for E. Let K_∞/K be the anticyclotomic $\mathbb{Z}_p$-extension of K. Let $\Gamma = \operatorname{Gal}(K_\infty/K)$ and let $\Lambda = \mathcal{O}_\wp[\![\Gamma]\!]$ be the corresponding Iwasawa algebra. Let the superscript ι denote the involution of Λ arising from inversion on Γ. Let X be the Pontrjagin dual of the discrete Selmer group $\mathcal{S} = \varinjlim_n \varinjlim_m \operatorname{Sel}_{\wp^m}(B_{/K_n})$ and let $S = \varprojlim_n \varprojlim_m \operatorname{Sel}_{\wp^m}(B_{/K_n})$, where K_n is the nth layer of the extension K_∞/K.

Let $\kappa_0 \in \operatorname{Sel}_{\wp^\infty}(B_{/K})$ be the Kummer image of the Heegner point $P_{g,\chi} \in B(K)$. A variant of the construction of $P_{g,\chi}$ over the layers K_n of the anticyclotomic extension K_∞/K leads to a norm-compatible system of generalized Heegner points and ultimately to a Heegner class $\kappa \in S$ deforming κ_0.

4.3.1. *The conjecture.* As essentially conjectured by Perrin-Riou [**53**], it is expected that

(a) the Heegner class $\kappa \in S$ is not Λ-torsion[9],
(b) $\operatorname{rank}_\Lambda S = \operatorname{rank}_\Lambda X = 1$,
(c) $\xi_\Lambda(S/\Lambda \cdot \kappa) \cdot \xi_\Lambda((S/\Lambda \cdot \kappa)^\iota) = \xi_\Lambda(X_{\mathrm{tor}})$, for $\xi_\Lambda(\cdot)$ the Λ-characteristic ideal and $(\cdot)_{\mathrm{tor}}$ the Λ-torsion submodule.

To deduce the implication (4.6) it suffices to know that the left-hand side of the conjectured equality in (c) divides the right-hand side:

$$\text{(4.7)} \qquad \xi_\Lambda(S/\Lambda \cdot \kappa) \cdot \xi_\Lambda((S/\Lambda \cdot \kappa)^\iota) \mid \xi_\Lambda(X_{\mathrm{tor}}).$$

---

[8]Everything in this section should be taken with a grain of salt, though the outline is correct in the large. To make the arguments work in practice, the Iwasawa algebras and the Selmer groups have to be suitably modified. In particular, it may not be possible to work over the integral Iwasawa algebras.

[9]a conjecture of Mazur (cf. [**49**], [**17**]).

By Iwasawa-theoretic descent to K for X, the hypothesis $\mathrm{corank}_{\mathcal{O}_\wp}\mathrm{Sel}_{\wp^\infty}(B_{/K}) = 1$ implies $\xi_\Lambda(X_{\mathrm{tor}})$ is not divisible by the augmentation ideal of Λ. The same is then true of $\xi_\Lambda(S/\Lambda \cdot \kappa)$, which then implies – again by descent to K – that κ_0 is non-torsion and hence that $P_{g,\chi} \neq 0$.

4.3.2. *Its proof.* The non-triviality of κ (part (a)) was proved in [**18**], and this implies part (b). The approach to proving part (c) in [**19**] is based on

- a p-adic Waldspurger formula and its Λ-adic analogue [**4**], [**48**], [**57**] and
- an anticyclotomic CM Iwasawa Main Conjecture [**56**], [**1**], [**2**].

The first of these should be seen as $\mathrm{GL}_{2/\mathbb{Q}}$-input and the second as $\mathrm{GL}_{1/K}$-input.

As p is an ordinary prime for E, p splits in K:

$$(p) = v\overline{v} \tag{ord}$$

with v determined via ι_p. Let $\Sigma = \{\iota_\infty\}$ be a CM type of K. The Heegner point main conjecture is then equivalent to yet another Iwasawa main conjecture:

(a') $\mathrm{rank}_\Lambda S_v = \mathrm{rank}_\Lambda X_v = 0$, and
(b') $\xi_\Lambda(X_v) = (\mathcal{L}_v^{ac}(B))$.

Here S_v is the compact $\wp$-adic Selmer group for B over K_∞ such that the local condition at the prime v (resp. $\overline{v}$) is relaxed (resp. strict), and X_v is the Pontrjagin dual of the analogous discrete Selmer group of B over K_∞. Furthermore, $\mathcal{L}_v^{ac}(B) \in \Lambda$ is the anticyclotomic p-adic L-function associated to the pair (g, χ^*) as in [**48**]. It interpolates the Rankin–Selberg central L-values associated to the self-dual pairs $(g, \chi^*\nu)$ for ν an arithmetic Hecke character of K with corresponding Galois character factoring through Γ and having Hodge–Tate weight at v at least 1.

The equivalence of the two main conjectures comes via the non-triviality of κ and the Λ-adic analogue of the p-adic Waldspurger formula. The latter expresses the p-adic L-function $\mathcal{L}_v^{ac}(B)$ in terms of the Λ-adic logarithm[10] of κ.

It remains to explain the proof of (a') and (b').

Much as the factorizations (4.3) and (4.5), the Λ-modules S_v and X_v and the p-adic L-function $\mathcal{L}_v^{ac}(B)$ decompose in terms of Selmer groups and p-adic L-functions for λ^* and $\lambda \cdot \frac{\chi^*}{\chi}$. In particular:

(i) $X_v \simeq X_\Sigma(\lambda^*) \oplus X_\Sigma(\lambda \cdot \frac{\chi^*}{\chi})$, and
(ii) $\mathcal{L}_v^{ac}(B) =_{\Lambda^\times} \mathcal{L}_\Sigma(\lambda^*) \cdot \mathcal{L}_\Sigma(\lambda \cdot \frac{\chi^*}{\chi})$.

Here the subscript Σ denotes that the anticyclotomic Selmer groups and p-adic L-functions of the Hecke characters are taken with respect to the CM type Σ. Also, '$=_{\Lambda^\times}$' denotes an equality up to an element in $\Lambda^\times$. Then Rubin's results [**56**] match the characteristic ideals of the right in (i) with the corresponding p-adic L-functions on the right in (ii). The equality (b') follows. The claim in (a') similarly follows from [**56**].

This completes the proof of the p-converse for the CM curve E.

REMARK 4.2. For an overview of the proof of a p-converse in a non-CM case see [**64**, §6].

[10] This interpolates the Bloch–Kato logarithm along the Iwasawa tower.

5. Some open problems

The works reported on herein include substantial progress towards Conjecture 2.5, at least in the $r \leq 1$ cases. However, a full resolution of the conjecture in this case requires:

- handling primes p of additive reduction,
- cases where $E[p]$ is reducible or the corresponding Galois representation has small image,
- allowing $p \leq 3$, especially $p = 2$.

At the primes of additive reduction even a formulation of some of the Iwasawa main conjectures (such as the Heegner point main conjecture) are missing. The same holds for many of the p-adic L-functions. As Iwasawa theory lies at the heart of much of the progress reported on, these are likely important problems to address. As always, the case of CM elliptic curves could be an instructive starting point.

Finally, there remains little to say in the $r > 1$ cases. It would be great to have

an example of an elliptic curve E over the rationals having rank at least 2 with $Ш(E)$ (proved) finite.

Acknowledgments

This paper was written at the request of Dinakar Ramakrishnan, and the authors are grateful to him for his suggestions and encouragement. They are also grateful to the referee for helpful comments. This paper was begun while C.S. was a Taussky-Todd distinguished visitor at Caltech, and he thanks the Caltech mathematics department for providing such a pleasant and inspiring environment[11] for beginning this and other projects.

References

[1] A. Agboola and B. Howard, *Anticyclotomic Iwasawa theory of CM elliptic curves* (English, with English and French summaries), Ann. Inst. Fourier (Grenoble) **56** (2006), no. 4, 1001–1048. MR2266884

[2] T. Arnold, *Anticyclotomic main conjectures for CM modular forms*, J. Reine Angew. Math. **606** (2007), 41–78, DOI 10.1515/CRELLE.2007.034. MR2337641

[3] A. Berti, M. Bertolini, and R. Venerucci, *Congruences between modular forms and the Birch and Swinnerton-Dyer conjecture*, Elliptic curves, modular forms and Iwasawa theory, Springer Proc. Math. Stat., vol. 188, Springer, Cham, 2016, pp. 1–31, DOI 10.1007/978-3-319-45032-2_1. MR3629647

[4] M. Bertolini, H. Darmon, and K. Prasanna, *Generalized Heegner cycles and p-adic Rankin L-series*, Duke Math. J. **162** (2013), no. 6, 1033–1148, DOI 10.1215/00127094-2142056. With an appendix by Brian Conrad. MR3053566

[5] M. Bertolini, H. Darmon, and V. Rotger, *Beilinson-Flach elements and Euler systems II: the Birch-Swinnerton-Dyer conjecture for Hasse-Weil-Artin L-series*, J. Algebraic Geom. **24** (2015), no. 3, 569–604, DOI 10.1090/S1056-3911-2015-00675-0. MR3344765

[6] M. Bhargava and C. Skinner, *A positive proportion of elliptic curves over $\mathbb{Q}$ have rank one*, J. Ramanujan Math. Soc. **29** (2014), no. 2, 221–242, DOI 10.1214/14-sts471. MR3237733

[7] A. Burungale, F. Castella, C. Skinner and Y. Tian, *p-converse to a theorem of Gross–Zagier and Kolyvagin: CM elliptic curves over totally real fields*, preprint 2021.

[8] A. Burungale, F. Castella, C. Skinner and Y. Tian, *p^∞-Selmer groups and rational points on CM elliptic curves*. Available at `https://web.math.ucsb.edu/~castella/CM-Q.pdf`, 2021.

[11]Especially the new Linde Hall!

[9] M. Bhargava and A. Shankar, *The average size of the 5-Selmer group of elliptic curves is* 6, *and the average rank is less than* 1, preprint, arXiv:1312.7859.

[10] M. Bhargava, C. Skinner and W. Zhang, *A majority of elliptic curves over* $\mathbb{Q}$ *satisfy the Birch and Swinnerton-Dyer conjecture*, preprint, arXiv:1407.1826.

[11] B. J. Birch, *Elliptic curves over Q: A progress report*, 1969 Number Theory Institute (Proc. Sympos. Pure Math., Vol. XX, State Univ. New York, Stony Brook, N.Y., 1969), Amer. Math. Soc., Providence, R.I., 1971, pp. 396–400. MR0314845

[12] B. J. Birch and H. P. F. Swinnerton-Dyer, *Notes on elliptic curves. I*, J. Reine Angew. Math. **212** (1963), 7–25, DOI 10.1515/crll.1963.212.7. MR146143

[13] B. J. Birch and H. P. F. Swinnerton-Dyer, *Notes on elliptic curves. II*, J. Reine Angew. Math. **218** (1965), 79–108, DOI 10.1515/crll.1965.218.79. MR179168

[14] C. Breuil, B. Conrad, F. Diamond, and R. Taylor, *On the modularity of elliptic curves over* **Q***: wild 3-adic exercises*, J. Amer. Math. Soc. **14** (2001), no. 4, 843–939, DOI 10.1090/S0894-0347-01-00370-8. MR1839918

[15] D. Bump, S. Friedberg, and J. Hoffstein, *Nonvanishing theorems for L-functions of modular forms and their derivatives*, Invent. Math. **102** (1990), no. 3, 543–618, DOI 10.1007/BF01233440. MR1074487

[16] Burungale, A. *The even parity Goldfeld conjecture: Congruent number elliptic curves*, J. Number Theory. Posted online June 29, 2021. DOI 10.1016/j.jnt.2021.05.001.

[17] A. A. Burungale, *Non-triviality of generalised Heegner cycles over anticyclotomic towers: a survey*, p-adic aspects of modular forms, World Sci. Publ., Hackensack, NJ, 2016, pp. 279–306. MR3587960

[18] A. A. Burungale and D. Disegni, *On the non-vanishing of p-adic heights on CM abelian varieties, and the arithmetic of Katz p-adic L-functions*, Ann. Inst. Fourier (Grenoble) **70** (2020), no. 5, 2077–2101. MR4245607

[19] A. Burungale, F. Castella, C. Skinner and Y. Tian, *p-converse to a theorem of Gross–Zagier and Kolyvagin: CM elliptic curves over totally real fields*, in preparation.

[20] A. Burungale, C. Skinner and Y. Tian, *Elliptic curves and Beilinson–Kato elements: rank one aspects*, preprint 2020.

[21] A. Burungale, C. Skinner and Y. Tian, *An Euler system for an elliptic newform over an imaginary qudratic field*, in progress.

[22] A. Burungale, C. Skinner and Y. Tian, *p-converse to a theorem of Gross–Zagier, Kolyvagin and Rubin, II, in progress.*

[23] A. A. Burungale and Y. Tian, *p-converse to a theorem of Gross-Zagier, Kolyvagin and Rubin*, Invent. Math. **220** (2020), no. 1, 211–253, DOI 10.1007/s00222-019-00929-7. MR4071412

[24] A. Burungale and Y. Tian, *A rank zero p-converse to a theorem of Gross–Zagier, Kolyvagin and Rubin*, preprint 2019.

[25] J. W. S. Cassels, *Diophantine equations with special reference to elliptic curves*, J. London Math. Soc. **41** (1966), 193–291, DOI 10.1112/jlms/s1-41.1.193. MR199150

[26] F. Castella, *p-adic heights of Heegner points and Beilinson-Flach classes*, J. Lond. Math. Soc. (2) **96** (2017), no. 1, 156–180, DOI 10.1112/jlms.12058. MR3687944

[27] F. Castella, *On the p-part of the Birch-Swinnerton-Dyer formula for multiplicative primes*, Camb. J. Math. **6** (2018), no. 1, 1–23, DOI 10.4310/CJM.2018.v6.n1.a1. MR3786096

[28] F. Castella and X. Wan, *Perrin-Riou's main conjecture for elliptic curves at supersingular primes*, preprint, arXiv:1607.02019.

[29] F. Castella, M. Çiperiani, C. Skinner and F. Sprung, *On the Iwasawa main conjectures for modular forms at non-ordinary primes*, preprint, arXiv:1804.10993.

[30] F. Castella, G. Grossi, J. Lee and C. Skinner, *On the Iwasawa theory of rational elliptic curves at Eisenstein primes*, preprint, arXiv:2008.02571.

[31] J. Coates and A. Wiles, *On the conjecture of Birch and Swinnerton-Dyer*, Invent. Math. **39** (1977), no. 3, 223–251, DOI 10.1007/BF01402975. MR463176

[32] J. Cremona, *The elliptic curve database for conductors to 130000*, Algorithmic number theory, Lecture Notes in Comput. Sci., vol. 4076, Springer, Berlin, 2006, pp. 11–29, DOI 10.1007/11792086_2. MR2282912

[33] H. Darmon and V. Rotger, *Diagonal cycles and Euler systems II: The Birch and Swinnerton-Dyer conjecture for Hasse-Weil-Artin L-functions*, J. Amer. Math. Soc. **30** (2017), no. 3, 601–672, DOI 10.1090/jams/861. MR3630084

[34] D. Goldfeld, *Sur les produits partiels eulériens attachés aux courbes elliptiques* (French, with English summary), C. R. Acad. Sci. Paris Sér. I Math. **294** (1982), no. 14, 471–474. MR679556

[35] D. Goldfeld, *Conjectures on elliptic curves over quadratic fields*, Number theory, Carbondale 1979 (Proc. Southern Illinois Conf., Southern Illinois Univ., Carbondale, Ill., 1979), Lecture Notes in Math., vol. 751, Springer, Berlin, 1979, pp. 108–118. MR564926

[36] R. Greenberg and V. Vatsal, *On the Iwasawa invariants of elliptic curves*, Invent. Math. **142** (2000), no. 1, 17–63, DOI 10.1007/s002220000080. MR1784796

[37] B. H. Gross and D. B. Zagier, *Heegner points and derivatives of L-series*, Invent. Math. **84** (1986), no. 2, 225–320, DOI 10.1007/BF01388809. MR833192

[38] B. H. Gross, *Lectures on the conjecture of Birch and Swinnerton-Dyer*, Arithmetic of L-functions, IAS/Park City Math. Ser., vol. 18, Amer. Math. Soc., Providence, RI, 2011, pp. 169–209, DOI 10.1090/pcms/018/08. MR2882691

[39] K. Heegner, *Diophantische Analysis und Modulfunktionen* (German), Math. Z. **56** (1952), 227–253, DOI 10.1007/BF01174749. MR53135

[40] B. Howard, *The Heegner point Kolyvagin system*, Compos. Math. **140** (2004), no. 6, 1439–1472, DOI 10.1112/S0010437X04000569. MR2098397

[41] D. Jetchev, C. Skinner, and X. Wan, *The Birch and Swinnerton-Dyer formula for elliptic curves of analytic rank one*, Camb. J. Math. **5** (2017), no. 3, 369–434, DOI 10.4310/CJM.2017.v5.n3.a2. MR3684675

[42] K. Kato, *Lectures on the approach to Iwasawa theory for Hasse-Weil L-functions via B_{dR}. I*, Arithmetic algebraic geometry (Trento, 1991), Lecture Notes in Math., vol. 1553, Springer, Berlin, 1993, pp. 50–163, DOI 10.1007/BFb0084729. MR1338860

[43] K. Kato, *p-adic Hodge theory and values of zeta functions of modular forms* (English, with English and French summaries), Astérisque **295** (2004), ix, 117–290. Cohomologies p-adiques et applications arithmétiques. III. MR2104361

[44] N. M. Katz and P. Sarnak, *Zeroes of zeta functions and symmetry*, Bull. Amer. Math. Soc. (N.S.) **36** (1999), no. 1, 1–26, DOI 10.1090/S0273-0979-99-00766-1. MR1640151

[45] G. Kings, D. Loeffler, and S. L. Zerbes, *Rankin-Eisenstein classes and explicit reciprocity laws*, Camb. J. Math. **5** (2017), no. 1, 1–122, DOI 10.4310/CJM.2017.v5.n1.a1. MR3637653

[46] S. Kobayashi, *The p-adic Gross-Zagier formula for elliptic curves at supersingular primes*, Invent. Math. **191** (2013), no. 3, 527–629, DOI 10.1007/s00222-012-0400-9. MR3020170

[47] V. A. Kolyvagin, *Euler systems*, The Grothendieck Festschrift, Vol. II, Progr. Math., vol. 87, Birkhäuser Boston, Boston, MA, 1990, pp. 435–483. MR1106906

[48] Y. Liu, S. Zhang, and W. Zhang, *A p-adic Waldspurger formula*, Duke Math. J. **167** (2018), no. 4, 743–833, DOI 10.1215/00127094-2017-0045. MR3769677

[49] B. Mazur, *Modular curves and the Eisenstein ideal*, Inst. Hautes Études Sci. Publ. Math. **47** (1977), 33–186 (1978). With an appendix by Mazur and M. Rapoport. MR488287

[50] B. Mazur, *Modular curves and arithmetic*, Proceedings of the International Congress of Mathematicians, Vol. 1, 2 (Warsaw, 1983), PWN, Warsaw, 1984, pp. 185–211. MR804682

[51] L. J. Mordell, *On the rational solutions of the indeterminate equations of the third and fourth degrees*, Proc. Cambridge Phil. Soc. (1922), 21, 179–192.

[52] J. Park, B. Poonen, J. Voight, and M. M. Wood, *A heuristic for boundedness of ranks of elliptic curves*, J. Eur. Math. Soc. (JEMS) **21** (2019), no. 9, 2859–2903, DOI 10.4171/JEMS/893. MR3985613

[53] B. Perrin-Riou, *Fonctions L p-adiques, théorie d'Iwasawa et points de Heegner* (French, with English summary), Bull. Soc. Math. France **115** (1987), no. 4, 399–456. MR928018

[54] D. E. Rohrlich, *On L-functions of elliptic curves and anticyclotomic towers*, Invent. Math. **75** (1984), no. 3, 383–408, DOI 10.1007/BF01388635. MR735332

[55] K. Rubin, *Tate-Shafarevich groups and L-functions of elliptic curves with complex multiplication*, Invent. Math. **89** (1987), no. 3, 527–559, DOI 10.1007/BF01388984. MR903383

[56] K. Rubin, *The "main conjectures" of Iwasawa theory for imaginary quadratic fields*, Invent. Math. **103** (1991), no. 1, 25–68, DOI 10.1007/BF01239508. MR1079839

[57] K. Rubin, *p-adic L-functions and rational points on elliptic curves with complex multiplication*, Invent. Math. **107** (1992), no. 2, 323–350, DOI 10.1007/BF01231893. MR1144427

[58] J. H. Silverman and J. T. Tate, *Rational points on elliptic curves*, 2nd ed., Undergraduate Texts in Mathematics, Springer, Cham, 2015, DOI 10.1007/978-3-319-18588-0. MR3363545

[59] C. Skinner, *Main conjectures and modular forms*, Current developments in mathematics, 2004, Int. Press, Somerville, MA, 2006, pp. 141–161. MR2459294

[60] C. Skinner and E. Urban, *The Iwasawa main conjectures for* GL_2, Invent. Math. **195** (2014), no. 1, 1–277, DOI 10.1007/s00222-013-0448-1. MR3148103
[61] C. Skinner, *Multiplicative reduction and the cyclotomic main conjecture for* GL_2, Pacific J. Math. **283** (2016), no. 1, 171–200, DOI 10.2140/pjm.2016.283.171. MR3513846
[62] C. Skinner, *A converse to a theorem of Gross, Zagier, and Kolyvagin*, Ann. of Math. (2) **191** (2020), no. 2, 329–354, DOI 10.4007/annals.2020.191.2.1. MR4076627
[63] C. Skinner and W. Zhang, *Indivisibility of Heegner points in the multiplicative case*, arXiv:1407.1099.
[64] C. Skinner, *Lectures on the Iwasawa theory of elliptic curves*, notes, 2018.
[65] A. Smith, *2^∞-Selmer groups, 2^∞-class groups, and Goldfeld's conjecture*, preprint, arXiv:1702.02325.
[66] J. Tate, *On the conjectures of Birch and Swinnerton-Dyer and a geometric analog [see MR1610977]*, Dix exposés sur la cohomologie des schémas, Adv. Stud. Pure Math., vol. 3, North-Holland, Amsterdam, 1968, pp. 189–214. MR3202555
[67] J. T. Tate, *The arithmetic of elliptic curves*, Invent. Math. **23** (1974), 179–206, DOI 10.1007/BF01389745. MR419359
[68] Y. Tian, *Congruent numbers and Heegner points*, Camb. J. Math. **2** (2014), no. 1, 117–161, DOI 10.4310/CJM.2014.v2.n1.a4. MR3272014
[69] X. Wan, *Iwasawa main conjecture for Rankin-Selberg p-adic L-functions*, Algebra Number Theory **14** (2020), no. 2, 383–483, DOI 10.2140/ant.2020.14.383. MR4195651
[70] X. Wan, *Heegner Point Kolyvagin System and Iwasawa Main Conjecture*, Acta Math. Sin. (Engl. Ser.) **37** (2021), no. 1, 104–120, DOI 10.1007/s10114-021-8355-7. MR4204538
[71] X. Wan, *Iwasawa Main Conjecture for Supersingular Elliptic Curves and BSD conjecture*, preprint, arXiv:1411.6352.
[72] A. Wiles, *The Birch and Swinnerton-Dyer conjecture*, The millennium prize problems, Clay Math. Inst., Cambridge, MA, 2006, pp. 31–41. MR2238272
[73] A. Wiles, *Foreword [In honour of John H. Coates on the occasion of his sixtieth birthday]*, Doc. Math. **Extra Vol.** (2006), 3–4. MR2290608
[74] A. Wiles, *Modular elliptic curves and Fermat's last theorem*, Ann. of Math. (2) **141** (1995), no. 3, 443–551, DOI 10.2307/2118559. MR1333035
[75] R. Venerucci, *On the p-converse of the Kolyvagin-Gross-Zagier theorem*, Comment. Math. Helv. **91** (2016), no. 3, 397–444, DOI 10.4171/CMH/390. MR3541715
[76] X. Yuan, S.-W. Zhang, and W. Zhang, *The Gross-Zagier formula on Shimura curves*, Annals of Mathematics Studies, vol. 184, Princeton University Press, Princeton, NJ, 2013. MR3237437
[77] W. Zhang, *Selmer groups and the indivisibility of Heegner points*, Camb. J. Math. **2** (2014), no. 2, 191–253, DOI 10.4310/CJM.2014.v2.n2.a2. MR3295917
[78] W. Zhang, *The Birch–Swinnerton-Dyer conjecture and Heegner points: a survey*, Current developments in mathematics 2013, Int. Press, Somerville, MA, 2014, pp. 169–203. MR3307716

California Institute of Technolology, 1200 E California Blvd, Pasadena California 91125; and The University of Texas at Austin, Austin, Texas 78712
Email address: ashayburungale@gmail.com

Department of Mathematics, Princeton University, Princeton New Jersey 08544-1000
Email address: cmcls@princeton.edu

Academy of Mathematics and Systems Science, Morningside center of Mathematics, Chinese Academy of Sciences, Beijing 100190; and School of Mathematical Sciences, University of Chinese Academy of Sciences, Beijing 10049
Email address: ytian@math.ac.cn

Proceedings of Symposia in Pure Mathematics
Volume **104**, 2021
https://doi.org/10.1090/pspum/104/01872

Bounding ramification by covers and curves

Hélène Esnault and Vasudevan Srinivas

Abstract. We prove that $\bar{\mathbb{Q}}_\ell$-local systems of bounded rank and ramification on a smooth variety X defined over an algebraically closed field k of characteristic $p \neq \ell$ are tamified outside of codimension 2 by a finite separable cover of bounded degree. In rank one, there is a curve which preserves their monodromy. There is a curve defined over the algebraic closure of a purely transcendental extension of k of finite degree which fulfills the Lefschetz theorem.

1. Introduction

The notions of fundamental group introduced by Poincaré and Riemann for topological manifolds and of Galois group of field extensions introduced by Galois were unified by Grothendieck's theory of étale fundamental groups. For complex varieties, by the Riemann existence theorem, the étale fundamental group is the profinite completion of the topological one, so both groups share many properties. It is no longer the case for a variety defined in characteristic $p > 0$, due to wild ramification. In this note we show how to give various 'upper bounds' of ramification.

Let X be a smooth connected variety of finite type over an algebraically closed field k of characteristic $p > 0$. Let $\mathcal{F}$ be a $\bar{\mathbb{Q}}_\ell$ local system. It is defined by a continuous representation $\rho : \pi_1^{\text{ét}}(X,x) \to GL_r(\bar{\mathbb{Q}}_\ell)$ from the étale fundamental group $\pi_1^{\text{ét}}(X,x)$ based in one geometric point x. Choosing a lattice stabilized by ρ defines a residual representation $\bar{\rho} : \pi_1^{\text{ét}}(X,x) \to GL_r(\bar{\mathbb{F}}_\ell)$ which by continuity has values in $GL_r(\mathbb{F})$ for a finite extension $\mathbb{F} \supset \mathbb{F}_\ell$. It is well defined modulo semi-simplification. The Galois étale cover of $X_{\bar{\rho}} \to X$ defined by $\bar{\rho}$ has the property that the pullback $\mathcal{F}|_{X_{\bar{\rho}}}$ of $\mathcal{F}$ to $X_{\bar{\rho}}$ is tame. We say that it *tamifies* $\mathcal{F}$. If we bound the ramification of $\mathcal{F}$ by an effective Cartier divisor D supported on $\bar{X} \setminus X$ where $j : X \hookrightarrow \bar{X}$ is a normal compactification in the sense of [**EK11**, Defn. 4.6], [**EK12**, 2.1], and we bound r, the degree of $X_{\bar{\rho}} \to X$ is not bounded. Indeed $D = 0$ is equivalent to ρ factoring over the tame quotient $\pi_1^{\text{ét,t}}(X,x)$ of $\pi_1^{\text{ét}}(X,x)$. Already in this case, the degree is not bounded. For example, the pro-ℓ-completion of the tame fundamental group of $\mathbb{P}^1 \setminus \{0,1,\infty\}$ is by Grothendieck's specialization

2020 *Mathematics Subject Classification.* Primary 14F35, 11S15.

The first author was supported during part of the preparation of the article by the Institute for Advanced Study, USA. The second author was supported by a J. C. Bose Fellowship of the Department of Science and Technology, India and the Department of Atomic Energy, Government of India, under project number 12-R&D-TFR-RTI4001.

theorem the pro-ℓ-completion of the free group on two letters, which is represented in $GL_2(\mathbb{F}_{\ell^n})$ for unbounded n.

We assume that X is quasi-projective. Let $\pi : Y \to X$ be a normal connected finite cover. We say that π *tamifies* $\mathcal{F}$ *outside of codimension* 2 if there is a normal compactification $Y \hookrightarrow \bar{Y}$ and a closed subset $\Sigma \subset \bar{Y}$ of codimension ≥ 2 such that $\mathcal{F}|_Y$ is tame along $(\bar{Y} \setminus Y) \setminus \Sigma$. We prove the following theorem.

THEOREM 1.1 (Boundedness theorem). *Given j, X, r, D, there is a natural number M such that for any $\mathcal{F}$ of rank and ramification bounded by (r, D), there is a finite generically étale cover $X_{\mathcal{F}} \to X$ of X of degree $\leq M$ which tamifies $\mathcal{F}$ outside of codimension* 2.

If k is a finite field, by L. Lafforgue's theorem in dimension 1 and Deligne's theorem [**EK12**, Thm.1.1] in general, up to twist by a character of k there are finitely many isomorphism classes of simple $\bar{\mathbb{Q}}_\ell$ local systems $\mathcal{F}$ in rank and ramification bounded by (r, D). This is analogous to Hermite-Minkowski theorem in number theory according to which there finitely many isomorphism classes of extensions of a given number field with bounded degree and bounded discriminant. As a consequence, there is a smooth curve $C \hookrightarrow X$ such that $\mathcal{F}|_C$ remains irreducible in bounded (r, D) ([**EK12**, Prop.B.1], see other references therein). Deligne asked in [**Del16**] whether over an algebraically closed field, there is a smooth curve $C \hookrightarrow X$ such that for any $\mathcal{F}$ with bounded (r, D), $\mathcal{F}|_C$ keeps the same monodromy group. To understand the question, recall that by Drinfeld's theorem [**Dri12**, Prop. C2] there is a full Lefschetz theorem for the tame fundamental group, that is there is such a curve for which the functoriality homomorphism $\pi_1^{\text{ét}}(C, x) \to \pi_1^{\text{ét,t}}(X, x)$ is surjective. On the other hand, as is well known, there is no Lefschetz theorem for $\pi_1^{\text{ét,t}}(X, x)$ (see e.g. [**Esn17**, Lem 5.4]). So Deligne's problem asks whether we can save the Lefschetz theorem bounding D, but also in addition r (this condition being unnecessary for $D = 0$) without any arithmeticity assumption. For $r = 1$ this is true if j is a good normal crossings compactification by Kerz-Saito's theorem, see [**KS14**, Thm. 1.1] and erratum to come. We give a complete answer to Deligne's question in rank one.

THEOREM 1.2 (Good curve in rank one). *There is a smooth $C \hookrightarrow X$ such that for any $\mathcal{F}$ of rank one with ramification bounded by D, $\mathcal{F}|_C$ keeps the same monodromy group.*

In fact, a slightly more precise statement is true, see Remark 5.9. Theorem 1.1 has the following corollary.

COROLLARY 1.3. *Given j, X, r, D, there is a natural number M such that for any $\mathcal{F}$ of rank and ramification bounded by (r, D), there is a smooth curve $C \hookrightarrow X$ and a finite generically étale cover $C' \to C$ of degree $\leq M$ such that $\mathcal{F}|_C$ has the same monodromy group as $\mathcal{F}$, C' is integral, and $\mathcal{F}|_{C'}$ is tame and has the same monodromy group as $\mathcal{F}|_{X_{\mathcal{F}}}$.*

Theorem 1.1, Corollary 1.3 and Theorem 1.2 give some evidence for a general positive answer to Deligne's problem.

The proof of Theorem 1.1 consists in globalizing the arguments used to prove [**EKS19**, Prop.2]. To this aim, by a standard argument we reduce the problem to $X \hookrightarrow \bar{X}$ being $\mathbb{A}^d \hookrightarrow \mathbb{P}^d$. There, one tool used is Harbater-Katz-Gabber local to global extension [**Kat86**, Thm.1.4.1], [**Har80**]. The proof of Corollary 1.3

relies on Drinfeld's tame Lefschetz theorem *loc. cit.* The proof of Theorem 1.2 relies on the same statement if $\bar{X}$ is smooth, together with some finiteness theorem on Frobenius invariant submodules of some local cohomology groups resting on [**Smi94**, Rmk. 4.4].

Finally we prove that a Lefschetz theorem without boundedness exists if we allow the wished curve C to be defined over the algebraic closure of a purely transcendental extension of k of finite degree, see Remark 6.1. The proof relies purely on the classical Bertini theorem and ought to be well known. We wrote it as we could not find a reference.

2. Elementary properties of $\mathcal{S}(j, r, D)$

Let k be an algebraically closed field of characteristic $p > 0$, X be a smooth connected variety of finite type over k, $j : X \hookrightarrow \bar{X}$ be a normal compactification, D be an effective Cartier divisor supported on $\bar{X} \setminus X$, r be a positive natural number. Recall (see [**EK11**, Defn. 4.6], used in [**EK12**, 2.1]) that an ℓ-adic local system $\mathcal{F}$ has *ramification bounded by* D if for any morphism of a smooth connected projective curve $\bar{\iota} : \bar{C} \to \bar{X}$ with $\iota^{-1}(X) \neq \emptyset$, the pullback to $\mathcal{F}$ to $\bar{\iota}^{-1}(X) =: C$ has Swan conductor bounded by $\bar{\iota}^{-1}D$ on $\bar{C}$.

NOTATION 2.1. We denote by $\mathcal{S}(j, r, D)$ the set of isomorphism classes of $\bar{\mathbb{Q}}_\ell$ local systems of rank $\leq r$ with ramification bounded by D.

We denote by $\pi_1^{\text{ét}}(X, x)$ the étale fundamental group based at a geometric point x. If there is no confusion, we simply write $\pi_1^{\text{ét}}(X)$.

2.1. Restriction to an open. Let $U \hookrightarrow X$ be a dense open subscheme, $j' : U \to \bar{X}$ the composition with j.

LEMMA 2.2. *Restriction to U induces an injective map*

$$\mathcal{S}(j, r, D) \hookrightarrow \mathcal{S}(j', r, D).$$

PROOF. This is an immediate consequence of $\pi_1^{\text{ét}}(U) \to \pi_1^{\text{ét}}(X)$ being surjective. □

2.2. Changing the compactification. Let $j_i : X \hookrightarrow \bar{X}_i$ be two normal compactifications, and D_1 be an effective Cartier divisor supported on $\bar{X}_1 \setminus X$.

LEMMA 2.3. *There is an effective divisor D_2 supported on $\bar{X}_2 \setminus X$ such that $\mathcal{S}(j_1, r, D_1) \subset \mathcal{S}(j_2, r, D_2)$.*

PROOF. Let $j_3 : X \hookrightarrow \bar{X}_3$ be a normal compactification which dominates j_i, $i = 1, 2$. Then for any curve $C \to X$, $\bar{C} \to \bar{X}_i$ factors through $\bar{C} \to \bar{X}_3 \to X_i$. We conclude that

$$\mathcal{S}(j_1, r, D_1) = \mathcal{S}(j_3, r, D_1 \times_{\bar{X}_1} \bar{X}_3)$$

thus for any D_2 such that $D_2 \times_{\bar{X}_2} \bar{X}_3 \supset D_1 \times_{\bar{X}_1} \bar{X}_3$ one has

$$\mathcal{S}(j_1, r, D_1) \subset \mathcal{S}(j_2, r, D_2).$$

□

2.3. Projection. Let $\bar{\pi} : \bar{X} \to \bar{Y}$ be a finite morphism of degree δ such that the restriction $\pi : X = \bar{\pi}^{-1}Y \to Y$ is étale, where $Y \hookrightarrow \bar{Y}$ is dense open. Let $j_Y : Y \hookrightarrow \bar{Y}$ be the open embedding.

LEMMA 2.4. 1) *There is an effective Cartier divisor D' supported on $\bar{Y} \setminus Y$ such that pushdown to Y induces a map*

$$\mathcal{S}(j, r, D) \to \mathcal{S}(j_Y, r\delta, D').$$

2) *Given an effective divisor D_Y supported on $\bar{Y} \setminus Y$, pullback induces a map $\mathcal{S}(j_Y, r, D_Y) \to \mathcal{S}(j, r, \bar{\pi}^* D_Y)$.*

PROOF. We prove 1). Let $\iota_Y : \bar{C}_Y \to \bar{Y}$ be a morphism of a smooth projective curve. Write $\pi^{-1}C_Y = C = \sqcup_i C_i$ for the union of irreducible components, which are disjoint as π is étale. This defines a commutative square

$$\begin{array}{ccc} C_i & \xrightarrow{\iota_i} & X \\ {\scriptstyle \pi|_{C_i}}\downarrow & & \downarrow{\scriptstyle \pi} \\ C_Y & \xrightarrow{\iota_Y} & Y \end{array}$$

where $\pi|_{C_i}$ has degree δ_i with $\sum_i \delta_i = \delta$. One has

$$\iota_Y^* \pi_* \mathcal{F} = \oplus_i (\pi|_{C_i})_* \iota_i^* \mathcal{F}$$

By [**EK12**, Lem.3.2,Prop.3.9], there is an effective Cartier divisor Δ supported on $\bar{Y} \setminus Y$ such that $\pi_* \mathbb{F}_\ell$ has ramification bounded by Δ and $(\pi|_{C_i})_* \mathbb{F}_\ell$ by $\iota_Y^* \Delta$. So by the Grothendieck-Ogg-Shafarevich formula applied to $\iota_i^* \mathcal{F}$ and $(\pi|_{C_i})_* \iota_i^* \mathcal{F}$, the Swan conductor of $(\pi|_{C_i})_* \iota_i^* \mathcal{F}$ is bounded by $r\delta_i \Delta + (\pi|_{C_i})_* D$. It is then enough to set $D' = \delta(r\Delta + \pi_* D)$ and prove 1). As for 2), this is immediate, just writing $\bar{C} \xrightarrow{\bar{\iota}} \bar{X} \xrightarrow{\bar{\pi}} \bar{Y}$. This finishes the proof of 2). □

3. Reduction of Theorem 1.1 to the case $X = \mathbb{A}^d$

The aim of this section is to prove

PROPOSITION 3.1. *If the theorem is true for $\mathbb{A}^d \hookrightarrow \mathbb{P}^d$ and any r, D, it is true for any $X \hookrightarrow \bar{X}$ with X of dimension d and any r, D.*

PROOF. By Lemma 2.2 we may assume that X is affine and admits an étale morphism $X \to \mathbb{A}^d$ as any nonsingular variety has a basis of such Zariski open subsets. We apply [**Ach17**, Prop.5.2] to conclude that there is then a finite étale map $\pi : X \to \mathbb{A}^d$. By Lemma 2.3 we may assume that $\bar{X}$ is the normalisation of $\mathbb{P}^d$ in $k(X)$. We apply Lemma 2.4 1) to j_Y being $\mathbb{A}^d \hookrightarrow \mathbb{P}^d$. Assume there is a natural number M such that for any $\mathcal{G} \in \mathcal{S}(j_Y, r\delta, D')$, there is a finite separable morphism $Y_\mathcal{G} \to Y = \mathbb{A}^d$ such that $\mathcal{G}|_{Y_\mathcal{G}}$ is tame outside of codimension 2. For $\mathcal{G} = \pi_* \mathcal{F}$, $\mathcal{F} \in \mathcal{S}(j, r, D)$, as $\pi^* \mathcal{G}$ surjects to $\mathcal{F}$, we conclude that $\mathcal{F}|_{Y_\mathcal{G} \times_{\mathbb{A}^d} X}$ is tame outside of codimension 2 as well. On the other hand, the morphism $Y_\mathcal{G} \times_{\mathbb{A}^d} X \to X$ is separable and of degree $\leq M$, so is a connected component of it. We choose a random one which we define to be $X_\mathcal{F}$. This finishes the proof. □

4. Proof of Theorem 1.1 and Corollary 1.3

The aim of this section is to prove Theorem 1.1 and Corollary 1.3. We start with a simple lemma. If G is a finite group we denote by $|G|$ its cardinality. For a natural number n, we set $n! = \prod_{i=1}^{n} i$. Let p be a prime number. If A, B are subgroups of G, where B normalizes A, we write $A \cdot B$ for the subgroup of G consisting of the image of $A \times B$ under $(a, b) \mapsto ab$.

LEMMA 4.1. *Let*

$$1 \to K \to G \to G/K \to 1$$

be an exact sequence of groups, and $H \subset G$ be a subgroup of G with the following properties:

(i) *The composite map $H \subset G \to G/K$ is an isomorphism (so $G = K \cdot H$).*
(ii) *K is finite, and Q is the unique p-Sylow subgroup of K.*
(iii) *K/Q is cyclic.*

Then there exists a normal subgroup $N \lhd G$ such that

1) *$N \cap Q = \{1\}$;*
2) *$|G/N| \leq (|Q|(|Q|!))!$.*

PROOF. By (ii) Q is a characteristic subgroup of K, thus also a normal subgroup of G, as K is normal in G. Let

$$Z_G(Q) = \mathrm{Ker}\big(G \to \mathrm{Aut}(Q),\ g \mapsto [x \mapsto gxg^{-1}]\big)$$

be the centraliser of Q in G. It is a normal subgroup of G of index $\leq |Q|!$. By (iii) applying the Schur-Zassenhaus theorem, $K = Q \cdot K'$ where K' is finite cyclic of order prime to p (the decomposition need not be unique). Thus

$$K \cap (Q \cdot Z_G(Q)) = Q \cdot (K \cap Z_G(Q)) = Q \times R,$$

where R is cyclic, of order prime to p, and now this direct product decomposition is unique (here $R \subset K \cap Z_G(Q)$ is isomorphic to the image of $K \cap Z_G(Q)$ in the cyclic group K/Q). As $Z_G(Q)$ is normal in G, $K \cap (Q \cdot Z_G(Q)) = Q \times R$ is normal in G, in particular H acts by conjugation on it, respecting the direct product decomposition. Hence $R \cdot H \subset G$ is a subgroup, by (i) of index equal to the index of $R \subset K$, which is at most $|Q|(|Q|!)$. Moreover $Q \cap R \cdot H = \{1\}$.

$$N = \bigcap_{g \in G} g(R \cdot H)g^{-1}$$

is a normal subgroup of G, of index at most $(|Q|(|Q|!))!$, and has trivial intersection with Q. □

PROOF OF THEOREM 1.1. By Proposition 3.1 we may assume that $j : X = \mathbb{A}^d \hookrightarrow \bar{X} = \mathbb{P}^d$. We set $Z = \mathbb{P}^d \setminus \mathbb{A}^d$. The local ring $\mathcal{O}_{\bar{X},Z}$ is isomorphic to the local ring $\mathcal{O}_{\mathbb{A}^1/k(Z),0}$. The choice of a parameter t on $\mathbb{A}^1$ identifies it with $k(Z)[t]_{(t)}$. The inclusion $\mathcal{K} \cong k(Z)(t) \subset \hat{\mathcal{K}} \cong k(Z)((t))$ of the field of fraction of $\mathcal{O}_{\bar{X},Z}$ into its completion defines $\hat{X} \to \mathbb{G}_m/k(Z)$ where $\hat{X} = \mathrm{Spec}\ \hat{\mathcal{K}}$.

Choosing an algebraic closure $\overline{\hat{\mathcal{K}}}$ defines a geometric point $x \to \hat{X} \to \mathbb{G}_m/k(Z)$. One has the diagram of exact sequences

$$\begin{array}{ccccccccc} & & & & 1 & & 1 & & \\ & & & & \downarrow & & \downarrow & & \\ 1 & \longrightarrow & P & \longrightarrow & I & \longrightarrow & I^{\text{tame}} & \longrightarrow & 1\\ & & \downarrow{\scriptstyle =} & & \downarrow & & \downarrow & & \\ 1 & \longrightarrow & P & \longrightarrow & \pi_1(\hat{X},x) & \longrightarrow & \pi_1^{\text{tame}}(\hat{X},x) & \longrightarrow & 1 \\ & & & & \downarrow & & \downarrow & & \\ & & & & \pi_1(k(Z),x) & \xrightarrow{=} & \pi_1(k(Z),x) & & \\ & & & & \downarrow & & \downarrow & & \\ & & & & 1 & & 1 & & \end{array}$$

where the groups on the right are the tame quotients of I and $\pi_1(\hat{X},x)$.

By [**Kat86**, Thm.1.4.1] one has a splitting

$$\begin{array}{ccc}\pi_1(\hat{X},x) & \longrightarrow & \pi_1(\mathbb{G}_m/k(Z),x)\\ & \searrow^{\cong} & \downarrow \\ & & \pi_1(\mathbb{G}_m/k(Z),x)[sp]\end{array} \tag{4.1}$$

where $[sp]$ stands for 'special', and is defined by the property that finite quotients have a unique p-Sylow. On the other hand, the choice of the rational point $\mathrm{Spec}(k(Z)) \to \mathbb{G}_m/k(Z)$ defined by $t = 1$ yields a splitting $\pi_1(k(Z),x) \to \pi_1(\mathbb{G}_m/k(Z),x)$, thus a splitting $\pi_1(k(Z),x) \to \pi_1(\mathbb{G}_m/k(Z),x)[sp]$, thus a splitting

$$\pi_1(k(Z),x) \xrightarrow{[t]} \pi_1(\hat{X},x).$$

For $\hat{\mathcal{F}}$ a $\bar{\mathbb{Q}}_\ell$-local system on $\hat{X}$, we choose a representative $\rho : \pi_1(\hat{X},x) \to GL_r(\bar{\mathbb{Q}}_\ell)$. The residual representation $\bar{\rho} : \pi_1(\hat{X},x) \to GL_r(\bar{\mathbb{F}}_\ell)$ is defined up to semi-simplification and isomorphism. We choose one. As $I \cap \mathrm{Ker}\bar{\rho}$ is normal in $\pi_1(\hat{X},x)$, setting

$$K = \bar{\rho}(I), \ \psi : \pi_1(\hat{X},x) \to G = \pi_1(\hat{X},x)/I \cap \mathrm{Ker}\bar{\rho},$$

one has a commutative diagram of exact sequences

$$\begin{array}{ccccccccc} 1 & \longrightarrow & I & \longrightarrow & \pi_1(\hat{X},x) & \longrightarrow & \pi_1(k(Z),x) & \longrightarrow & 1 \\ & & \downarrow{\scriptstyle \bar{\rho}} & & \downarrow{\scriptstyle \psi} & \searrow{\scriptstyle \bar{\rho}} & \downarrow{\scriptstyle =} & & \\ 1 & \longrightarrow & K & \longrightarrow & G & \longrightarrow & \pi_1(k(Z),x) & \longrightarrow & 1 \\ & & & & \downarrow & & & & \\ & & & & GL_k(\bar{\mathbb{F}}_\ell) & & & & \end{array}$$

The bottom exact sequence is split by $\psi \circ [t] : \pi_1(k(Z), x) \to G$. We set

$$\bar{\rho}(P) = Q,\ H = \big(\psi \circ [t](\pi_1(k(Z), x))\big) \subset G.$$

We apply Lemma 4.1. Let $N \subset G$ be the normal subgroup of G defined in *loc.cit.*. By 2) the finite Galois cover $\hat{X}_N \to \hat{X}$ defined by the composite $\pi_1(\hat{X}, x) \xrightarrow{\bar{\rho}} G \to G/N$ has degree $\leq (|Q|(|Q|!)!$ and by 1) it has the property that the pullback of $\hat{\mathcal{F}}$ to $\hat{X}_N$ is tame. On the other hand, via (4.1), there is a Galois cover $(\mathbb{G}_m/k(Z))_N \to \mathbb{G}_m/k(Z)$ which restricts to $\hat{X}_N \to \hat{X}$, of the same degree.

We now globalize the construction on $\mathbb{A}^d$. Let $(x_0 = t, x_1, \dots, x_d) \in H^0(\mathbb{P}^d, \mathcal{O}(1))$ be a system of coordinates so $t = 0$ defines $\mathbb{P}^d \setminus \mathbb{A}^d$. The choice of this system of coordinates defines the factorization

$$\hat{X} \to \mathbb{G}_m/k(Z) \to \mathbb{A}^d.$$

The right morphism induces an isomorphism on the field of fractions $\mathcal{K}$. We define $\mathcal{K}_N$ to be the function field of $(\mathbb{G}_m/k(Z))_N$, so that $\mathcal{K} \subset \mathcal{K}_N$ is a finite Galois extension. Let $X_{\mathcal{F}}$ be the normalisation of X in $\mathcal{K}_N$. Then $X_{\mathcal{F}} \to X$ has degree $\leq (|Q||Q|!)!$ and for $\hat{\mathcal{F}}$ being the restriction of $\mathcal{F} \in \mathcal{S}(j, r, D)$, the pullback of $\mathcal{F}$ to $X_{\mathcal{F}}$ is tame. It remains to show that $|Q|$ is bounded if $\hat{\mathcal{F}}$ is the restriction to $\hat{X}$ of $\mathcal{F} \in \mathcal{S}(j, r, D)$.

Let $C \hookrightarrow X$ be a curve in good position with respect to $Z \times_{\bar{X}} \bar{Y}$ in the sense of [**EKS19**, Defn.7] where $\bar{Y} \to \bar{X}$ is the cover associated to the choice of a residual representation of $\mathcal{F}$. Then by Section 4 ($\star$) of *loc. cit.* the image of the inertia at the points of $\bar{C} \setminus C$ of $\bar{\rho}|_C$ is equal to K, thus the image of the wild inertia is equal to Q. We now apply Proposition 2 of *loc.cit.*. This finishes the proof. $\square$

Proof of Corollary 1.3. Let $\mathcal{F} \in \mathcal{S}(j, r, D)$ and $\rho : \pi_1^{\text{ét}}(X, x) \to GL_r(\mathcal{O})$ representing it, where $\mathcal{O}^{\oplus r} \subset \bar{\mathbb{Q}}_\ell$ is a lattice stabilized by ρ. Let $\mathfrak{m} \subset \mathcal{O}$ be the maximal ideal. Then by [**BOU**, Cor. I.6.3.4], for any smooth irreducible $C \hookrightarrow X$ through x, the induced representation $\rho|_C : \pi_1^{\text{ét}}(C, x) \to \pi_1^{\text{ét}}(X, x) \xrightarrow{\rho} GL_r(\mathcal{O})$ has the same image if and only if it has after post-composing with $\pi : GL_r(\mathcal{O}) \to GL_r(\mathcal{O}/\mathfrak{m}^2)$ (see [**EK12**, Lem.B.2, Proof]). By [**Jou83**, Thm.6.3 (iv)], fixing an ample $|\mathcal{H}|$ linear system on $\bar{X}$, there is a dense open subset of $|\mathcal{H}|^{d-1}$ such that for any closed point $(f_1, \dots, f_{d-1})$ in it, the corresponding subscheme $\bar{C} = (f_1) \cap \ldots \cap (f_{d-1})$ is a smooth complete intersection curve in good position with respect to $\bar{X} \setminus X$, its pullback to $W \to X$ is connected, its pullback C' to $\bar{X}_{\mathcal{F}}$ is integral and in good position with respect to $\bar{X}_{\mathcal{F}} \setminus X_{\mathcal{F}}$. Here $W \to X$ is the finite étale cover associated to $\pi \circ \rho$. Thus by [**EK12**, Lem.B.2], $\mathcal{F}|_C$ has the same monodromy group as $\mathcal{F}$. By [**Dri12**, Lem.C.2] applied to C', $\mathcal{F}|_{C'}$, which is tame, has the same monodromy as $\mathcal{F}|_{X_{\mathcal{F}}}$. This finishes the proof. $\square$

5. Rank one

Definition 5.1. If $\mathcal{S}$ is a given family of $\bar{\mathbb{Q}}_\ell$-local systems on X, a smooth connected projective curve $C \subset X$ which is the complete intersection of generically smooth very ample divisors in $\bar{X}$ in good position with respect to $\bar{X} \setminus X$ is called *good for* $\mathcal{S}$ if the restriction map $\mathcal{F} \to \mathcal{F}|_C$ induces an isomorphism on the monodromy groups for any $\mathcal{F} \in \mathcal{S}$.

We rephrase Theorem 1.2:

Theorem 5.2. *Given j, D, there is a good curve for $\mathcal{S}(j, 1, D)$.*

Let n be a positive natural number, and $\mathcal{S}(j,1,D,n) \subset \mathcal{S}(j,1,D)$ be the subset of isomorphism classes of rank one $\mathcal{F}$ with monodromy group $\mathbb{Z}/p^n$.

LEMMA 5.3. *Given j, D, a good curve for $\cup_{\mathbb{N}\ni n\geq 1}\mathcal{S}(j,1,D,n)$ is a good curve for $\mathcal{S}(j,1,D)$.*

PROOF. Let $\mathcal{F} \in \mathcal{S}(j,1,D)$ with underlying continuous representation $\rho : \pi_1^{\text{ét}}(X) \to \mathcal{O}_E^\times$ where $\mathcal{O}_E$ is a finite normal extention of $\mathbb{Z}_\ell$. Let $\mathfrak{m} \subset \mathcal{O}_E$ be the maximal ideal. Let $\bar\rho$ be the residual representation $\bar\rho : \pi_1^{\text{ét}}(X) \to \mathcal{O}_E^\times \to (\mathcal{O}_E/\mathfrak{m})^\times$. Then $\bar\rho(\pi_1^{\text{ét}}(X)) = \mathbb{Z}/p^n\mathbb{Z} \times \mathbb{Z}/N\mathbb{Z}$ for n, N positive natural numbers with $(N,p) = 1$. So $\rho(\pi_1^{\text{ét}}(X)) = \mathbb{Z}/p^n \times G \subset \mathcal{O}_E^\times$ where G is a profinite group of pro-order prime to p. By [**Dri12**, Prop.C2], any smooth connected projective curve $C \subset X$ which is the complete intersection of very ample divisors in $\bar X$ in good position with respect to $\bar X \setminus X$ is good for all tame $\bar{\mathbb{Q}}_\ell$-local systems. Thus any such C is good for $\rho^t : \pi_1^{\text{ét}}(X) \xrightarrow{\rho} \mathbb{Z}/p^n\mathbb{Z} \times G \xrightarrow{\text{projection}} G \subset \mathcal{O}_E^\times$. This proves the lemma. $\square$

LEMMA 5.4. *Given j, D, a good curve for $\mathcal{S}(j,1,D,1)$ is a good curve for $\cup_{\mathbb{N}\ni n\geq 1}\mathcal{S}(j,1,D,n)$.*

PROOF. Let $n \geq 2$ and $\rho : \pi_1^{\text{ét}}(X) \to \mathbb{Z}/p^n\mathbb{Z}$ a surjective representation. Let C be a good curve for $\bar\rho : \pi_1^{\text{ét}}(X) \to \mathbb{Z}/p^n\mathbb{Z} \to \mathbb{Z}/p\mathbb{Z}$. As a homomorphism $K \to \mathbb{Z}/p^n\mathbb{Z}$, where K is any group, is surjective if and only if the composed homomorphism $K \to \mathbb{Z}/p^n\mathbb{Z} \to \mathbb{Z}/p\mathbb{Z}$ is, the lemma follows. $\square$

In order to prove Theorem 5.2 we may assume that X is affine. Let us denote by $F : X \to X$ the absolute Frobenius. For an effective Cartier divisor Δ with support in $\bar X \setminus X$, we define $A(\bar X, \Delta)$ to be the image of $H^0(\bar X, \mathcal{O}(\Delta))$ in $H^1_{\text{ét}}(X, \mathbb{Z}/p\mathbb{Z}) = H^0(X,\mathcal{O})/(F-1)H^0(X,\mathcal{O})$ where this description of $H^1_{\text{ét}}(X,\mathbb{Z}/p\mathbb{Z})$ follows from Artin-Schreier theory.

LEMMA 5.5. *Given a natural number $m \geq 1$, there is a normal generically smooth very ample divisor $\bar H \subset \bar X$ in good position with respect $\bar X \setminus X$ such that the restriction homomorphism $A(\bar X, npD) \to A(\bar H, \bar H \cap npD)$ is an isomorphism for all natural number $1 \leq n \leq m$. There is curve $\bar C \to \bar X$ which is a complete intersection of such H such that $A(\bar X, pnD) \to A(\bar C, \bar C \cdot pnD)$ is an isomorphism for all natural number $1 \leq n \leq m$.*

PROOF. We choose $\bar H$ a normal generically smooth very ample divisor $\bar H \subset \bar X$ in good position with respect $\bar X \setminus X$ such that

$$H^i(\bar X, \mathcal{O}_{\bar X}(-H + nD')) = 0, \text{ for } D' = D \text{ or } D' = pD,\ i = 0,1, \quad \text{and for all } 1 \leq n \leq m.$$

The existence of a normal very ample $\bar H$ in good position follows from [**Sei50**], and Bertini's theorem. The cohomology vanishing property holds for $i = 0,1$ by the Enriques-Severi-Zariski lemma [**Har77**, Cor.7.8], as there are finitely many such n, and the invertible sheaves $\mathcal{O}_{\bar X}(nD')$ for $D' = D$ or pD are S_2. It then follows that the restriction maps

$$H^0(\bar X, \mathcal{O}(nD')) \xrightarrow{\cong} H^0(\bar H, \mathcal{O}(nD' \cap \bar H))$$
$$(F-1)H^0(\bar X, \mathcal{O}(nD)) \xrightarrow{\cong} (F-1)H^0(\bar H, \mathcal{O}(nD \cap \bar H))$$

are isomorphisms for $1 \le n \le m$, and thus

$$A(\bar{X}, npD) \to A(\bar{H}, npD \cap \bar{H})$$

as well. This finishes the quest for $\bar{H}$. For $\bar{C}$ we iterate the argument with $\bar{X}$ replaced by $\bar{H}$ etc. This finishes the proof.

□

Let Y be a smooth connected variety defined over k, $j_Y : Y \hookrightarrow \bar{Y}$ be a good semi-compactification, that is, $\bar{Y}$ is smooth connected and $\bar{Y} \setminus Y$ is a normal crossings divisor. Given an effective divisor D supported on $\bar{Y} \setminus Y$, we define $\mathcal{S}(j_Y, r, D)$, $\mathcal{S}(j_Y, 1, D, n)$ etc. analogously. We denote by $\lfloor \frac{D}{p} \rfloor$ the floor of the $\mathbb{Q}$-divisor $\frac{D}{p}$, that is the largest Cartier divisor Δ on Y such that $p\Delta \subset D$. Following [**KS14**, 3-2] we define (with slightly simplified notation) the complex $\mathbb{Z}/p\mathbb{Z}_D$ on $\bar{Y}_{\text{ét}}$ by the formula

$$\mathbb{Z}/p\mathbb{Z}_D = \operatorname{cone}(\mathcal{O}(\lfloor \frac{D}{p} \rfloor) \xrightarrow{F-1} \mathcal{O}(D))[-1].$$

LEMMA 5.6. *One has* $\mathcal{S}(j_Y, 1, D, 1) = H^1_{\text{ét}}(\bar{Y}, \mathbb{Z}/p\mathbb{Z}_D)$.

PROOF. If $\bar{Y}$ is affine, then

$$H^1_{\text{ét}}(\bar{Y}, \mathbb{Z}/p\mathbb{Z}_D) = H^0(\bar{Y}, \mathcal{O}(D))/(F-1)H^0(\bar{Y}, \mathcal{O}(\lfloor \frac{D}{p} \rfloor),$$

so it follows from [**KS14**, Prop.2.5]. In general, let $\mathcal{H}^i$ be the Zariski sheaf associated to $\bar{U} \to H^i(\bar{Y} \cap \bar{U}, \mathbb{Z}/p\mathbb{Z}_D)$. For $i = 0$ it is equal to the constant sheaf $\mathbb{Z}/p\mathbb{Z}$ thus

$$H^1_{\text{ét}}(\bar{Y}, \mathbb{Z}/p\mathbb{Z}_D) = H^0_{\text{Zar}}(\bar{Y}, \mathcal{H}^1).$$

This finishes the proof. □

We come back to $j : X \hookrightarrow \bar{X}$ a projective normal connected compactification of a smooth variety and denote by $\bar{Y}$ its smooth locus. We set $S = \bar{X} \setminus \bar{Y}$ for the singular locus, defining the factorization $j : X \to \bar{X} \setminus S \xrightarrow{\alpha} \bar{X}$. We extend to $\bar{X}_{\text{ét}}$ the definition of $\mathbb{Z}/p\mathbb{Z}_D$ on $\bar{Y}_{\text{ét}}$ for p-divisible Cartier divisors supported on $\bar{X} \setminus X$, as follows. Let D be an effective Cartier divisor supported on $\bar{X} \setminus X$. For any natural number m, we define the complex $\mathbb{Z}/p\mathbb{Z}_{mpD}$ on $\bar{X}_{\text{ét}}$ by the formula

$$\mathbb{Z}/p\mathbb{Z}_{mpD} = \operatorname{cone}(\mathcal{O}(mD) \xrightarrow{F-1} \mathcal{O}(mpD))[-1].$$

For an effective divisor D' on $\bar{Y}$ and any m, we have the maps

$$\text{on } \bar{X}_{\text{ét}} : \mathbb{Z}/p\mathbb{Z}_{mpD} \to \mathbb{Z}/p\mathbb{Z}_{(m+1)pD}, \text{ on } \bar{Y}_{\text{ét}} : \mathbb{Z}/p\mathbb{Z}_D \to \mathbb{Z}/p\mathbb{Z}_{D+D'}$$

raising the level stemming from the natural inclusions $\mathcal{L} \to \mathcal{L}(\Delta)$ for a line bundle $\mathcal{L}$ and an effective Cartier divisor Δ.

LEMMA 5.7. *For any* $m \ge 2$, *the level raising maps*

$$H^1_{\text{ét}}(\bar{X}, \mathbb{Z}/p\mathbb{Z}_{(m-1)pD}) \to H^1_{\text{ét}}(\bar{X}, \mathbb{Z}/p\mathbb{Z}_{mpD}),\ H^1_{\text{ét}}(\bar{Y}, \mathbb{Z}/p\mathbb{Z}_D) \to H^1_{\text{ét}}(\bar{Y}, \mathbb{Z}/p\mathbb{Z}_{D+D'})$$

are injective, and so are the restriction maps

$$H^1_{\text{ét}}(\bar{X}, \mathbb{Z}/p\mathbb{Z}_{mpD}) \to H^1_{\text{ét}}(\bar{Y}, \mathbb{Z}/p\mathbb{Z}_{mpD}) \to H^1_{\text{ét}}(X, \mathbb{Z}/p\mathbb{Z})$$

for any $m \ge 0$.

PROOF. For the level raising map we compute $H^1_{et}(-, \mathbb{Z}/p\mathbb{Z}_?)$ in the Zariski topology as $H^1(-, \mathcal{O}(\lfloor \frac{\Delta}{p} \rfloor) \xrightarrow{F-1} \mathcal{O}(\Delta))$ for an effective Cartier divisor Δ and argue that if $(F-1)\psi = \varphi$ for local sections $\varphi \in \mathcal{O}(\Delta)$ and $\psi \in \mathcal{O}(*\Delta)$ then $\psi \in \mathcal{O}(\lfloor \frac{\Delta}{p} \rfloor)$. Here $\frac{\Delta}{p}$ is Cartier in the $\bar{X}$-version, on the smooth $\bar{Y}$ we allow is to be a $\mathbb{Q}$-divisor. For the restriction map this is the same proof noting that $S \subset \bar{X}$ has codimension ≥ 2 thus $\bar{Y} \cap Y$ hits all the codimension 1 points of $\bar{X} \setminus X$. □

PROPOSITION 5.8. *There is a natural number $m \geq 1$ such that in the diagram*

$$\begin{array}{ccc} & & H^1_{\text{ét}}(\bar{Y}, \mathbb{Z}/p\mathbb{Z}_D) \\ & & \big\downarrow \\ H^1_{\text{ét}}(\bar{X}, \mathbb{Z}/p\mathbb{Z}_{mpD}) & \hookrightarrow & H^1_{\text{ét}}(\bar{Y}, \mathbb{Z}/p\mathbb{Z}_{mpD}) \end{array}$$

the image of $H^1_{\text{ét}}(\bar{Y}, \mathbb{Z}/p\mathbb{Z}_D)$ by the level raising map falls in the image of $H^1_{\text{ét}}(\bar{X}, \mathbb{Z}/p\mathbb{Z}_{mpD})$ by the restriction map.

PROOF. By Lemma 5.7 we may replace $H^1_{\text{ét}}(\bar{Y}, \mathbb{Z}/p\mathbb{Z}_D)$ by $H^1_{\text{ét}}(\bar{Y}, \mathbb{Z}/p\mathbb{Z}_{pD})$ in the statement. For any $m \geq 1$ we have an exact sequence

$$0 \to H^1_{\text{ét}}(\bar{X}, \mathbb{Z}/p\mathbb{Z}_{pmD}) \to H^1_{\text{ét}}(\bar{Y}, \mathbb{Z}/p\mathbb{Z}_{pmD}) \to$$
$$Q_m = \text{Ker}(H^0(\bar{X}, R^1\alpha_*\mathcal{O} \otimes (mD)) \xrightarrow{F-1} H^0(\bar{X}, R^1\alpha_*\mathcal{O} \otimes (pmD)))$$

so it is enough to show that the level raising map $Q_1 \to Q_m$ is zero for m large. Let $W \subset S$ be the non Cohen-Macaulay locus of $\bar{X}$. It is closed and $S \setminus W \hookrightarrow S$ is dense ([**EGA IV(2)**, Cor.6.11.3]). This defines the factorization

$$\alpha : \bar{X} \setminus S \xrightarrow{\beta} \bar{X} \setminus W \xrightarrow{\gamma} \bar{X}$$

inducing the exact sequence

$$(\star) \quad 0 \to R^1\gamma_*\mathcal{O} \to R^1\alpha_*\mathcal{O} \to \gamma_*R^1\beta_*\mathcal{O}.$$

For any variety Z, we denote by $Z^{(i)}$, resp. $Z^{(\geq i)}$ the set of codimension i, resp $\geq i$ points. By [**SGA2**, VIII, Cor.2.3], since

$$\inf_{x \in \bar{X} \setminus W} \{\text{depth}\, \mathcal{O}_x + \dim \overline{\{x\}} - \dim W \cap \overline{\{x\}}\} > 2,$$

which holds as $\bar{X} \setminus W$ is Cohen-Macaulay, and W has codimension ≥ 3 in $\bar{X}$, the Zariski sheaf $R^1\gamma_*\mathcal{O} = \mathcal{H}^2_W(\bar{X}, \mathcal{O})$ is coherent. So it is annihilated by a power $\mathcal{I}^n_W$ of the ideal sheaf of W, thus by $\mathcal{O}_{\bar{X}}(-mD)$ with m large chosen so $\mathcal{I}^n_W \supset \mathcal{O}_{\bar{X}}(-mD)$. The level raising map on Q_a for some $a \geq 1$ respects the filtration defined by $(\star)$. Thus for $a \geq m$ it sends Q_1 to

$$Q_a^{CM} = \text{Ker}(H^0(\bar{X}, R^1\gamma_*\mathcal{O} \otimes (aD)) \xrightarrow{F-1} H^0(\bar{X}, R^1\gamma_*\mathcal{O} \otimes (paD))) \subset Q_a$$

and we have to show that for b large, the level raising map $Q_a^{CM} \to Q_{a+b}^{CM}$ is zero. Points $x \in \bar{X}^{(\geq c)} \cap (S \setminus W)$ have depth $\geq c$, thus by [**SGA2**, III, Lem. 3.1, Prop. 3.3] $H^i_x(\bar{X}, \mathcal{L}) = 0$ for $i < c$, and any invertible $\mathcal{L}$ on $\bar{X}$. Thus the restriction map to the finitely many points in $(\bar{X})^{(2)} \cap (S \setminus W)$

$$R^1\beta_*\mathcal{O} \hookrightarrow \oplus_{x \in (\bar{X})^{(2)} \cap (\bar{X} \setminus W)} H^2_x(\bar{X}, \mathcal{O})$$

is injective. Denoting by $\mathcal{O}_x$ the local ring of $\bar{X}$ at x, in which D is defined by $g \in \mathcal{O}_x$, the vertical arrows define isomorphism of complexes

$$\begin{array}{ccc} \mathcal{O}(aD)_x & \xrightarrow{F-1} & \mathcal{O}(paD)_x \\ {\scriptstyle g^a}\downarrow & & \downarrow{\scriptstyle g^{ap}} \\ \mathcal{O}_x & \xrightarrow[F-g^{ap-a}\cdot 1]{} & \mathcal{O}_x \end{array}$$

We set

$$Q_a(x) = \mathrm{Ker}(H^2_x(\mathcal{O}) \otimes \mathcal{O}(aD) \xrightarrow{F-1} H^2_x(\mathcal{O}) \otimes \mathcal{O}(paD))$$
$$Q'_a(x) = \mathrm{Ker}(H^2_x(\mathcal{O}_x) \xrightarrow{F-g^{ap-a}\cdot 1} H^2_x(\mathcal{O}_x)).$$

For every $b \geq 0$ we have a commutative diagram

$$\begin{array}{ccc} Q_a(x) & \xrightarrow{1} & Q_{a+b}(x) \\ {\scriptstyle g^a}\downarrow & & \downarrow{\scriptstyle g^{a+b}} \\ \mathcal{O}_x \cdot Q'_a(x) & \xrightarrow{g^b} & \mathcal{O}_x \cdot Q'_{a+b}(x) \end{array}$$

where the upper horizontal arrow is the level raising map, the vertical maps are injective and $\mathcal{O}_x \cdot Q'_m(x)$ is the $\mathcal{O}_x$-submodule of $H^2_x(\mathcal{O}_x)$ generated by $Q'_m(x)$. By [**Smi94**, Rmk.4.4], $H^2_x(\mathcal{O}_x)$ has one maximal Frobenius invariant proper $\mathcal{O}_x$-submodule K_x, the tight closure of 0 in $H^2_x(\mathcal{O}_x)$, which in addition is finite. On the other hand, $F(\lambda y) = (\lambda^p g^{ap-a})y$ for $\lambda \in \mathcal{O}_x, y \in Q'_a(x)$, so that $\mathcal{O}_x \cdot y$ has finite $\mathcal{O}_x$-length, and is Frobenius invariant, for each such y; thus $\mathcal{O}_x \cdot y \subset K_x$ for each such y. Hence $\mathcal{O}_x \cdot Q'_a(x) \subset K_x$ for any a. Hence $\mathcal{O}_x \cdot Q'_a(x)$ is annihilated by a power of the maximal ideal $\mathfrak{m}_x^n \subset \mathcal{O}_x$, thus by g^b for b so large that $\mathfrak{m}_x^n \supset \mathcal{O}_{\bar{X}}(-bD)$. Thus the level raising map is zero for b large. This finishes the proof. □

PROOF OF THEOREM 5.2. By enlarging D we may assume that X is affine. By Lemma 5.3 and Lemma 5.4 it is enough to find a good curve for $\mathcal{S}(j, 1, \Delta, 1)$ with $\Delta = mpD$ as in Proposition 5.8. For $\bar{C}$ as in Lemma 5.5, $A(\bar{X}, \Delta) \to A(\bar{C}, \bar{C} \cdot \Delta)$ is injective and by Proposition 5.8 $A(\bar{X}, \Delta) \supset \mathcal{S}(j, 1, D, 1)$. This finishes the proof. □

REMARK 5.9. We denote by $\mathcal{S}(j_Y, 1, D \cap \bar{Y}) \supset \mathcal{S}(j, 1, D)$ the set of rank 1 $\bar{\mathbb{Q}}_\ell$ local systems with ramification bounded by $D \cap \bar{Y}$ on $\bar{Y}$ where $\bar{Y}$ denotes the smooth locus of $\bar{X}$ as in Proposition 5.8, and D is a Cartier divisor on $\bar{X}$. Then in fact the curve C constructed in Theorem 5.2 is good for $\mathcal{S}(j_Y, 1, D \cap \bar{Y})$ as well. This enables to sharpen Deligne's question as to whether in higher rank there is a curve good for $\mathcal{S}(j_Y, r, D \cap \bar{Y}) \supset \mathcal{S}(j, r, D)$.

6. Remarks

Let $k, X, \bar{X}$ be as in Section 1, such that $\bar{X}$ is projective. For any field extension $K \supset k$, if C_K is a smooth geometrically connected curve over K, we denote by $\bar{\eta} \mapsto C_K$ a geometric point, and by $\bar{K}$ an algebraic closure of K in $K(\bar{\eta})$. We make the following elementary remark.

REMARK 6.1. There is a purely transcendental extension K/k of finite type and a morphism $C_K \to X$ of a smooth geometrically connected curve over K such that the composed homomorphism

$$\pi_1^{\text{ét}}(C_{\bar{K}}, \bar{\eta}) \to \pi_1^{\text{ét}}(C_K, \bar{\eta}) \to \pi_1^{\text{ét}}(X, \bar{\eta})$$

is surjective.

In other words, the Lefschetz theorem in the strong form is true if we allow finite type field extensions, even only purely transcendental ones, and then the algebraic closure. So Deligne's problem whether or not the Lefschetz theorem is true after bounding r and D is really a question over the original algebraically closed field of definition.

PROOF. We may assume X is affine, by passing to a dense open subset if necessary. Let $f = (f_1, \dots, f_n) : X \to \mathbb{A}_k^n$ be an embedding. In particular, f is unramified. As in [**Jou83**, Section 6.5], we consider

$$Z_X^{(d-1)} \subset X \times_k (\mathbb{A}^{n+1})^{d-1}$$

defined by

$$u_{i0} + \sum_{j=1}^{n} u_{ij} f_j(x) = 0, \quad 1 \le i \le d-1,$$

where $(u_{ij})_{0 \le j \le n}$ are the coordinate functions on the i-th copy of $\mathbb{A}^{n+1}$. By [**Jou83**, Thm. 6.6], the projection $\pi_X : Z_X^{(d-1)} \to \mathbb{A}_k^{(n+1)(d-1)}$ is dominant. By [**Jou83**, 6.5.3], the other projection $Z_X^{(d-1)} \to X$ is a trivial fibre bundle with fibre $(\mathbb{A}^n)^{d-1}$. In particular $Z_X^{(d-1)}$ is smooth over k, thus the generic fibre C_K of π_X is smooth over $K = k(\mathbb{A}_k^{(n+1)(d-1)})$. As $\dim f(X) \ge (d-1)+1 = d$, C_K is geometrically integral by [**Jou83**, Thm. 6.6 (3)], of dimension $\dim Z_X^{(d-1)} - (n+1)(d-1) = 1$. By construction $C_K \subset X_K$ is a closed embedding.

To prove that C_K has the desired property, we have to show that if $\varphi : Y \to X$ is any connected finite étale covering, then $Y \times_X C_{\bar{K}}$ is connected. Let $f_Y = f \circ \varphi : Y \to \mathbb{A}_k^n$ be the composition, which is again unramified. Then the construction of [**Jou83**, Section 6.5] applied to f_Y gives rise to a similarly defined $Z_Y^{(d-1)} \subset Y \times_k (\mathbb{A}^{n+1})^{d-1}$ and a corresponding smooth and geometrically integral curve $C(Y)_K$, which is the geometric generic fibre of $Z_Y^{(d-1)} \to (\mathbb{A}^{n+1})^{d-1}$. By definition

$$Z_Y^{(d-1)} = Z_X^{(d-1)} \times_X Y, \; Y_K \supset C(Y)_K = C_K \times_X Y = C_K \times_{X_K} Y_K.$$

Hence as claimed, the inverse image $C(Y)_{\bar{K}}$ of $C_{\bar{K}}$ under $\varphi \times_k \bar{K} : Y_{\bar{K}} \to X_{\bar{K}}$ is integral, in particular connected. This finishes the proof. □

Acknowledgments

The first author thanks Pierre Deligne for sending the email [**Del16**]. We thank Haoyu Hu, Takeshi Saito and Enlin Yang for promptly answering our questions on the various notions of Swan conductor, and Moritz Kerz for a discussion on [**KS14**].

References

[Ach17] P. Achinger, *Wild ramification and $K(\pi, 1)$ spaces*, Invent. Math. **210** (2017), no. 2, 453–499, DOI 10.1007/s00222-017-0733-5. MR3714509

[Del16] P. Deligne: *Email to Hélène Esnault*, dated Feb. 29th, 2016.

[Dri12] V. Drinfeld, *On a conjecture of Deligne* (English, with English and Russian summaries), Mosc. Math. J. **12** (2012), no. 3, 515–542, 668, DOI 10.17323/1609-4514-2012-12-3-515-542. MR3024821

[Esn17] H. Esnault, *Survey on some aspects of Lefschetz theorems in algebraic geometry*, Rev. Mat. Complut. **30** (2017), no. 2, 217–232, DOI 10.1007/s13163-017-0223-8. MR3642032

[EK11] H. Esnault and M. Kerz: *Notes on Deligne's letter to Drinfeld dated March 5, 2007*, `http://page.mi.fu-berlin.de/esnault/preprints/helene/103-110617.pdf`

[EK12] H. Esnault and M. Kerz, *A finiteness theorem for Galois representations of function fields over finite fields (after Deligne)*, Acta Math. Vietnam. **37** (2012), no. 4, 531–562. MR3058662

[EKS19] H. Esnault, L. Kindler, and V. Srinivas, *A note on fierce ramification*, J. Algebra **528** (2019), 250–259, DOI 10.1016/j.jalgebra.2019.03.020. MR3933258

[Har80] D. Harbater, *Moduli of p-covers of curves*, Comm. Algebra **8** (1980), no. 12, 1095–1122, DOI 10.1080/00927878008822511. MR579791

[Har77] R. Hartshorne, *Geometry: Euclid and beyond*, Undergraduate Texts in Mathematics, Springer-Verlag, New York, 2000, DOI 10.1007/978-0-387-22676-7. MR1761093

[Jou83] J.-P. Jouanolou, *Théorèmes de Bertini et applications* (French), Progress in Mathematics, vol. 42, Birkhäuser Boston, Inc., Boston, MA, 1983. MR725671

[Kat86] N. M. Katz, *Local-to-global extensions of representations of fundamental groups* (English, with French summary), Ann. Inst. Fourier (Grenoble) **36** (1986), no. 4, 69–106. MR867916

[KS14] M. Kerz and S. Saito, *Lefschetz theorem for abelian fundamental group with modulus*, Algebra Number Theory **8** (2014), no. 3, 689–701, DOI 10.2140/ant.2014.8.689. MR3218806

[Sei50] A. Seidenberg, *The hyperplane sections of normal varieties*, Trans. Amer. Math. Soc. **69** (1950), 357–386, DOI 10.2307/1990364. MR37548

[Smi94] K. E. Smith, *Tight closure of parameter ideals*, Invent. Math. **115** (1994), no. 1, 41–60, DOI 10.1007/BF01231753. MR1248078

[BOU] N. Bourbaki, *Éléments de mathématique* (French), Masson, Paris, 1980. Algèbre. Chapitre 10. Algèbre homologique. [Algebra. Chapter 10. Homological algebra]. MR610795

[EGA IV(2)] Éléments de Géométrie Algébrique *Étude locale des schémas et des morphismes de schémas*, Seconde partie, Publ.math.I.H.É.S. **24** (1965), 5–231.

[SGA2] Séminaire de Géométrie Algébrique *Cohomologie locale des faisceaux cohérents et théorèmes locaux et globaux*, North-Holland Publishing Company (1968).

Freie Universität Berlin, Arnimallee 3, 14195, Berlin, Germany
Email address: esnault@math.fu-berlin.de

TIFR, School of Mathematics, Homi Bhabha Road, 400005 Mumbai, India
Email address: srinivas@math.tifr.res.in

Proceedings of Symposia in Pure Mathematics
Volume **104**, 2021
https://doi.org/10.1090/pspum/104/01877

The Lieb–Thirring inequalities: Recent results and open problems

Rupert L. Frank

CONTENTS

This review celebrates the generous gift by Ronald and Maxine Linde for the remodeling of the Caltech mathematics department and the author is very grateful to the editors of this volume for the invitation to contribute. We attempt to survey recent results and open problems connected to Lieb–Thirring inequalities. In view of several excellent existing reviews [**17, 95, 112, 117, 136**] as well as highly recommended textbooks [**138, 140**], we sometimes put our focus on developments during the past decade.

The author would like to thank all his collaborators on the topic of Lieb–Thirring inequalities and, in particular, A. Laptev, S. Larson, M. Lewin, E. H. Lieb, P. T. Nam and T. Weidl for helpful remarks on a preliminary version of this review.

1. The Lieb–Thirring problem

1.1. A Sobolev inequality for orthonormal functions. In 1975, Lieb and Thirring proved the following theorem [**142**], see also [**143**].

2020 *Mathematics Subject Classification.* Primary 35P15; Secondary 81Q10.

Partial support through U.S. National Science Foundation grants DMS-1363432 and DMS-1954995 and through the Deutsche Forschungsgemeinschaft (DFG, German Research Foundation) through Germany's Excellence Strategy EXC - 2111 - 390814868 is acknowledged.

THEOREM 1.1. *Let $d \geq 1$. There is a constant $K_d > 0$ such that for all $N \in \mathbb{N}$ and all functions $u_1, \ldots, u_N \in H^1(\mathbb{R}^d)$ that are orthonormal in $L^2(\mathbb{R}^d)$ one has*

$$\sum_{n=1}^{N} \int_{\mathbb{R}^d} |\nabla u_n|^2 \, dx \geq K_d \int_{\mathbb{R}^d} \left(\sum_{n=1}^{N} |u_n|^2 \right)^{1+\frac{2}{d}} dx \,. \tag{1.1}$$

The main point in this theorem is that the constant K_d is independent of the number N of functions. Clearly, if the orthonormality requirement is dropped, then the constant on the right side would decrease like $N^{-\frac{2}{d}}$, as can be seen by taking all u_n to be equal.

The work of Lieb and Thirring was motivated by giving a new proof of stability of matter and the constant K_3 enters into their stability estimate. We will discuss this in more detail in Section 2 below and the reader who wants to see the Lieb–Thirring inequality 'in action', before studying its more theoretic aspects, might wish to jump directly to that section. For various other applications related to stability of matter we refer to [**136, 140**].

Lieb–Thirring bounds are closely connected to justifications of density functional theories, see, for instance, [**124, 134**]. In addition, they have proved useful to bound the dimension of attractors for the Navier–Stokes flow [**135**]; see also [**27, 190**]. They also appear in the context of spectral theory of Jacobi matrices and one-dimensional Schrödinger operators, see, e.g., [**104, 105**].

The Lieb–Thirring theorem leads naturally to the following challenge, which is a famous open problem in the field.

OPEN PROBLEM 1.2. Find the optimal constant K_d in (1.1).

Lieb and Thirring suggested two possible scenarios for optimality that lead to different constants and conjectured that the optimal constant K_d is given by the lesser of the two constants in these scenarios. Let us describe this in more detail.

1.2. The one-particle constant. A well-known Sobolev interpolation inequality, sometimes called Gagliardo–Nirenberg or Moser inequality, states that for any $d \geq 1$ there is a constant $K_d^{(1)} > 0$ such that for all $u \in H^1(\mathbb{R}^d)$ one has

$$\int_{\mathbb{R}^d} |\nabla u|^2 \, dx \geq K_d^{(1)} \int_{\mathbb{R}^d} |u|^{2+\frac{4}{d}} \, dx \left(\int_{\mathbb{R}^d} |u|^2 \, dx \right)^{-\frac{2}{d}} . \tag{1.2}$$

In connection with Lieb–Thirring inequalities, the constant $K_d^{(1)}$ is called the *one-particle constant.*

Clearly, choosing $N = 1$ in (1.1), we obtain an inequality of the form (1.2) and therefore the optimal constants in these inequalities satisfy

$$K_d \leq K_d^{(1)} \, . \tag{1.3}$$

Let us summarize what is known about the optimal constant $K_d^{(1)}$ and optimizers in (1.2). For detailed proofs and references we refer to [**53**]. In dimension $d = 1$, one has [**155**]

$$K_1^{(1)} = \frac{\pi^2}{4} \, .$$

and equality in (1.2) is attained if and only if u coincides, up to translation, dilation and multiplication by a constant, with $Q(x) = (\cosh x)^{-1/2}$.

In dimensions $d \geq 2$ it is known that (1.2) has an optimizer (one method of proof is suggested in [**143**] and another one is carried out in [**195**]) and that this optimizer is positive, radial [**84**] and unique up to translation, dilation and multiplication by a constant [**110**] (see also [**152**] and references therein). Clearly, the optimizer Q of (1.2) can be normalized such that it satisfies the Euler–Lagrange equation in the form

$$-\Delta Q - Q^{1+4/d} = -Q \qquad \text{in } \mathbb{R}^d, \tag{1.4}$$

and in this normalization, the optimal constant is related to the L^2 norm of Q by

$$K_d^{(1)} = \frac{d}{d+2} \|Q\|_2^{4/d}. \tag{1.5}$$

This follows by integrating (1.4) against Q and $x \cdot \nabla Q$. Moreover, Q is not only the unique minimizer up to symmetries, but also the unique positive solution of (1.4) [**84, 110**]. Therefore, Q can be computed numerically using the shooting method and then $K_d^{(1)}$ can be evaluated using (1.5). This computation appears in the appendix of [**143**] by Barnes for $d = 2, 3$.

1.3. The semiclassical constant. To get a different upper bound on K_d, we want to choose the functions u_n in (1.1) as plane waves. In order to make them belong to $H^1(\mathbb{R}^d)$ we need to multiply them by a cut-off function. Concerning their normalization, the following lemma is useful. It says that a certain relaxation of the problem does not change the optimal constant.

LEMMA 1.3. *Let $(u_n) \subset H^1(\mathbb{R}^d)$ be a sequence of functions that are orthonormal in $L^2(\mathbb{R}^d)$ and let $(\nu_n) \subset [0,1]$ be a sequence of numbers. Then*

$$\sum_{n=1}^{\infty} \nu_n \int_{\mathbb{R}^d} |\nabla u_n|^2 \, dx \geq K_d \int_{\mathbb{R}^d} \left(\sum_{n=1}^{\infty} \nu_n |u_n|^2 \right)^{1+\frac{2}{d}} dx,$$

where K_d is the optimal constant in (1.1).

PROOF. By monotone convergence, we may assume that only finitely many of the ν_n's are nonzero. We write

$$\sum_{n=1}^{\infty} \nu_n \int_{\mathbb{R}^d} |\nabla u_n|^2 \, dx = \int_0^1 \left(\sum_{\nu_n > \tau} \int_{\mathbb{R}^d} |\nabla u_n|^2 \, dx \right) d\tau$$

and use the bound (1.1) for fixed τ. In this way, we obtain

$$\sum_{n=1}^{\infty} \nu_n \int_{\mathbb{R}^d} |\nabla u_n|^2 \, dx \geq K_d \int_{\mathbb{R}^d} \left(\int_0^1 \left(\sum_{\nu_n > \tau} |u_n|^2 \right)^{1+\frac{2}{d}} d\tau \right) dx.$$

For fixed $x \in \mathbb{R}^d$, we apply Hölder's inequality in the τ integral,

$$\int_0^1 \left(\sum_{\nu_n > \tau} |u_n|^2 \right)^{1+\frac{2}{d}} d\tau \geq \left(\int_0^1 \sum_{\nu_n > \tau} |u_n|^2 \, d\tau \right)^{1+\frac{2}{d}} = \left(\sum_n \nu_n |u_n|^2 \right)^{1+\frac{2}{d}}.$$

Inserting this into the above integral leads to the claimed inequality. □

Let χ be a real-valued, compactly supported, Lipschitz function satisfying $|\chi| \leq 1$ and $\chi(0) = 1$. For two parameters $L, \mu > 0$ we consider the integral operator γ in $L^2(\mathbb{R}^d)$ with integral kernel

$$\gamma(x,y) = \chi(x/L) \int_{|\xi|^2<\mu} e^{i\xi\cdot(x-y)} \frac{d\xi}{(2\pi)^d} \chi(y/L) , \qquad x, y \in \mathbb{R}^d .$$

We have for any $\psi \in L^2(\mathbb{R}^d)$

$$\langle \psi, \gamma\psi\rangle = \int_{|\xi|^2<\mu} \left| (2\pi)^{-d/2} \int_{\mathbb{R}^d} e^{-i\xi\cdot y} \chi(y/L)\psi(y)\, dy \right| d\xi .$$

This shows that the operator γ is selfadjoint and nonnegative and, since the right side does not exceed

$$\int_{\mathbb{R}^d} \left| (2\pi)^{-d/2} \int_{\mathbb{R}^d} e^{-i\xi\cdot y} \chi(y/L)\psi(y)\, dy \right| d\xi = \int_{\mathbb{R}^d} |\chi(y/L)\psi(y)|^2\, dy \leq \|\psi\|_2^2 ,$$

its norm is ≤ 1. It is straightforward to see that γ is compact in $L^2(\mathbb{R}^d)$ and that its range is contained in $H^1(\mathbb{R}^d)$. Therefore there are functions $u_n \in H^1(\mathbb{R}^d)$ that are orthonormal in $L^2(\mathbb{R}^d)$ and numbers $\nu_n \in (0,1]$ such that

$$\gamma = \sum_n \nu_n |u_n\rangle\langle u_n| .$$

We want to apply the inequality in Lemma 1.3 with these u_n and ν_n. We have

$$\sum_n \nu_n |u_n(x)|^2 = \gamma(x,x) = \chi(x/L)^2 \int_{|\xi|^2<\mu} \frac{d\xi}{(2\pi)^d} = \frac{\omega_d}{(2\pi)^d} \mu^{\frac{d}{2}} \chi(x/L)^2 ,$$

where ω_d is the measure of the unit ball in $\mathbb{R}^d$. Thus,

$$\int_{\mathbb{R}^d} \left(\sum_n \nu_n |u_n(x)|^2 \right)^{1+\frac{2}{d}} dx = \left(\frac{\omega_d}{(2\pi)^d} \right)^{1+\frac{2}{d}} \mu^{\frac{d}{2}+1} L^d \int_{\mathbb{R}^d} \chi(y)^{1+\frac{2}{d}}\, dy .$$

On the other hand,

$$\begin{aligned}
\sum_n \nu_n \int_{\mathbb{R}^d} |\nabla u_n|^2\, dx &= \int_{|\xi|^2<\mu} \int_{\mathbb{R}^d} \left| \nabla \left(\chi(x/L) e^{i\xi\cdot x} \right) \right|^2 dx \frac{d\xi}{(2\pi)^d} \\
&= \int_{|\xi|^2<\mu} \int_{\mathbb{R}^d} \left(|\xi|^2 \chi(x/L)^2 + L^{-2} |\nabla\chi(x/L)|^2 \right) dx \frac{d\xi}{(2\pi)^d} \\
&= \frac{d}{d+2} \frac{\omega_d}{(2\pi)^d} \mu^{\frac{d}{2}+1} L^d \int_{\mathbb{R}^d} \chi(y)^2\, dy + \frac{\omega_d}{(2\pi)^d} \mu^{\frac{d}{2}} L^{d-2} \int_{\mathbb{R}^d} |\nabla\chi(y)|^2\, dy.
\end{aligned}$$

Thus, Lemma 1.3 with these u_n and ν_n yields the inequality

$$\frac{d}{d+2} \int_{\mathbb{R}^d} \chi(y)^2\, dy + \mu^{-1} L^{-2} \int_{\mathbb{R}^d} |\nabla\chi(y)|^2\, dy \geq K_d \left(\frac{\omega_d}{(2\pi)^d} \right)^{\frac{2}{d}} \int_{\mathbb{R}^d} \chi(y)^{1+\frac{2}{d}}\, dy .$$

In this inequality, we pass to the limit $\mu L^2 \to \infty$. This removes the gradient term on the left side and we can let χ approximate a characteristic function. In this way we finally arrive at

$$K_d \leq \frac{d}{d+2} \frac{(2\pi)^2}{\omega_d^{\frac{2}{d}}} =: K_d^{\mathrm{cl}} . \tag{1.6}$$

For an alternative proof of (1.6) which does not use Lemma 1.3, see [**87**, Lemma 10].

The constant on the right side of (1.6) is called the *classical (or semiclassical) constant.* The reason for this name will become clear later when discussing Weyl asymptotics for eigenvalues of Schrödinger operators. The discussion in this subsection is closely related to the Thomas–Fermi approximation in density functional theory, see [**124, 131, 134, 141**] and references therein.

As an aside, we mention that an analogue of the Lieb–Thirring inequality is valid for measures on phase space and is frequently used in kinetic theory; see, e.g., [**146**, Eq. (14)]. This analogue is valid with the classical constant.

1.4. The Lieb–Thirring conjecture. Lieb and Thirring [**143**] conjectured that the best constant K_d in (1.1) is given by the worst one in the previous two scenarios.

CONJECTURE 1.4. *For any $d \geq 1$,*

$$K_d = \min\left\{K_d^{(1)}, K_d^{\mathrm{cl}}\right\}. \tag{1.7}$$

Using the numerical computations of $K_d^{(1)}$ (see also [**56**] for an analytic argument), one sees that this is equivalent to

$$K_d = \begin{cases} K_d^{(1)} & \text{if } d = 1, 2, \\ K_d^{\mathrm{cl}} & \text{if } d \geq 3. \end{cases}$$

1.5. The currently best bound. In [**142**], Lieb–Thirring proved $K_3 \geq (4\pi)^{-\frac{2}{3}} K_3^{\mathrm{cl}}$ in dimension $d = 3$. Note that $(4\pi)^{-\frac{2}{3}} \approx 0.185001$. Since then there have been many contributions devoted to improving the lower bounds on K_d [**17, 36, 39, 97, 135**]. In particular, Dolbeault, Loss and Laptev showed in [**36**] that $K_d \geq (4/\pi^2)^{\frac{1}{d}} K_d^{\mathrm{cl}}$ for all $d \geq 1$. Note that $4/\pi^2 \approx 0.405285$ and $(4/\pi^2)^{\frac{1}{3}} \approx 0.740037$. The following bound, which is currently the best one in all dimensions, was obtained in [**59**].

THEOREM 1.5. *For all $d \geq 1$,*

$$K_d \geq (0.471851)^{\frac{1}{d}} K_d^{\mathrm{cl}}. \tag{1.8}$$

Note that in $d = 3$ the excess factor is $(0.471851)^{\frac{1}{3}} \approx 0.7785137$. Moreover, in $d = 1$ it is natural to compute the excess factor with respect to the conjectured constant $K_1^{(1)} = \pi^2/4$ instead of $K_1^{\mathrm{cl}} = \pi^2/3$. We have

$$K_1 \geq (0.471851)\,\tfrac{4}{3}\,K_1^{(1)} \approx 0.629134\,K_1^{(1)}.$$

The fact that the dimension enters the excess factor in the form of an exponent $1/d$ is due to the Laptev–Weidl method of lifting the dimension [**116**]. We will explain this technique in more detail in Subsection 6.1 below. The essence is that one tries to prove an inequality similar to (1.1) in dimension $d = 1$, but allows the functions to take values in $\mathbb{C}^q$ and seeks for constants independent of the 'internal dimension' q.

An interesting bound was proved in [**157**], namely, for any $d \geq 1$ there is a $C_d < \infty$ such that inequality (1.1) holds with the right side replaced by

$$(1-\varepsilon)K_d^{\mathrm{cl}} \int_{\mathbb{R}^d} \left(\sum_{n=1}^N |u_n|^2\right)^{1+\frac{2}{d}} dx - C_d\,\varepsilon^{-3-\frac{4}{d}} \int_{\mathbb{R}^d} \left|\nabla\left(\sum_{n=1}^N |u_n|^2\right)^{\frac{1}{2}}\right|^2 dx$$

for any $\varepsilon > 0$. In many applications the additional gradient term is of lower order and therefore the inequality is 'almost as good' as the one with the semiclassical constant. For a relevant discussion, see [**134**]. It is an interesting open question whether the inequality holds without the factor $1-\varepsilon$ in front of the first term and with a finite constant in front of the gradient term.

2. Application: Stability of Matter

We now recall the original motivation of the Lieb–Thirring inequality, namely the problem of stability of matter. This is the statement that the energy of a system of N electrons and K nuclei interacting through Coulomb forces is bounded from below by a (negative) constant times $N+K$. The constant is only allowed to depend on the charge of the nuclei.

The fact that such a bound holds was first shown by Dyson and Lenard [**37,38**]. In [**142**] Lieb and Thirring provided a new proof, which is shorter, gives a better bound on the involved constant and identified the Lieb–Thirring inequality as the crucial analytic ingredient in this argument. For more on the problem of stability of matter and, in particular, for further references and extensions to other physical systems, we refer to the book [**140**].

The state of a system of N quantum particles in $\mathbb{R}^d$ is described by a function $\psi \in L^2(\mathbb{R}^{dN})$, typically normalized such that $\|\psi\|_2 = 1$. We ignore here spin for the sake of simplicity. The corresponding one-particle density is the nonnegative function ρ_ψ on $\mathbb{R}^d$, defined by

$$\rho_\psi(x) = \sum_{n=1}^N \int_{\mathbb{R}^{d(N-1)}} |\psi(x_1,\ldots,x_{n-1},x,x_{n+1},\ldots,x_N)|^2 \, dx_1 \cdots dx_{n-1} \, dx_{n+1} \cdots dx_N .$$

Note that

$$\int_{\mathbb{R}^d} \rho_\psi(x)\, dx = N \, \|\psi\|_2^2 . \tag{2.1}$$

If the quantum particles are fermions, the function ψ satisfies an additional requirement. Let $\mathcal{S}_N$ be the symmetric group of $\{1,\ldots,N\}$. A function $\psi : \mathbb{R}^{dN} \to \mathbb{C}$ is called *antisymmetric* if for all $\sigma \in \mathcal{S}_N$

$$\psi(x_{\sigma(1)},\ldots,x_{\sigma(N)}) = (\operatorname{sgn}\sigma)\, \psi(x_1,\ldots,x_N) \quad \text{for all } x=(x_1,\ldots,x_N) \in \mathbb{R}^{dN} = (\mathbb{R}^d)^N .$$

The following is a Sobolev inequality for antisymmetric functions, which is a consequence of and, in fact, equivalent to the Lieb–Thirring inequality in Theorem 1.1.

COROLLARY 2.1. *Let $d \geq 1$, $N \geq 1$ and let $\psi \in H^1(\mathbb{R}^{dN})$ be antisymmetric. Then*

$$\int_{\mathbb{R}^{dN}} |\nabla\psi|^2 \, dx_1 \cdots dx_N \geq K_d \, \|\psi\|_2^{-\frac{4}{d}} \int_{\mathbb{R}^d} \rho_\psi(x)^{1+\frac{2}{d}} \, dx ,$$

where K_d is the optimal constant in (1.1).

We emphasize that both the constant K_d and the exponent $1+\frac{2}{d}$ on the right side are independent of N. Note that if $u_1,\ldots,u_N \in H^1(\mathbb{R}^d)$ are orthonormal in $L^2(\mathbb{R}^d)$, then for

$$\psi(x) = \frac{1}{\sqrt{N!}} \det\left(u_n(x_m)\right)_{n,m=1}^N ,$$

the inequality in Corollary 2.1 reduces to that in Theorem 1.1.

PROOF. We may assume that $\|\psi\|_2 = 1$. We define an integral operator γ_ψ on $L^2(\mathbb{R}^d)$ via its integral kernel

$$\gamma_\psi(x,y) := \sum_{n=1}^{N} \int_{\mathbb{R}^{d(N-1)}} \psi(x_1,\ldots,x_{n-1},x,x_{n+1},\ldots,x_N)\overline{\psi(x_1,\ldots,x_{n-1},x,x_{n+1},\ldots,x_N)} \, dx_1\cdots dx_{n-1}\, dx_{n+1}\cdots dx_N\,.$$

Thus, for any $f \in L^2(\mathbb{R}^d)$,

$$\langle f, \gamma_\psi f\rangle = \sum_{n=1}^{N} \int_{\mathbb{R}^{d(N-1)}} \left| \int_{\mathbb{R}^d} \overline{\psi(x_1,\ldots,x_{n-1},x,x_{n+1},\ldots,x_N)} f(x)\, dx \right|^2 dx_1\cdots dx_{n-1}\, dx_{n+1}\cdots dx_N\,. \tag{2.2}$$

This implies that the operator γ_ψ is selfadjoint and nonnegative. Moreover, since

$$\operatorname{Tr}\gamma_\psi = \sum_{n=1}^{N} \int_{\mathbb{R}^{dN}} |\psi(x)|^2\, dx = N\,,$$

the operator is, in particular, compact and we can write

$$\gamma_\psi = \sum_k \nu_k |u_k\rangle\langle u_k|$$

with orthonormal functions u_k and positive numbers ν_k. In terms of these functions and numbers we can write

$$\int_{\mathbb{R}^{dN}} |\nabla\psi|^2\, dx_1\cdots dx_N = \sum_k \nu_k \int_{\mathbb{R}^d} |\nabla u_k|^2\, dx$$

and

$$\rho_\psi(x) = \sum_k \nu_k |u_k(x)|^2 \qquad \text{for a.e. } x \in \mathbb{R}^d\,.$$

Therefore, the inequality in the corollary follows immediately from the inequality in Lemma 1.3, provided we can show that $\nu_k \leq 1$ for all k. This, in turn, is equivalent to the inequality

$$\langle f, \gamma_\psi f\rangle \leq \|f\|_2^2 \qquad \text{for all } f \in L^2(\mathbb{R}^d)\,, \tag{2.3}$$

which we shall show now. The antisymmetry of ψ enters in the proof of (2.3).

For the proof of (2.3) we may assume that $\|f\|_2 = 1$. Then we may choose an orthonormal basis $(e_\alpha)_{\alpha\in\mathbb{N}}$ of $L^2(\mathbb{R}^d)$ with $e_1 = f$. For $j \in \mathbb{N}^N$ let

$$\psi_j^\sharp = \int_{\mathbb{R}^{dN}} \overline{e_{j_1}(x_1)\cdots e_{j_N}(x_N)}\psi(x)\, dx_1\cdots dx_N\,,$$

so that

$$\psi = \sum_{j\in\mathbb{N}^N} \psi_j^\sharp\, e_{j_1}\otimes\cdots\otimes e_{j_N}$$

and therefore, by (2.2) and orthonormality,

$$\langle f, \gamma_\psi f\rangle = \sum_{n=1}^{N} \sum_{j,j'\in\mathbb{N}^N} \psi_j^\sharp \overline{\psi_{j'}^\sharp} \delta_{j_1,j_1'}\cdots\delta_{j_n,1}\delta_{j_n',1}\cdots\delta_{j_N,j_N'} = \sum_{n=1}^{N}\sum_{j\in\mathbb{N}^N} \left|\psi_j^\sharp\right|^2 \delta_{j_n,1}\,.$$

The antisymmetry of ψ implies that

$$\psi^{\sharp}_{(\sigma(j_1),\ldots,\sigma(j_N))} = (\operatorname{sgn}\sigma)\, \psi^{\sharp}_{(j_1,\ldots,j_N)} \qquad \text{for all } \sigma \in \mathcal{S}_N\,,\ j \in \mathbb{N}^N\,.$$

Thus,

$$\sum_{n=1}^{N} \sum_{j\in\mathbb{N}^N} \left|\psi^{\sharp}_{j}\right|^2 \delta_{j_n,1} = N! \sum_{1=j_1<j_2<\ldots<j_N} \left|\psi^{\sharp}_{j}\right|^2 \leq \sum_{j\in\mathbb{N}^n} \left|\psi^{\sharp}_{j}\right|^2 = \|\psi\|_2^2 = 1\,.$$

This proves (2.3) and completes the proof. □

Note that $\langle f, \gamma_\psi f\rangle = \langle \psi, K_f \psi\rangle$, where $K_f = \sum_{n=1}^N P_n$ and P_n has integral kernel $f(x_n)\overline{f(x_n')}$. Thus, (2.3) is equivalent to $\|K_f\| \leq \|f\|_2^2$ and, indeed, one can show that the spectrum of K_f, acting on antisymmetric functions in $L^2(\mathbb{R}^{dN})$, consists only of 0 and $\|f\|_2^2$, which again yields (2.3).

The energy of a Coulomb system consisting of N electrons in a state described by an antisymmetric $\psi \in L^2(\mathbb{R}^{3N})$ and of K classical nuclei at positions $R = (R_1, \ldots, R_K) \in \mathbb{R}^{3K}$ and with charges $Z = (Z_1, \ldots, Z_K) \in [0,\infty)^K$ is

$$\mathcal{E}_{R,Z}[\psi] := \int_{\mathbb{R}^{3N}} \left(|\nabla\psi|^2 + V_{R,Z}(x)|\psi|^2\right) dx$$

with the Coulomb potential

$$V_{R,Z}(x) = -\sum_{n=1}^{N}\sum_{k=1}^{K} \frac{Z_k}{|x_n - R_k|} + \sum_{1\leq n<m\leq N} \frac{1}{|x_n - x_m|} + \sum_{1\leq k<\ell\leq K} \frac{Z_k Z_\ell}{|R_k - R_\ell|}\,.$$

The ground state energy is

$$E_{R,Z}(N) := \inf\left\{\mathcal{E}_{R,Z}[\psi] :\ \psi \in H^1(\mathbb{R}^{3N}) \text{ antisymmetric, } \|\psi\|_2 = 1\right\}.$$

The following theorem states that matter is stable.

THEOREM 2.2. *For all $R_1, \ldots, R_K \in \mathbb{R}^3$ and $Z_1, \ldots, Z_K \in [0,\infty)$,*

$$E_{R,Z}(N) \geq -\frac{3\,\pi^{\frac{4}{3}}}{2^{\frac{2}{3}}\,5} K_3^{-1}\,(2z+1)^2\,(N+K)\,,$$

where K_3 is the optimal constant in (1.1) *with $d = 3$ and $z = \max_{1\leq k\leq K} Z_k$.*

We outline a proof of this theorem due to Solovej [**185**]. For another proof, which gives a slightly better value of the constant, we refer to [**140**, Section 7.1].

To bound the kinetic energy contribution to $\mathcal{E}_{R,Z}[\psi]$ we use the Lieb–Thirring inequality in the form of Corollary 2.1. To bound the potential energy contribution to $\mathcal{E}_{R,Z}[\psi]$ we use the following pointwise inequality due to Baxter [**5**].

LEMMA 2.3. *For any $x_1, \ldots, x_N, R_1, \ldots, R_K \in \mathbb{R}^3$ and $Z_1, \ldots, Z_K \in [0,\infty)$,*

$$V_{R,Z}(x) \geq -\sum_{n=1}^{N} \frac{2z+1}{\delta_R(x_n)}\,,$$

where $\delta_R(x_n) = \min\{|x_n - R_k| :\ 1 \leq k \leq K\}$ and $z = \max_{1\leq k\leq K} Z_k$.

For the proof of this lemma we refer to [**140**, Theorem 5.4]. (There the inequality is only stated in the case where all Z_k's are equal, but since for given k, $Z_k \mapsto V_{R,Z}(x)$ is affine linear and so attains its minimum on the interval $[0,z]$ either at 0 or at z, we can reduce the general case to this special case.)

PROOF OF THEOREM 2.2. According to Corollary 2.1 and Lemma 2.3 we have for all antisymmetric $\psi \in H^1(\mathbb{R}^{3N})$ with $\|\psi\|_2 = 1$,

$$\mathcal{E}_{R,Z}[\psi] \geq K_3 \int_{\mathbb{R}^3} \rho_\psi^{\frac{5}{3}}\, dx - \int_{\mathbb{R}^3} \frac{2z+1}{\delta_R} \rho_\psi \, dx\,.$$

In the second integral on the right side we add and subtract a number $\mu > 0$ from $(2z+1)\delta_R^{-1}$ and we recall (2.1). By Hölder's inequality, we obtain

$$\mathcal{E}_{R,Z}[\psi] \geq K_3 T - \left(\int_{\mathbb{R}^3} \left(\frac{2z+1}{\delta_R} - \mu \right)_+^{\frac{5}{2}} dx \right)^{\frac{2}{5}} T^{\frac{3}{5}} - \mu N \qquad \text{with } T = \int_{\mathbb{R}^3} \rho_\psi^{\frac{5}{3}}\, dx\,.$$

Optimizing the right side with respect to T, we obtain

$$\mathcal{E}_{R,Z}[\psi] \geq -\frac{2\cdot 3^{\frac{3}{2}}}{5^{\frac{5}{2}}} K_3^{-\frac{3}{2}} \int_{\mathbb{R}^3} \left(\frac{2z+1}{\delta_R} - \mu \right)_+^{\frac{5}{2}} dx - \mu N\,.$$

By scaling, with $\tilde{R}_k = \mu(2z+1)^{-1} R_k$,

$$\int_{\mathbb{R}^3} \left(\frac{2z+1}{\delta_R} - \mu \right)_+^{\frac{5}{2}} dx = (2z+1)^3 \mu^{-\frac{1}{2}} \int_{\mathbb{R}^3} \left(\frac{1}{\delta_{\tilde{R}}} - 1 \right)_+^{\frac{5}{2}} dx\,.$$

Optimizing the resulting bound in μ, we obtain

$$\mathcal{E}_{R,Z}[\psi] \geq -\frac{3^2}{5^{\frac{5}{3}}} K_3^{-1} (2z+1)^2 \left(\int_{\mathbb{R}^3} \left(\frac{1}{\delta_{\tilde{R}}} - 1 \right)_+^{\frac{5}{2}} dx \right)^{\frac{2}{3}} N^{\frac{1}{3}}\,.$$

To complete the proof, it suffices to bound the integral in terms of K. Using

$$\left(\frac{1}{\delta_{\tilde{R}}(x)} - 1 \right)_+^{\frac{5}{2}} = \max_{1\leq k\leq K} \left(\frac{1}{|x-\tilde{R}_k|} - 1 \right)_+^{\frac{5}{2}} \leq \sum_{k=1}^{K} \left(\frac{1}{|x-\tilde{R}_k|} - 1 \right)_+^{\frac{5}{2}}$$

and

$$\int_{\mathbb{R}^3} \left(\frac{1}{|y|} - 1 \right)_+^{\frac{5}{2}} dy = 4\pi \int_0^1 (r^{-1}-1)^{\frac{5}{2}} r^2\, dr = \frac{5\pi^2}{4}\,,$$

we find

$$\int_{\mathbb{R}^3} \left(\frac{1}{\delta_{\tilde{R}}} - 1 \right)_+^{\frac{5}{2}} dx \leq \sum_{k=1}^{K} \int_{\mathbb{R}^3} \left(\frac{1}{|x-\tilde{R}_k|} - 1 \right)_+^{\frac{5}{2}} dx = \frac{5\pi^2}{4} K\,.$$

Thus, we obtain

$$\mathcal{E}_{R,Z}[\psi] \geq -\frac{3^2\, \pi^{\frac{4}{3}}}{2^{\frac{4}{3}}\, 5} K_3^{-1} (2z+1)^2 K^{\frac{2}{3}} N^{\frac{1}{3}}\,.$$

Since $K^{\frac{2}{3}} N^{\frac{1}{3}} \leq (2^{\frac{2}{3}}/3)(K+N)$, this yields the inequality in the theorem. □

Besides being of independent interest, the stability of matter theorem is used in the proof of the existence of the thermodynamic limit for Coulomb systems [**120, 137**]. In some applications, for instance, to thermodynamics for systems where the number of nuclei can fluctuate or even be completely random [**16**], it is useful to have a bound as in Theorem 2.2 but with a right side independent of K; see [**47**, Theorem 8] or [**91**, Lemma 8].

3. The Lieb–Thirring inequality for Schrödinger operators

3.1. Duality. In this subsection we discuss an equivalent formulation of the Lieb–Thirring inequality in terms of eigenvalues of Schrödinger operators. By 'equivalence' of the two inequalities we mean that the best constants in these two inequalities are in one-to-one correspondence.

We denote by $E_n(-\Delta+V)$, $n\in\mathbb{N}$, the negative eigenvalues of the Schrödinger operator $-\Delta+V$ in $L^2(\mathbb{R}^d)$, repeated according to multiplicities and arranged in nondecreasing order with the convention that $E_n(-\Delta+V)=0$ if $-\Delta+V$ has less than n negative eigenvalues. More precisely, for real $V\in L^{1+\frac{d}{2}}(\mathbb{R}^d)$ one can use the Sobolev interpolation inequality (1.2) to show that the quadratic form $\int_{\mathbb{R}^d}(|\nabla u|^2+V|u|^2)\,dx$ with form domain $H^1(\mathbb{R}^d)$ is lower bounded and closed and therefore induces a selfadjoint, lower bounded operator $-\Delta+V$ in $L^2(\mathbb{R}^d)$. The $E_n(-\Delta+V)$ are defined as the negative eigenvalues of this operator. Alternatively, and without any recourse to spectral theory, the $E_n(-\Delta+V)$ could be defined by the variational principle, which only involves the above quadratic form, but no operator [**138**].

The Lieb–Thirring inequality for Schrödinger operators reads as follows, where we use the notation $a_-=\max\{-a,0\}$ for the negative part.

THEOREM 3.1. *Let $d\geq 1$. There is a constant $L_d<\infty$ such that for any real $V\in L^{1+\frac{d}{2}}(\mathbb{R}^d)$,*

$$\sum_n |E_n(-\Delta+V)|\leq L_d\int_{\mathbb{R}^d}V(x)_-^{1+\frac{d}{2}}\,dx\,. \tag{3.1}$$

Moreover, this inequality is equivalent to inequality (1.1) *in the sense that the best constants satisfy*

$$\left((1+\tfrac{d}{2})L_d\right)^{1+\frac{2}{d}}\left((1+\tfrac{2}{d})K_d\right)^{1+\frac{d}{2}}=1\,. \tag{3.2}$$

This equivalence is due to Lieb and Thirring [**142, 143**]. In fact, their original proof of (1.1) proceeded via proving (3.1). The first direct proof of (1.1) was found in [**39**] in $d=1$ and in [**170**] for general d. Further direct proofs appeared in [**150**] (see also [**157**]) and [**171**]. The advantage of the reformulation (3.1) in terms of Schrödinger operators is that one can use what is called the Birman–Schwinger principle [**13, 176**], which reduces the problem of estimating negative eigenvalues of an unbounded operator to the problem of estimating eigenvalues of a compact operator. This compact operator has the structure of a product of multiplication operators in position and in momentum space. As an aside, we mention that in their paper [**143**], in order to bound operators of this form, Lieb and Thirring derived a matrix inequality that has proved very useful in many other applications as well.

PROOF. The proof is based on the variational principle for sums of eigenvalues, which says that

$$\sum_{n=1}^N\int_{\mathbb{R}^d}\left(|\nabla u_n|^2+V|u_n|^2\right)dx\geq\sum_{n=1}^N E_n(-\Delta+V) \tag{3.3}$$

for all N and all functions $u_1,\ldots,u_N\in H^1(\mathbb{R}^d)$ that are orthonormal in $L^2(\mathbb{R}^d)$. Moreover, equality in this inequality is attained if $-\Delta+V$ has at least N negative eigenvalues and the u_n are the eigenfunctions corresponding to the $E_n(-\Delta+V)$ for $n=1,\ldots,N$.

Thus, if (3.1) holds, it follows from (3.3) that for all N and all functions $u_1,\ldots,u_N \in H^1(\mathbb{R}^d)$ that are orthonormal in $L^2(\mathbb{R}^d)$,

$$\sum_{n=1}^{N} \int_{\mathbb{R}^d} |\nabla u_n|^2\,dx \geq -\int_{\mathbb{R}^d} V \sum_{n=1}^{N} |u_n|^2\,dx - L_d \int_{\mathbb{R}^d} V(x)_-^{1+\frac{d}{2}}\,dx\,.$$

Optimizing this inequality with respect to $V \in L^{1+\frac{d}{2}}(\mathbb{R}^d)$ leads to the choice

$$V(x) = -\left(\tfrac{d+2}{2}\,L_d\right)^{-\frac{2}{d}} \left(\sum_{n=1}^{N} |u_n|^2\right)^{\frac{2}{d}}$$

and to the bound

$$\sum_{n=1}^{N} \int_{\mathbb{R}^d} |\nabla u_n|^2\,dx \geq \tfrac{d}{d+2}\left(\tfrac{2+d}{2}\,L_d\right)^{-\frac{2}{d}} \int_{\mathbb{R}^d} \left(\sum_{n=1}^{N} |u_n|^2\right)^{1+\frac{2}{d}} dx\,.$$

This proves (1.1) with optimal constant satisfying $K_d \geq \frac{d}{d+2}\left(\frac{2+d}{2}\,L_d\right)^{-\frac{2}{d}}$.

Conversely, if $V \in L^{1+\frac{d}{2}}(\mathbb{R}^d)$ is given and if $-\Delta+V$ has at least N negative eigenvalues $E_n(-\Delta+V)$ with corresponding orthonormal eigenfunctions u_n, then by (1.1),

$$\sum_{n=1}^{N} |E_n(-\Delta+V)| = -\sum_{n=1}^{N} \int_{\mathbb{R}^d} \left(|\nabla u_n|^2 + V|u_n|^2\right) dx \leq -K_d \int_{\mathbb{R}^d} \rho^{1+\frac{2}{d}}\,dx + \int_{\mathbb{R}^d} V_-\rho\,dx$$

where $\rho = \sum_{n=1}^{N} |u_n|^2$. Optimizing the right side with respect to all functions ρ, we obtain

$$\sum_{n=1}^{N} |E_n(-\Delta+V)| \leq \tfrac{2}{2+d}\left(\tfrac{d+2}{d} K_d\right)^{-\frac{d}{2}} \int_{\mathbb{R}^d} V(x)_-^{1+\frac{d}{2}}\,dx\,.$$

This proves (3.1) with optimal constant satisfying $L_d \leq \frac{2}{2+d}\left(\frac{d+2}{d} K_d\right)^{-\frac{d}{2}}$. (Note that we used a similar argument already in the proof of Theorem 2.2.) □

In Subsections 1.2 and 1.3 we have presented two possible scenarios for the best constant K_d in (1.1). Let us discuss what this means for the best constant L_d in (3.1).

3.2. The one-particle constant. The same argument as in the proof of Theorem 3.1 shows that there is a constant $L_d^{(1)} < \infty$ such that for all $V \in L^{1+\frac{d}{2}}(\mathbb{R}^d)$,

$$|E_1(-\Delta+V)| \leq L_d^{(1)} \int_{\mathbb{R}^d} V(x)_-^{1+\frac{d}{2}}\,dx\,,$$

and the best constant in this inequality is related to the best constant $K_d^{(1)}$ in inequality (1.2) by

$$\left((1+\tfrac{d}{2})L_d^{(1)}\right)^{1+\frac{2}{d}} \left((1+\tfrac{2}{d})K_d^{(1)}\right)^{1+\frac{d}{2}} = 1\,. \tag{3.4}$$

Thus, the analogue of the constant $K_d^{(1)}$ in the dual picture is the best constant in the problem of minimizing the eigenvalue of a Schrödinger operator under the constraint of a fixed $L^{1+\frac{d}{2}}$-norm. This is a particular case of a question raised by Keller [**102**].

3.3. The semiclassical constant. We recall that for any real $V \in L^{1+\frac{d}{2}}(\mathbb{R}^d)$, one has

$$\lim_{\hbar\to 0} \hbar^d \sum_n |E_n(-\hbar^2\Delta + V)| = L_d^{\mathrm{cl}} \int_{\mathbb{R}^d} V(x)_-^{1+\frac{d}{2}}\, dx \tag{3.5}$$

with

$$L_d^{\mathrm{cl}} = \frac{2}{d+2} \frac{\omega_d}{(2\pi)^d}\,.$$

Indeed, one can prove asymptotics (3.5) for continuous, compactly supported V using Dirichlet–Neumann bracketing and then one can use (3.1) to extend these asymptotics to all real $V \in L^{1+\frac{d}{2}}(\mathbb{R}^d)$.

The asymptotics (3.5) become more intuitive if one notices that

$$\sum_n |E_n(-\hbar^2\Delta + V)| = \operatorname{Tr}\left(-\hbar^2\Delta + V\right)_-$$

and that

$$\hbar^{-d} L_d^{\mathrm{cl}} \int_{\mathbb{R}^d} V(x)_-^{1+\frac{d}{2}}\, dx = \iint_{\mathbb{R}^d\times\mathbb{R}^d} \left(|\hbar\xi|^2 + V(x)\right)_- \frac{dx\, d\xi}{(2\pi)^d}\,.$$

Thus, (3.5) says that

$$\operatorname{Tr}\left(-\hbar^2\Delta + V\right)_- \sim \iint_{\mathbb{R}^d\times\mathbb{R}^d} \left(|\hbar\xi|^2 + V(x)\right)_- \frac{dx\, d\xi}{(2\pi)^d} \qquad \text{as } \hbar \to 0\,.$$

On the left side, one applies the function $E \mapsto E_-$ to the operator $-\hbar^2\Delta + V$ and then takes the trace, and on the right side, one applies the same function to the symbol $|\hbar\xi|^2 + V(x)$ and then integrates over phase space. Asymptotics (3.5) say that this agrees to leading order in the semiclassical limit $\hbar \to 0$.

Note that the constants L_d^{cl} and K_d^{cl} in (1.6) are related by

$$\left((1+\tfrac{d}{2})L_d^{\mathrm{cl}}\right)^{1+\frac{2}{d}} \left((1+\tfrac{2}{d})K_d^{\mathrm{cl}}\right)^{1+\frac{d}{2}} = 1\,. \tag{3.6}$$

Thus, the analogue of K_d^{cl} in the dual picture are Weyl asymptotics.

3.4. The Lieb–Thirring conjecture. Because of (3.2), (3.4) and (3.6), Conjecture 1.4 can be equivalently stated as

$$L_d = \max\left\{L_d^{(1)}, L_d^{\mathrm{cl}}\right\},$$

that is, $L_d = L_d^{(1)}$ if $d = 1, 2$ and $L_d = L_d^{\mathrm{cl}}$ if $d \geq 3$.

3.5. The currently best bound. Because of (3.2) and (3.6), Theorem 1.5 can be equivalently stated as

$$L_d \leq 1.456\, L_d^{\mathrm{cl}} \qquad \text{for all } d \geq 1\,. \tag{3.7}$$

4. Lieb–Thirring inequalities for Schrödinger operators. II

4.1. A family of inequalities. In [**143**] Lieb and Thirring observed that (3.1) is only a special case of a whole scale of inequalities. One has

THEOREM 4.1. *Let* $\gamma \geq \frac{1}{2}$ *if* $d = 1$, $\gamma > 0$ *if* $d = 2$ *and* $\gamma \geq 0$ *if* $d \geq 3$. *There is a constant* $L_{\gamma,d} < \infty$ *such that for any real* $V \in L^{\gamma+\frac{d}{2}}(\mathbb{R}^d)$,

$$\sum_n |E_n(-\Delta + V)|^\gamma \leq L_{\gamma,d} \int_{\mathbb{R}^d} V(x)_-^{\gamma+\frac{d}{2}}\, dx\,. \tag{4.1}$$

Here, as before, $E_n(-\Delta+V)$ denote the negative eigenvalues of the Schrödinger operator $-\Delta+V$, which can be shown to be well-defined under the conditions on V in the theorem. Moreover, we use the convention that for $\gamma = 0$ the left side of (4.1) denotes the number of negative eigenvalues of $-\Delta + V$, counting multiplicities.

The inequality in Theorem 4.1 is due to Lieb and Thirring [**143**] in the non-endpoint cases. The case $\gamma = 0$ in dimensions $d \geq 3$ is due to Cwikel [**29**], Lieb [**129, 130**] and Rozenblum [**166, 167**] and often referred to as the Cwikel–Lieb–Rozenblum (CLR) inequality. The case $\gamma = \frac{1}{2}$ in dimension $d = 1$ is due to Weidl [**192**].

The work of Rozenblum on the case $\gamma = 0$ completed earlier extensive work by Birman and Solomyak (summarized, for instance, in [**15**]); see also [**180**] for a close-to optimal result and a conjecture. Alternative proofs of the CLR bound were obtained in [**26, 47, 52, 96, 128**].

According to the duality Theorem 3.1, Open Problem 1.2 is equivalent to finding the best constant $L_d = L_{1,d}$ in (3.1) for $\gamma = 1$. This naturally leads to

OPEN PROBLEM 4.2. Find the optimal constants $L_{\gamma,d}$ in (4.1).

Motivated by the case $\gamma = 1$ it is natural to look at two particular scenarios.

4.2. The one-particle constant. Theorem 4.1 implies, in particular, that for all γ and d as in that theorem there is a constant $L^{(1)}_{\gamma,d} < \infty$ such that for all $V \in L^{\gamma+\frac{d}{2}}(\mathbb{R}^d)$,

$$|E_1(-\Delta + V)|^\gamma \leq L^{(1)}_{\gamma,d} \int_{\mathbb{R}^d} V(x)_-^{\gamma+\frac{d}{2}}\, dx\,. \tag{4.2}$$

Finding the optimal $L^{(1)}_{\gamma,d}$ is Keller's problem [**102**] of minimizing the lowest eigenvalue of a Schrödinger operator $-\Delta + V$ in $\mathbb{R}^d$ under the constraint of a fixed L^p norm of the potential V.

For $\gamma = 0$ (which is allowed in dimensions $d \geq 3$) the interpretation of inequality (4.2) is that $-\Delta + V$ has no negative eigenvalue if $L^{(1)}_{0,d} \int_{\mathbb{R}^d} V(x)_-^{\frac{d}{2}}\, dx < 1$.

For all γ such that $\gamma + \frac{d}{2} \geq 1$, the same duality argument as in the case $\gamma = 1$ shows that the optimal constant $L^{(1)}_{\gamma,d}$ is related to the optimal constant $K^{(1)}_{p',d}$ in the Sobolev interpolation inequality

$$\int_{\mathbb{R}^d} |\nabla u|^2\, dx \geq K^{(1)}_{p,d} \left(\int_{\mathbb{R}^d} |u|^{2p}\, dx \right)^{\frac{2}{d(p-1)}} \|u\|_2^{-\frac{4p}{d(p-1)}+2} \tag{4.3}$$

by

$$L^{(1)}_{p'-d/2,d} \left(K^{(1)}_{p,d}\right)^{\frac{d}{2}} = \left(\frac{d}{2p'}\right)^{\frac{d}{2}} \left(\frac{2p'-d}{2p'}\right)^{\frac{2p'-d}{2}}. \tag{4.4}$$

Note that as γ runs through the range in Theorem 4.1, the integrability exponent $2p = 2(\gamma + \frac{d}{2})/(\gamma + \frac{d}{2} - 1)$ in (4.3) runs through the range $2 < 2p \leq \infty$ if $d = 1$, $2 < 2p < \infty$ if $d = 2$ and $2 < 2p \leq \frac{2d}{d-2}$ if $d \geq 3$. It is well-known that (4.3) holds precisely in this range of parameters. In particular, the nonvalidity of (4.3) for $2p = \infty$ if $d = 2$ implies that (4.2), and consequently (4.1), are not valid for $\gamma = 0$ if $d = 2$.

In dimensions $d = 1$, the value of the optimal constant $K_{p,1}^{(1)}$ and the set of optimizers was determined by Nagy [**155**]. Via (4.4) this leads to the formula

$$L_{\gamma,1}^{(1)} = \frac{1}{\sqrt{\pi}} \frac{\Gamma(\gamma+1)}{\Gamma(\gamma+\frac{1}{2})} \frac{(\gamma-\frac{1}{2})^{\gamma-\frac{1}{2}}}{(\gamma+\frac{1}{2})^{\gamma+\frac{1}{2}}}. \tag{4.5}$$

In dimensions $d \geq 3$, the optimal constant in (4.3) is known for the exponent $2p = \frac{2d}{d-2}$, corresponding to $\gamma = 0$, and the optimizers are known [**3, 164, 165, 189**].

In the remaining cases $2 < 2p < \frac{2d}{d-2}$ for $d \geq 2$ the optimal constants $K_{p,d}^{(1)}$ are not known explicitly. However, just like for (1.2), one can show that there is an optimizer and that this optimizer is unique up to translation, dilation and multiplication by a constant. For proofs we refer to the references mentioned in Subsection 1.2.

The duality used to derive (4.4) also shows that optimizers u of (4.3) are in one-to-one correspondence (modulo symmetries) with optimal potentials V in (4.2). In particular, we infer that there is a unique (up to symmetries) optimizing potential of (4.2). A stability estimate for (4.2) was proved in [**25**].

4.3. The semiclassical constant. Analogously to (3.5) one can show that for all γ and d as in Theorem 4.1 and all $V \in L^{\gamma+\frac{d}{2}}(\mathbb{R}^d)$,

$$\lim_{\hbar \to 0} \hbar^d \sum_n |E_n(-\hbar^2\Delta + V)|^\gamma = L_{\gamma,d}^{\mathrm{cl}} \int_{\mathbb{R}^d} V(x)_-^{\gamma+\frac{d}{2}}\, dx \tag{4.6}$$

with

$$L_{\gamma,d}^{\mathrm{cl}} = (4\pi)^{-\frac{d}{2}} \frac{\Gamma(\gamma+1)}{\Gamma(\gamma+1+\frac{d}{2})}. \tag{4.7}$$

As in the special case $\gamma = 1$, the relation to semiclassics becomes clearer if one writes the expression on the right side of (4.6) as

$$L_{\gamma,d}^{\mathrm{cl}} \int_{\mathbb{R}^d} V(x)_-^{\gamma+\frac{d}{2}}\, dx = \iint_{\mathbb{R}^d\times\mathbb{R}^d} (|\xi|^2 + V(x))_-^\gamma \frac{dx\, d\xi}{(2\pi)^d}. \tag{4.8}$$

As an aside, we mention that the validity of the Lieb–Thirring inequality is crucial for establishing asymptotics (4.6) under the assumption $V \in L^{\gamma+\frac{d}{2}}(\mathbb{R}^d)$. We refer to [**14**] for examples of $V \in L^{\gamma+\frac{d}{2}}(\mathbb{R}^d)$ where (4.6) fails for $\gamma = 0$ in $d = 2$.

4.4. Known results. It follows from the previous two subsections that the optimal constant $L_{\gamma,d}$ in (4.1) satisfies

$$L_{\gamma,d} \geq \max\left\{L_{\gamma,d}^{(1)}, L_{\gamma,d}^{\mathrm{cl}}\right\}.$$

It was originally conjectured by Lieb and Thirring [**143**] that one has equality in this inequality. This is still believed to be the case in dimension $d = 1$.

CONJECTURE 4.3. *For $d = 1$ and $\gamma \geq \frac{1}{2}$,*

$$L_{\gamma,1} = \max\left\{L^{(1)}_{\gamma,1}, L^{\mathrm{cl}}_{\gamma,1}\right\}.$$

Using the explicit expressions for $L^{(1)}_{\gamma,1}$ and $L^{\mathrm{cl}}_{\gamma,1}$ in (4.5) and (4.7) one sees that Conjecture 4.3 is equivalent to

$$L_{\gamma,1} = \begin{cases} L^{(1)}_{\gamma,1} & \text{if } \gamma \leq \frac{3}{2}, \\ L^{\mathrm{cl}}_{\gamma,1} & \text{if } \gamma \geq \frac{3}{2}. \end{cases}$$

In the following theorem we will collect all the known optimal results about the constants $L_{\gamma,d}$.

THEOREM 4.4. *The best constant $L_{\gamma,d}$ in (4.1) is given by*

(a) $L_{\gamma,d} = L^{\mathrm{cl}}_{\gamma,d}$ *if $\gamma \geq \frac{3}{2}$ and $d \geq 1$.*

(b) $L_{\frac{1}{2},1} = L^{(1)}_{\frac{1}{2},1}$ *if $\gamma = \frac{1}{2}$ and $d = 1$.*

In the dimension $d = 1$, part (a) is due to Gardner, Greene, Kruskal and Miura [**83**, Eq. (3.27)] for $\gamma = 3/2$ and due to Lieb and Thirring [**143**] for $\gamma = \frac{2k+1}{2}$, $k \in \mathbb{N}$, and due to Aizenman and Lieb for all $\gamma \geq \frac{3}{2}$. The higher dimensional case in (a) is due to Laptev and Weidl [**116**]; see also [**6**] for a partially alternate proof. Part (b) is due to Hundertmark, Lieb and Thomas [**143**] with a partially alternate proof in [**97**].

In view of Theorem 4.4, Conjecture 4.3 is proved for $\gamma = \frac{1}{2}$ and for $\gamma \geq \frac{3}{2}$. The case $\gamma \in (\frac{1}{2}, \frac{3}{2})$ is open.

Later, in Subsection 6.1 we will discuss the Laptev–Weidl argument [**116**] used in the proof of part (a) of Theorem 4.4 in $d \geq 2$.

Here, let us briefly explain the Aizenman–Lieb argument [**2**] that is used in the proof of part (a) of Theorem 4.4. Clearly, for any $0 \leq \gamma < \sigma$ there is a positive constant $C_{\gamma,\sigma}$ such that

$$E_-^\sigma = C_{\gamma,\sigma} \int_0^\infty (E+\tau)_-^\gamma \tau^{\sigma-\gamma-1}\, d\tau\,.$$

Thus,

$$\sum_n |E_n(-\Delta+V)|^\sigma = C_{\gamma,\sigma} \int_0^\infty \sum_n |E_n(-\Delta+V+\tau)|^\gamma \tau^{\sigma-\gamma-1}\, d\tau$$

and, in view of (4.8),

$$L^{\mathrm{cl}}_{\sigma,d} \int_{\mathbb{R}^d} V(x)_-^{\sigma+\frac{d}{2}}\, dx = C_{\gamma,\sigma} \int_0^\infty \left(L^{\mathrm{cl}}_{\gamma,d} \int_{\mathbb{R}^d} (V(x)+\tau)_-^{\gamma+\frac{d}{2}}\, dx \right) \tau^{\sigma-\gamma-1}\, d\tau\,.$$

It follows from these two equations that the optimal constants $L_{\gamma,d}$ and $L_{\sigma,d}$ satisfy

$$\frac{L_{\sigma,d}}{L^{\mathrm{cl}}_{\sigma,d}} \leq \frac{L_{\gamma,d}}{L^{\mathrm{cl}}_{\gamma,d}} \qquad \text{if } \sigma > \gamma\,.$$

In particular, if $L_{\gamma,d} = L^{\mathrm{cl}}_{\gamma,d}$ for some γ, then $L_{\sigma,d} = L^{\mathrm{cl}}_{\sigma,d}$ for all $\sigma > \gamma$.

Next, let us discuss the maximum between $L^{(1)}_{\gamma,d}$ and $L^{\mathrm{cl}}_{\gamma,d}$, which appears in the original form of the Lieb–Thirring conjecture. A small variation of the Aizenman–Lieb argument together with some facts about the optimizing potential for $L^{(1)}_{\gamma,d}$ implies [**56**] that $\gamma \mapsto L^{(1)}_{\gamma,d}/L^{\mathrm{cl}}_{\gamma,d}$ is strictly decreasing. In Subsection 4.6 we will note that $L^{(1)}_{0,d} < L^{\rm cl}_{0,d}$ for $d \geq 8$ and therefore $L^{(1)}_{\gamma,d} < L^{\rm cl}_{\gamma,d}$ for all $\gamma \geq 0$. On the other hand, for $1 \leq d \leq 7$ there is a unique $\gamma_c(d)$ where the functions $\gamma \mapsto L^{(1)}_{\gamma,d}$ and $\gamma \mapsto L^{\rm cl}_{\gamma,d}$ intersect. Indeed, for $d = 1$ this is explicit and for $2 \leq d \leq 7$ see [**56**]. According to the numerics from [**143**], one has

$$\gamma_c(d) = \begin{cases} \frac{3}{2} & \text{if } d = 1\,, \\ 1.165 & \text{if } d = 2\,, \\ 0.8627 & \text{if } d = 3\,. \end{cases}$$

Let us complement the 'positive' results in Theorem 4.4 by 'negative' results.

PROPOSITION 4.5. *The best constant* $L_{\gamma,d}$ *in* (4.1) *satisfies*

(a) $L_{\gamma,d} > L^{\rm cl}_{\gamma,d}$ *if* $\gamma < \frac{3}{2}$ *in* $d = 1$ *and* $\gamma < 1$ *in* $d \geq 2$.

(b) $L_{\gamma,d} > L^{(1)}_{\gamma,d}$ *if* $\gamma > \max\{2 - d/2, 0\}$ *in* $1 \leq d \leq 6$ *and* $\gamma \geq 0$ *in* $d \geq 7$.

We have repeated the results for $d = 1$ for the sake of completeness. Part (a) in $d \geq 2$ is due to Helffer and Robert [**93**]. Part (b) for $\gamma = 0$ is due to Glaser, Grosse and Martin [**85**] and will be discussed in Subsection 4.6. Part (b) for $\gamma > 0$ is from [**56**]; see Subsection 6.2 below for some details of the argument.

Let us discuss the state of the original Lieb–Thirring conjecture and some possible modifications, which take into accound the negative results from Proposition 4.5 as well as the numerical experiments from [**123**].

Dimension $d = 1$. Conjecture 4.3 is generally believed to be true. The only remaining case is the range $1/2 < \gamma < 3/2$, where the optimal constant should be $L^{(1)}_{\gamma,1}$.

Dimension $d = 2$. In the range $0 < \gamma \leq 1$, it is conceivable that $L_{\gamma,2} = L^{(1)}_{\gamma,2}$, as originally conjectured by Lieb and Thirring. (This is suggested by numerics in the appendix of [**143**] and in [**123**]. However, both these computations missed the phenomena described next for $\gamma > 1$, so it is not clear how reliable they are.)

The situation in the range $1 < \gamma < 3/2$ is rather unclear. By part (b) of Proposition 4.5, one has $L_{\gamma,2} > L^{(1)}_{\gamma,2}$ for $\gamma > 1$. Since $L^{(1)}_{\gamma,2} > L^{\rm cl}_{\gamma,2}$ for $\gamma < \gamma_c(2) = 1.165$, this shows that the original Lieb–Thirring conjecture fails in the range $(1, \gamma_c(2)]$. Moreover, it is shown in [**56**] that, if there is an optimizing potential for some $\gamma > 1$, then this potential has infinitely many negative eigenvalues. Instead of (or besides) the existence of such an optimal potential, it is conceivable that the optimal potentials in the bound for the first N eigenvalues (which exist [**56**]) converge, when suitably normalized, as $N \to \infty$ to a potential that does not belong to $L^{\gamma+1}$ like, for instance, a periodic potential; see [**57**] for numerics in this direction, as well as an analytic result for $d = 1$ and $\gamma = 3/2$.

Dimensions $d \geq 3$. It is conceivable that $L_{\gamma,3} = L^{(1)}_{\gamma,3}$ for $0 \leq \gamma \leq 1/2$ if $d = 3$, as originally conjectured by Lieb and Thirring. (This is suggested by numerics in [**123**], however, with the same caveat as before.) On the other hand, according to Proposition 4.5 one has $L_{\gamma,d} > \max\{L^{(1)}_{\gamma,d}, L^{\rm cl}_{\gamma,d}\}$ for $1/2 < \gamma < 1$ if $d = 3$, for $0 < \gamma < 1$ if $4 \leq d \leq 6$ and for $0 \leq \gamma < 1$ if $d \geq 7$, so in all these cases the original Lieb–Thirring conjecture fails. Moreover, by [**58**] if there is a optimizing potential for $L_{\gamma,d}$ for these γ, then this potential has infinitely many negative eigenvalues.

According to Conjecture 1.4 and the Aizenman–Lieb argument, it is believed that $L_{\gamma,d} = L^{\rm cl}_{\gamma,d}$ for $\gamma \geq 1$ and $d \geq 3$.

4.5. Currently best bounds. Let us summarize bounds on the optimal constant $L_{\gamma,d}$ for $\gamma < 3/2$. The best bounds in the literature are

$$L_{\gamma,d} \leq \begin{cases} 1.456\ L_{\gamma,d}^{\mathrm{cl}} & \text{if } 1 \leq \gamma < 3/2\,, \\ 2\ L_{\gamma,1}^{\mathrm{cl}} & \text{if } 1/2 \leq \gamma < 1 \text{ and } d = 1\,, \\ 2.912\ L_{\gamma,d}^{\mathrm{cl}} & \text{if } 1/2 \leq \gamma < 1 \text{ and } d \geq 2\,. \end{cases}$$

By the Aizenman–Lieb argument, these bounds follow from the corresponding bounds at the smallest value at γ. Thus, the first bound follows from (3.7), the second one from [**98**] and the third one by the Laptev–Weidl lifting argument from [**97**] and (3.7). This lifting argument yields, more generally, the bound

$$L_{1/2,d} \leq 2\ L_{1,d-1}\,.$$

Bounds for the range $0 \leq \gamma < 1/2$ in $d \geq 3$ follow by the Aizenman–Lieb argument from corresponding bounds for $\gamma = 0$. The best value for $L_{0,3}$ in $d = 3$ is due to Lieb in [**129, 130**],

$$L_{0,3} \leq 6.86924\ L_{0,3}^{\mathrm{cl}}$$

and is to be compared with the lower bound from the Sobolev inequality $L_{0,3} \geq (8/\sqrt{3})\ L_{0,3}^{\mathrm{cl}} \approx 4.6188\ L_{0,3}^{\mathrm{cl}}$. Lieb's proof uses a new formula for Wiener integrals, called Lieb's formula, which is further discussed in [**181**, Theorem 8.2]. The best bounds for $d = 4$ and for $d \geq 5$ are in [**129, 130**] and [**96**], respectively.

Bounds for the range $0 < \gamma < 1/2$ in $d = 2$ have received relatively little attention in the literature. In particular, we are not aware of an investigation of the asymptotic behavior of $L_{\gamma,2}$ as $\gamma \to 0$. Probably, both $L_{\gamma,2}$ and $L_{\gamma,2}^{(1)}$ behave like a constant times γ^{-1}. Are the two constants the same? The asymptotics of $L_{\gamma,2}^{(1)}$ can be obtained via (4.4) from arguments similar to those in [**163**]. A logarithmic endpoint type inequality is shown in [**107**].

4.6. The number of negative eigenvalues. Let us discuss in more detail the (open) problem of finding the optimal constant $L_{0,d}$ for $d \geq 3$, that is, to maximize the quotient between the number of negative eigenvalues of $-\Delta + V$ and $\int_{\mathbb{R}^d} V_-^{\frac{d}{2}}\, dx$.

It is convenient to introduce the notation $N_{\leq}(-\Delta+V)$ to denote the number of nonpositive eigenvalues of $-\Delta + V$, counting multiplicities, plus the number of zero energy resonances, corresponding to solutions $u \in \dot{H}^1(\mathbb{R}^d) \setminus L^2(\mathbb{R}^d)$ of $(-\Delta+V)u = 0$. This definition appears naturally in this context since $N_{\leq}(-\Delta+V)$ is the limit of the number of negative eigenvalues of $-\Delta + V_+ - (1+\varepsilon)V_-$ as $\varepsilon \to 0+$, so inequality (4.1), even if the left side only counts negative eigenvalues, implies

$$N_{\leq}(-\Delta+V) \leq L_{0,d} \int_{\mathbb{R}^d} V(x)_-^{\frac{d}{2}}\, dx\,.$$

We begin by presenting the example of [**85**] that shows $L_{0,d} > \max\{L_{0,d}^{\mathrm{cl}}, L_{0,d}^{(1)}\}$ for $d \geq 7$. Our presentation is somewhat different from theirs and fills in some details. The basis is the following computation, which we explain later in this subsection.

LEMMA 4.6. *Let $d \geq 3$ and, for $L \in \mathbb{N}_0$,*

$$V^{(L)}(x) = -\left(L + \frac{d-2}{2}\right)\left(L + \frac{d}{2}\right)\left(\frac{2}{1+|x|^2}\right)^2, \qquad x \in \mathbb{R}^d\,.$$

Then

$$N_{\leq}(-\Delta + V^{(L)}) = \frac{2}{d!} \frac{(L+d-1)!\,(L+\frac{d}{2})}{L!}$$

and

$$\int_{\mathbb{R}^d} \left(V^{(L)}\right)_-^{\frac{d}{2}} dx = \left((L+\tfrac{d-2}{2})(L+\tfrac{d}{2})\right)^{\frac{d}{2}} |\mathbb{S}^d| \,.$$

As a consequence of this lemma,

$$L_{0,d} \geq \sup_{L\in\mathbb{N}_0} \frac{N_{\leq}(-\Delta + V^{(L)})}{\int_{\mathbb{R}^d} \left(V^{(L)}\right)_-^{\frac{d}{2}} dx} = \frac{2}{d!\,|\mathbb{S}^d|} \sup_{L\in\mathbb{N}_0} a_L = L_{0,d}^{\mathrm{cl}} \sup_{L\in\mathbb{N}_0} a_L$$

with

$$a_L := \frac{(L+d-1)!\,(L+\frac{d}{2})}{L!\,\left((L+\frac{d-2}{2})(L+\frac{d}{2})\right)^{d/2}} \,.$$

Note that, by the form of optimizers in the Sobolev inequality [**3**, **164**, **165**, **189**],

$$L_{0,d}^{(1)} = L_{0,d}^{\mathrm{cl}}\, a_0 \,.$$

On the other hand, since $a_L \to 1$ as $L \to \infty$,

$$L_{0,d}^{\mathrm{cl}} = L_{0,d}^{\mathrm{cl}} \lim_{L\to\infty} a_L \,.$$

Thus, in order to show that $L_{0,d} > \max\{L_{0,d}^{\mathrm{cl}}, L_{0,d}^{(1)}\}$, it suffices to show that one has $\sup_{L\in\mathbb{N}_0} a_L > \max\{\lim_{L\to\infty} a_L, a_0\}$. This is possible if $d \geq 7$. Indeed, as suggested to me by S. Larson, to whom I am grateful, using $\ln(1+x) = x + \mathcal{O}(x^2)$ as $x \to 0$, one sees that

$$\ln a_L = \tfrac{d}{2}\, L^{-1} + \mathcal{O}(L^{-2}) \qquad \text{as } L \to \infty \,,$$

so $a_L > 1 = \lim_{L'\to\infty} a_{L'}$ for all sufficiently large L. On the other hand, $a_1 > a_0$ if $d = 7$. Since $a_0 < 1 = \lim_{L\to\infty} a_L$ if $d \geq 8$, we have indeed shown that $\sup_{L\in\mathbb{N}_0} a_L > \max\{\lim_{L\to\infty} a_L, a_0\}$ for all $d \geq 7$.

Glaser, Grosse and Martin [**85**] make the following conjecture.

CONJECTURE 4.7. *Let $d \geq 3$ and $\gamma = 0$. Then*

$$L_{0,d} = L_{0,d}^{\mathrm{cl}} \sup_{L\in\mathbb{N}_0} a_L \,.$$

In particular, it is conjectured that $L_{0,d} = L_{0,d}^{(1)}$ if $d \leq 6$. The CLR bound with the conjectured constant $L_{0,4}^{(1)}$ holds for radial potentials in $d = 4$ [**85**]. Moreover, the Lieb–Thirring conjecture for $\gamma = 1$ in $d = 1$ would imply the CLR bound with the conjectured constant $L_{0,3}^{(1)}$ for radial potentials in $d = 3$ [**85**].

Further evidence for Conjecture 4.7 comes from the following observation, which is analogous to one made in a related context in [**52**], namely that the problem of computing the optimal $L_{0,d}$ is conformally invariant. More precisely, if h is a conformal transformation of $\mathbb{R}^d \cup \{\infty\}$ with Jacobian denoted by J_h and if

$$V_h(x) = J_h(x)^{2/d}\, V(h(x)) \,,$$

then

$$\int_{\mathbb{R}^d} V_h(x)_-^{\frac{d}{2}}\, dx = \int_{\mathbb{R}^d} V(x)_-^{\frac{d}{2}}\, dx \qquad \text{and} \qquad N_{\leq}(-\Delta + V_h) = N_{\leq}(-\Delta + V) \,.$$

The first equality is clear and the second one follows from the variational principle in the form (sometimes called Glazman's lemma)

$$\begin{aligned}&N_{\leq}(-\Delta+V)\\&=\sup\left\{\dim\mathcal{M}:\ \mathcal{M}\subset\dot{H}^1(\mathbb{R}^d)\,,\ \int_{\mathbb{R}^d}\left(|\nabla u|^2+V|u|^2\right)dx\leq 0\ \forall u\in\mathcal{M}\right\},\end{aligned}$$

if we note that for $v(x)=J_h(x)^{(d-2)/(2d)}u(h(x))$ one has

$$\int_{\mathbb{R}^d}|\nabla v|^2\,dx=\int_{\mathbb{R}^d}|\nabla u|^2\,dx\,,\qquad\int_{\mathbb{R}^d}V_h|v|^2\,dx=\int_{\mathbb{R}^d}V|u|^2\,dx\,.$$

(Here, the first equality is verified by noting that any conformal transformation of $\mathbb{R}^d\cup\{\infty\}$ is a composition of a translation, a dilation, a rotation, a reflection and an inversion.)

In view of the conformal invariance it is natural to consider the optimzation problem on the sphere. We will use this procedure to prove Lemma 4.6. We consider the inverse stereographic projection $\mathcal{S}:\mathbb{R}^d\to\mathbb{S}^d$,

$$\mathcal{S}_j(x)=\frac{2x_j}{1+|x|^2}\,,\qquad j=1,\ldots d\,,\qquad\mathcal{S}_{d+1}(x)=\frac{1-|x|^2}{1+|x|^2}\,.$$

Then, by a similar argument as before, if

$$V(x)=\left(\frac{2}{1+|x|^2}\right)^2W(\mathcal{S}(x))\,,$$

then

$$\int_{\mathbb{R}^d}V(x)_-^{\frac{d}{2}}\,dx=\int_{\mathbb{S}^d}W(\omega)_-^{\frac{d}{2}}\,d\omega\qquad\text{and}\qquad N_{\leq}(-\Delta+V)=N_{\leq}(-\Delta_{\mathbb{S}^d}+\tfrac{d(d-2)}{4}+W)\,.$$

Here $-\Delta_{\mathbb{S}^d}$ is the Laplace–Beltrami operator on $\mathbb{S}^d$. Its eigenvalues are given by $\ell(\ell+d-1)$, $\ell\in\mathbb{N}_0$, with multiplicity

$$\nu_\ell=\frac{(2\ell+d-1)\ (\ell+d-2)!}{(d-1)!\ \ell!}\,.$$

Note that the potential $V^{(L)}$ in Lemma 4.6 corresponds to the constant potential $W^{(L)}=-\left(L+\frac{d-2}{2}\right)\left(L+\frac{d}{2}\right)$ on $\mathbb{S}^d$. We have

$$\int_{\mathbb{R}^d}V^{(L)}(x)_-^{\frac{d}{2}}\,dx=\int_{\mathbb{S}^d}W^{(L)}(\omega)_-^{\frac{d}{2}}\,d\omega=\left((L+\tfrac{d-2}{2})(L+\tfrac{d}{2})\right)^{\frac{d}{2}}|\mathbb{S}^d|$$

and, since $\ell(\ell+d-1)+\frac{d(d-2)}{4}-\left(L+\frac{d-2}{2}\right)\left(L+\frac{d}{2}\right)\leq 0$ iff $\ell\leq L$,

$$N_{\leq}(-\Delta_{\mathbb{S}^d}+\tfrac{d(d-2)}{4}+W^{(L)})=\sum_{\ell=0}^{L}\nu_\ell=\frac{2}{d!}\ \frac{(L+d-1)!\,(L+\frac{d}{2})}{L!}\,.$$

This completes the proof of Lemma 4.6.

To summarize, Conjecture 4.7 says that the optimal constant in the CLR inequality is given, after mapping the problem conformally to the sphere, by a constant potential. This would be similar to other optimization problems with conformal invariance, both for single functions [**132**] and for functions of eigenvalues [**154**].

5. Further directions of study

After the overview over the standard Lieb–Thirring inequalities in the previous sections, we now address some extensions and generalizations. Our presentation emphasizes, probably unjustly, developments in the last decade and/or developments in which the author was involved. The overall focus is on open problems, some major, some minor, and it is hoped that the presentation stimulates further progress.

5.1. Pólya's conjecture. A classical question in the field of spectral estimates concerns the best value of the constant $L_{\gamma,d}^{\rm dom}$ in the inequality

$$\sum_n (E_n(-\Delta_\Omega)-\mu)_-^\gamma \le L_{\gamma,d}^{\rm dom}\,|\Omega|\,\mu^{\gamma+\frac d2} \qquad \text{for all } \mu\ge 0$$

valid for all open sets $\Omega\subset\mathbb{R}^d$ of finite measure. Here $-\Delta_\Omega$ denotes the Dirichlet Laplacian in Ω and $E_n(-\Delta_\Omega)$ its eigenvalues in nondecreasing order, counted according to multiplicity.

Clearly, Weyl asymptotics imply that $L_{\gamma,d}^{\rm dom}\ge L_{\gamma,d}^{\rm cl}$ for all $\gamma\ge 0$. A famous conjecture by Polya states that $L_{\gamma,d}^{\rm dom}=L_{\gamma,d}^{\rm cl}$ for all $\gamma\ge 0$. (Strictly speaking, Polya only considered $\gamma=0$. By the Aizenman–Lieb argument, equality $L_{\gamma,d}^{\rm dom}=L_{\gamma,d}^{\rm cl}$ for some $\gamma=\gamma_0$ implies equality for any $\gamma>\gamma_0$. So Polya's conjecture for $\gamma=0$ implies the conjecture as stated.)

Polya has given an elegant proof of his conjectured bound in the special case of tiling domains [**161**]. Further results for product domains can be found in [**111**].

The connection between Polya's conjecture and the Lieb–Thirring problem is that $L_{\gamma,d}^{\rm dom}\le L_{\gamma,d}$. This follows from the variational principle by taking $V(x)=-\mu$ for $x\in\Omega$ and $V(x)\ge 0$ for $x\notin\Omega$ in the Lieb–Thirring inequality.

Berezin [**8**] and Li and Yau [**128**] (the latter in an equivalent, dual form) proved that $L_{\gamma,d}^{\rm dom}=L_{\gamma,d}^{\rm cl}$ for $\gamma\ge 1$.

There has been relatively little progress on Pólya's conjecture. In particular, it is still unknown whether the inequality holds with the semiclassical constant in the special case where Ω is a disc in $d=2$.

Some recent work concerns the analogue of Pólya's conjecture in the presence of a homogeneous magnetic field. While the analogue of the Berezin–Li–Yau bound continues to hold in this setting [**42**], the analogue of Pólya's conjecture fails for any $0\le\gamma<1$ [**72**]. Also, in [**109**] it was shown that the analogue of Pólya's conjecture fails for the fractional Laplacian $(-\Delta)_\Omega^s$ in $d=1$, as well as for most s in $d=2$.

Evidence for Pólya's conjecture comes from the sign of the subleading term in Weyl's asymptotic law [**100**]. Bounds that capture lower order correction terms appear, typically for $\gamma\ge 3/2$ or $\gamma\ge 1$, in [**63, 64, 81, 82, 108, 194**] and references therein; see also [**118**] for an application of these ideas to shape optimization problems.

5.2. Magnetic Lieb–Thirring inequalities. The Lieb–Thirring inequality in the presence of a magnetic field reads

$$\sum_n |E_n((-i\nabla+A)^2+V)|^\gamma \le L_{\gamma,d}^{\rm mag}\int_{\mathbb{R}^d} V(x)_-^{\gamma+\frac d2}\,dx\,,$$

where γ is as in Theorem 4.1. By definition, $L_{\gamma,d}^{\rm mag}$ is independent of $A\in L^2_{\rm loc}(\mathbb{R}^d,\mathbb{R}^d)$.

Several of the proofs of Theorem 4.1 extend to the magnetic case with the same constant. It is an open problem, however, whether the optimal constant $L^{\mathrm{mag}}_{\gamma,d}$ coincides with the optimal constant $L_{\gamma,d}$.

This is trivially the case if $d = 1$, where every magnetic field can be gauged away. Moreover, Laptev and Weidl [**116**] showed that $L^{\mathrm{mag}}_{\gamma,d} = L^{\mathrm{cl}}_{\gamma,d} = L_{\gamma,d}$ for $\gamma \geq 3/2$ in any dimension d. All bounds that are obtained using their method starting from a one-dimensional inequality remain valid in the magnetic case, including the current best bound (3.7). There is a semi-abstract result [**50**], which says that $L^{\mathrm{mag}}_{\gamma,d}$ does not exceed $L_{\gamma,d}$ by more than a factor depending only on γ and d. This result is also applicable to spectral inequalities of a more complicated form than Lieb–Thirring inequalities.

The case of the Pauli operator, that is, $(\sigma\cdot(-i\nabla+A))^2$ instead of $(-i\nabla+A)^2$, is considerably more difficult and we refer to [**23, 41, 43–45, 139, 183**] and references therein.

5.3. Lieb–Thirring inequalities for powers of the Laplacian. The Lieb–Thirring inequality for powers $s > 0$ of the Laplacian reads

$$\sum_n |E_n((-\Delta)^s + V)|^\gamma \leq L_{\gamma,d,s} \int_{\mathbb{R}^d} V(x)_-^{\gamma+\frac{d}{2s}}\, dx\,,$$

where

$$\begin{cases} \gamma \geq 1 - \frac{d}{2s} & \text{if } d < 2s\,, \\ \gamma > 1 - \frac{d}{2s} & \text{if } d = 2s\,, \\ \gamma \geq 0 & \text{if } d > 2s\,. \end{cases}$$

The inequality in the cases $\gamma > (1 - d/2s)_+$ and $\gamma = 0$ can be proved using the methods from [**143**] and [**29, 167**], respectively. The inequality for $\gamma = 1 - d/2s > 0$ appears in [**158, 192**] for integer s and in [**55**] for $s < 1$. The proof for noninteger $s > 1$ should follow along the same lines.

While some of the above proofs yield reasonably good constants $L_{\gamma,d,s}$, nothing seems to be known about their optimal values for $s \neq 1$. In particular, one might wonder whether $L_{\gamma,d,s}$ is given by its semiclassical analogue for sufficiently large γ.

On the other hand, for $d = 1$ and any integer $s \geq 2$ it is shown in [**48**] that in the critical case $\gamma = 1 - d/2s$, the optimal constant $L_{\gamma,d,s}$ is strictly larger than the corresponding one-particle constant, contrary to a conjecture in [**158**]. One might wonder whether it is equal to the one-particle constant in $d = 1$ for $s \in (1/2, 3/2)$.

5.4. Lieb–Thirring inequalities for discrete Schrödinger operators. Results for Jacobi matrices and discrete Schrödinger operators can be found, for instance, in [**4, 99, 104, 168, 173, 174**] and in the references therein. Due to the lack of scaling invariance the form of the inequality and therefore also the question of optimal constants is less clear in this setting.

5.5. The oval problem. The Lieb–Thirring Conjecture 4.3 would imply, in particular, that $|E_1(-\frac{d^2}{dx^2}+V)|+|E_2(-\frac{d^2}{dx^2}+V)|$ is bounded by $L^{(1)}_{1,1}\int_{\mathbb{R}} V_-^{\frac{3}{2}}\, dx$. Benguria and Loss [**7**] reformulated this weaker conjecture as an isoperimetric problem for certain planar curves and proved an initial result. Further progress is contained in [**9, 24, 34, 144**], but the problem is still open.

5.6. Semiclassical monotonicity. Remarkably, in [**187**] it was shown that the function $\hbar \mapsto \hbar^{-d} \sum_n |E_n(-\hbar^2\Delta + V)|^\gamma$ is nonincreasing for $\gamma \geq 2$ and $d \geq 1$. Moreover, taking $V(x) = |x|^2 - 1$ and $\hbar$ near $(d+2)^{-2}$, one sees that the assumption $\gamma \geq 2$ is necessary.

5.7. Reverse Lieb–Thirring inequalities. In [**85, 175**] the Lieb–Thirring bound for $\gamma = 1/2$ in $d = 1$ is complemented by the lower bound

$$\sum_n |E_n(-\tfrac{d^2}{dx^2} + V)|^{1/2} \geq -L^{\rm cl}_{1/2,1} \int_{\mathbb{R}} V(x)\, dx$$

with optimal constant $L^{\rm cl}_{1/2,1} = 1/4$. Similar bounds for $V \leq 0$ were proved for $0 < \gamma < 1/2$ if $d = 1$ [**30**] and for $\gamma = 0$ if $d = 2$ [**90, 106**]; see also [**158, 177**]. While for most of these bounds, optimal (or almost optimal) values of the constants have not been investigated, remarkably, for $\gamma = 0$ in $d = 2$ one has the optimal inequality

$$N_{\leq}(V) \geq 1 + \left\lfloor \left(\frac{1}{8\pi} \int_{\mathbb{R}^3} V(x)\, dx \right)_- \right\rfloor$$

For $V \leq 0$ this follows by conformal invariance as in Subsection 4.6 from the corresponding result on $\mathbb{S}^2$ in [**101**], which also contains references to earlier partial results. As in [**89**] the bound extends to not necessarily nonpositive V. In particular,

$$N_{\leq}(V) \geq \left(\frac{1}{8\pi} \int_{\mathbb{R}^3} V(x)\, dx \right)_- ,$$

which, in the radial case, goes back to [**85**],

A completely unrelated form of a reverse Lieb–Thirring inequality is shown in [**35**], namely, the inequality in Theorem 4.1 for $\gamma < -d/2$. The constant is the classical one. This follows by integrating the Golden–Thompson inequality [**86, 188, 191**].

5.8. Bounds on the number of negative eigenvalues in 2D. The CLR inequality does not hold for $\gamma = 0$ in $d = 2$ and there have been many attempts of finding suitable analogues. Phenomena one has to deal with are the existence of weakly coupled bound states [**179**] as well the existence of L^1 potentials with non-Weyl asymptotics [**14**].

Contributions to this area include [**60, 88, 103, 107, 114, 115, 153, 178, 184, 186, 193**]. In particular, the paper [**114**] raises the question of characterizing all $V \in L^1(\mathbb{R}^2)$ (or all $0 \geq V \in L^1(\mathbb{R}^2)$) such that either $\limsup_{\alpha\to\infty} \alpha^{-1} N(-\Delta + \alpha V) < \infty$ or such that (4.6) with $d = 2$ and $\gamma = 0$ holds. This problem was solved in the radial case in [**114**], but is still open in general. The eigenvalue bounds in [**88, 103, 115, 178, 184**] can be understood as sufficient conditions for an asymptotically linear bound.

5.9. Hardy–Lieb–Thirring inequalities. These are bounds where the operator $-\Delta$ is replaced by an operator $-\Delta - w$ with a function (Hardy weight) $w \geq 0$ such that $-\Delta - w \geq 0$. For the case $w(x) = (d-2)^2/(4|x|^2)$, as well as its extensions to powers of the Laplacian and magnetic fields, we refer to [**40, 49, 69**] and, for applications to the problem of stability of relativistic matter in magnetic fields, to [**68**]. For bounds on domains where w blows up at the boundary, see [**71, 82**], and for the fractional Pauli operator, see [**18**].

5.10. Equivalence of Sobolev and Lieb–Thirring inequalities. While it is clear that Lieb–Thirring inequalities imply Sobolev (interpolation) inequalities, it is quite remarkable that, in an abstract setting under certain assumptions, the converse implication holds as well. This was shown in [**122**] for CLR inequalities and extended in [**69, 70**] to LT inequalities. The analogue of Weidl's result for $\gamma = 1/2$ [**192**] is missing in this abstract framework.

5.11. Lieb–Thirring inequalities at positive density. In [**65, 66**] Lieb–Thirring inequalities were extended to the case of a positive, constant background density or, equivalently, to the case of potentials that tend to a positive constant at infinity. Informally, the inequalities can be written as

$$\begin{aligned}&\operatorname{Tr}\left((-\Delta+V-\mu)_-^\gamma-(-\Delta-\mu)_-^\gamma+\gamma(-\Delta-\mu)_-^{\gamma-1}V\right)\\&\qquad\leq L'_{\gamma,d}\int_{\mathbb{R}^d}\left((V-\mu)_-^{\gamma+\frac{d}{2}}-\mu^{\gamma+\frac{d}{2}}+\left(\gamma+\tfrac{d}{2}\right)\mu^{\gamma+\frac{d}{2}-1}V\right)dx\end{aligned}$$

with $\mu > 0$. With a suitable interpretation of the left side, these inequalities were shown in [**66**] for $\gamma \geq 1$ in dimensions $d \geq 2$. Conditions under which the difference $(-\Delta+V-\mu)_-^\gamma - (-\Delta-\mu)_-^\gamma$ is trace class where given in [**73**], see also [**74, 75**].

These Lieb–Thirring inequalities have found applications in the study of quantum many body systems at positive density, for instance, in [**125–127**].

The optimal values of the constants $L'_{\gamma,d}$ are not known. Are they semiclassical for $\gamma \geq 3/2$?

Moreover, for $\gamma = 1$ in $d = 1$ the inequality holds only with a logarithmic correction term. Does the inequality hold without this term for $\gamma > 1$?

Are there similar inequalities for $\gamma < 1$? In particular, for $\gamma = 0$ this is related to bounds for the spectral shift functions; see, e.g., [**162, 182**].

5.12. Lieb–Thirring inequalities for interacting systems. The paper [**150**] initiated the study of Lieb–Thirring inequalities for interacting systems. These inequalities generalize the form of the Lieb–Thirring inequality in Corollary 2.1, but the left side now takes into account interactions between the particles and the antisymmetry requirement is modified. Some works on this topic are [**79, 119, 147–149, 151**] and the references therein.

5.13. Lieb–Thirring inequalities for complex potentials. There has been some interest recently in extending Lieb–Thirring inequalities to the case of complex-valued potentials. It is known (see, e.g. [**54**]) that if $V \in L^{\gamma+d/2}(\mathbb{R}^d)$ with γ as in Theorem 4.1, then $-\Delta+V$ can be defined as an m-sectorial operator and its spectrum in $\mathbb{C}\setminus[0,\infty)$ consists of isolated eigenvalues of finite algebraic multiplicity.

Keller-type inequalities, that is, bounds on eigenvalues in terms of the L^p norm of the potential appeared first in [**1**]. Bounds on sums of eigenvalues outside a cone around the positive axis were proved in [**61**]. The Laptev–Safronov conjecture [**113**] concerns the optimal range of Keller-type inequalities and is still open; see [**51, 80**] for some results. For Keller-type bounds with other norms than L^p norms, see, for instance, [**28, 31, 46, 121, 172**] and references therein.

In [**19**] it is shown that for any $\gamma > d/2$ there is a bounded $V \in L^{\gamma+d/2}(\mathbb{R}^d)$ such that $-\Delta+V$ has infinitely many eigenvalues in the lower halfplane that accumulate at every point in $[0,\infty)$. Whether such V exist even for $\gamma > 1/2$ in $d \geq 2$ is open.

Bounds on sums of powers of eigenvalues are typically obtained either by identifying eigenvalues with zeros of an analytic function and then using tools from complex analysis, or by operator theoretic techniques. We refer to [**22, 32, 33, 54, 62, 76, 92**] and references therein. Despite these works, the natural form of the Lieb–Thirring inequality in the complex case seems to be unclear; see [**20**] for a counterexample in $d = 1$ to one possible form.

Much earlier, Pavlov [**159, 160**] has shown that the threshold between finitely and infinitely many eigenvalues, which is a $|x|^{-2}$ decay in the real case, becomes a $\exp(-c\sqrt{|x|})$ decay in the case of a complex potential. For a bound on the number of negative eigenvalues in the analogous problem for Jacobi matrices, see [**21**].

5.14. Inequalities for orthonormal systems. Lieb [**133**] showed that if $0 < \alpha < d/2$ and if $f_1, \ldots, f_N$ are orthonormal in $L^2(\mathbb{R}^d)$, then

$$\left\| \sum_{n=1}^N \left|(-\Delta)^{-\frac{\alpha}{2}} f_n\right|^2 \right\|_{\frac{d}{d-2\alpha}} \leq C_{d,\alpha}\, N^{\frac{d-2\alpha}{d}}.$$

This is a strengthening of the Hardy–Littlewood–Sobolev inequality, which concerns the case $N = 1$. The important feature of this bound is that N appears on the right side with an (optimal) exponent $\frac{d-2\alpha}{d} < 1$. Without orthogonality, the exponent would be 1. For an alternative proof, see [**169**], and for a conjecture about the optimal constant, see [**52**]. For related inequalities, see [**87, 94**].

In [**67, 76, 77**] a similar extension of the Strichartz inequality to orthonormal functions was proved. For instance, if $p, q \geq 1$ satisfy $\frac{2}{p} + \frac{d}{q} = d$ and $1 \leq q < 1 + \frac{2}{d-1}$ and if $f_1, \ldots, f_N$ are orthonormal in $L^2(\mathbb{R}^d)$, then

$$\left\| \sum_n \nu_n \left|e^{it\Delta} f_n\right|^2 \right\|_{L^p_t L^q_x} \leq C_{d,q} \|\nu\|_{\ell^{\frac{2q}{q+1}}}.$$

For further results and open problems, see [**10–12, 156**]. For applications of these bounds to the dynamics of quantum many-body systems, see, for instance, [**125, 126**].

The papers [**54, 76, 78**] also contain further bounds on orthonormal systems related to Fourier restriction estimates. These have applications to bounds on eigenvalues of Schrödinger operators with complex potentials.

6. Some proofs

6.1. Proof of Theorem 1.5. The following result is due to [**59**].

THEOREM 6.1. *Let $d \geq 1$ and $q \geq 1$. Let $N \in \mathbb{N}$ and let $u_1, \ldots, u_N \in H^1(\mathbb{R}^d, \mathbb{C}^q)$ be orthonormal in $L^2(\mathbb{R}^d, \mathbb{C}^q)$. Then*

$$\sum_{n=1}^N \int_{\mathbb{R}^d} |\nabla u_n(x)|^2_{\mathbb{C}^q}\, dx \geq \tilde{K}_d \int_{\mathbb{R}^d} \operatorname{Tr}_{\mathbb{C}^q} \left(R(x)^{1+2/d} \right) dx,$$

where $R(x)$ is the Hermitian nonnegative $q \times q$ matrix given by

$$R(x) = \sum_{n=1}^N u_n(x) u_n(x)^*$$

and where

$$\tilde{K}_d = \frac{2^{6/d}\, d^2\, (2\pi)^2}{(d+2)^{2+4/d}\, |\mathbb{S}^{d-1}|^{2/d}}\, \mathcal{I}_d^{-2/d} = \frac{2^{6/d}\, d^{1-2/d}}{(d+2)^{1+4/d}}\, \mathcal{I}_d^{-2/d}\, K_d^{\mathrm{cl}}$$

with

$$\mathcal{I}_d = \inf\left\{ \left(\int_0^\infty w(s)^2\, ds\right)^{d/2} \int_0^\infty \frac{(1-g(t))^2}{t^{1+d/2}}\, dt :\ f,\, w \geq 0\,,\ \int_0^\infty f(s)^2\, ds = 1, \right.$$
$$\left. g(t) = \int_0^\infty w(s) f(st)\, ds \right\}.$$

PROOF. *Step 1.* Let f be a nonnegative function on $(0,\infty)$ with $\int_0^\infty f(s)^2\, ds = 1$. For any $E>0$ we define functions $u_1^E, \ldots, u_N^E$ by

$$\widehat{u_n^E}(\xi) = f(E/|\xi|^2)\widehat{u_n}(\xi) \qquad \text{for all } \xi \in \mathbb{R}^d\,.$$

Then, since

$$|\xi|^2 = \int_0^\infty f(E/|\xi|^2)^2\, dE \qquad \text{for all } \xi \in \mathbb{R}^d\,,$$

we have

$$\sum_{n=1}^N \int_{\mathbb{R}^d} |\nabla u_n(x)|_{\mathbb{C}^q}^2\, dx = \sum_{n=1}^N \int_{\mathbb{R}^d} |\xi|^2 |\widehat{u_n}(\xi)|_{\mathbb{C}^q}^2\, d\xi = \sum_{n=1}^N \int_{\mathbb{R}^d} \int_0^\infty |\widehat{u_n^E}(\xi)|_{\mathbb{C}^q}^2\, dE\, d\xi$$
$$= \int_{\mathbb{R}^d} \sum_{n=1}^N \int_0^\infty |u_n^E(x)|_{\mathbb{C}^q}^2\, dE\, dx\,. \tag{6.1}$$

Our goal will be to bound $\sum_{n=1}^N \int_0^\infty |u_n^E(x)|_{\mathbb{C}^q}^2\, dE$ from below pointwise in x. We note that in Rumin's original argument [**170**], $\sum_{n=1}^N |u_n^E(x)|_{\mathbb{C}^q}^2$ is bounded pointwise in x and E. The additional integration in E, however, allows us to improve the constant.

Step 2. Let $R^E(x)$ be the Hermitian nonnegative $q\times q$ matrix given by

$$R^E(x) = \sum_{n=1}^N u_n^E(x) u_n^E(x)^*\,.$$

Moreover, let w be a nonnegative, square-integrable function on $(0,\infty)$ and let

$$g(t) := \int_0^\infty w(s) f(st)\, ds\,.$$

The crucial step in the proof will be to show that for any $x\in\mathbb{R}^d$, $\varepsilon>0$ and $\mu>0$ one has, in the sense of matrices,

$$R(x) \leq (1+\varepsilon)\mu^{-1}\|w\|_2^2 \int_0^\infty R^E(x)\, dE + (1+\varepsilon^{-1}) A \mu^{d/2}\,, \tag{6.2}$$

where

$$A = 2^{-1}(2\pi)^{-d}|\mathbb{S}^{d-1}| \int_0^\infty (1-g(t))^2 t^{-1-d/2}\, dt\,.$$

In order to prove (6.2), for any $E>0$ let $v_1^E, \ldots, v_N^E \in L^2(\mathbb{R}^d, \mathbb{C}^q)$ be defined by

$$\widehat{v_n^E}(\xi) = g(E/|\xi|^2)\widehat{u_n}(\xi) \qquad \text{for all } \xi \in \mathbb{R}^d\,.$$

For $e \in \mathbb{C}^q$, $n \in \{1,\ldots,N\}$ and $\mu > 0$ we bound

$$
\begin{aligned}
|e^* u_n(x)|^2 &= |e^* v_n^\mu(x)|^2 + 2\operatorname{Re}\overline{e^* v_n^\mu(x)}\ e^*\left(u_n(x) - v_n^\mu(x)\right) + |e^*(u_n(x) - v_n^\mu(x))|^2 \\
&\leq (1+\varepsilon)\,|e^* v_n^\mu(x)|^2 + (1+\varepsilon^{-1})\,|e^*(u_n(x) - v_n^\mu(x))|^2 . \qquad (6.3)
\end{aligned}
$$

For the first term on the right side we have, by the Schwarz inequality,

$$
\begin{aligned}
|e^* v_n^\mu(x)|^2 &= (2\pi)^{-d} \left| \int_{\mathbb{R}^d} \int_0^\infty e^{i\xi\cdot x} w(s) f(s\mu/|\xi|^2) e^* \widehat{u_n}(\xi)\, ds\, d\xi \right|^2 \\
&\leq \|w\|_2^2\, (2\pi)^{-d} \int_0^\infty \left| \int_{\mathbb{R}^d} e^{i\xi\cdot x} f(s\mu/|\xi|^2) e^* \widehat{u_n}(\xi)\, d\xi \right|^2 ds \\
&= \|w\|_2^2 \int_0^\infty |e^* u_n^{s\mu}(x)|^2\, ds \\
&= \mu^{-1} \|w\|_2^2 \int_0^\infty |e^* u_n^E(x)|^2\, dE .
\end{aligned}
$$

Inserting this into (6.3) and summing over n, we obtain

$$
\begin{aligned}
e^* R(x) e &= \sum_n |e^* u_n(x)|^2 \\
&\leq (1+\varepsilon)\mu^{-1}\|w\|_2^2 \int_0^\infty \sum_{n=1}^N |e^* u_n^E(x)|^2\, dE \\
&\quad + (1+\varepsilon^{-1}) \sum_{n=1}^N |e^*(u_n(x) - v_n^\mu(x))|^2 \\
&= (1+\varepsilon)\mu^{-1}\|w\|_2^2 \int_0^\infty e^* R^E(x) e\, dE + (1+\varepsilon^{-1}) \sum_{n=1}^N |e^*(u_n(x) - v_n^\mu(x))|^2 .
\end{aligned}
$$

To bound the second term on the right side, we write

$$
e^*(u_n(x) - v_n^\mu(x)) = (2\pi)^{-d/2} \int_{\mathbb{R}^d} e^{i\xi\cdot x} (1 - g(\mu/|\xi|^2)) e^* \widehat{u_n}(\xi)\, d\xi = (\chi_{x,\mu} e, \widehat{u_n}) ,
$$

where the last inner product is in $L^2(\mathbb{R}^d, \mathbb{C}^q)$ and where

$$
\chi_{x,\mu}(\xi) := (2\pi)^{-d/2} e^{-i\xi\cdot x} (1 - g(\mu/|\xi|^2)) \qquad \text{for all } \xi \in \mathbb{R}^d .
$$

Since the $\widehat{u_n}$ are orthonormal, we obtain by Bessel's inequality

$$
\sum_{n=1}^N |e^*(u_n(x) - v_n^\mu(x))|^2 \leq \|\chi_{x,\mu} e\|^2 = \tilde{A}\, \mu^{d/2}\, |e|_{\mathbb{C}^q}^2 ,
$$

where

$$
\tilde{A} = (2\pi)^{-d} \int_{\mathbb{R}^d} (1 - g(1/|\eta|^2))^2\, d\eta = A .
$$

To summarize, we have shown that

$$
e^* R(x) e \leq (1+\varepsilon)\mu^{-1}\|w\|_2^2 \int_0^\infty e^* R^E(x) e\, dE + (1+\varepsilon^{-1})\, A\, \mu^{d/2}\, |e|_{\mathbb{C}^q}^2 ,
$$

which is the same as (6.2).

Step 3. We denote by $\lambda_j(H)$, $j = 1, \ldots, q$, the eigenvalues of a Hermitian $q \times q$ matrix H, arranged in nonincreasing order and counted according to multiplicities. Then, by the variational principle, the matrix inequality (6.2) implies that

$$\lambda_j(R(x)) \leq (1+\varepsilon)\mu^{-1}\|w\|_2^2\lambda_j\left(\int_0^\infty R^E(x)\,dE\right)+(1+\varepsilon^{-1})\,A\,\mu^{d/2} \quad \text{for } j = 1, \ldots, q.$$

Optimizing with respect to $\varepsilon > 0$ and $\mu > 0$ for each j, we obtain

$$\lambda_j(R(x)) \leq \left(\frac{2}{d}\right)^{\frac{2d}{d+2}} \left(1+\frac{d}{2}\right)^2 \|w\|_2^{\frac{2d}{d+2}} A^{\frac{2}{d+2}} \left(\lambda_j\left(\int_0^\infty R^E(x)\,dE\right)\right)^{\frac{d}{d+2}},$$

which is the same as

$$\lambda_j\left(\int_0^\infty R^E(x)\,dE\right) \geq \left(\frac{d}{2}\right)^2 \left(1+\frac{d}{2}\right)^{-\frac{2(d+2)}{d}} \|w\|_2^{-2} A^{-\frac{2}{d}} \left(\lambda_j(R(x))\right)^{1+\frac{2}{d}}.$$

Thus,

$$\begin{aligned}\int_0^\infty \sum_{n=1}^N |u_n^E(x)|_{\mathbb{C}^q}^2\,dE &= \operatorname{Tr}_{\mathbb{C}^q} \int_0^\infty R^E(x)\,dE = \sum_{j=1}^q \lambda_j\left(\int_0^\infty R^E(x)\,dE\right) \\ &\geq \left(\frac{d}{2}\right)^2 \left(1+\frac{d}{2}\right)^{-\frac{2(d+2)}{d}} \|w\|_2^{-2} A^{-\frac{2}{d}} \sum_{j=1}^q \left(\lambda_j(R(x))\right)^{1+\frac{2}{d}} \\ &= \left(\frac{d}{2}\right)^2 \left(1+\frac{d}{2}\right)^{-\frac{2(d+2)}{d}} \|w\|_2^{-2} A^{-\frac{2}{d}} \operatorname{Tr}_{\mathbb{C}^q}\left(R(x)^{1+\frac{2}{d}}\right).\end{aligned}$$

Inserting this bound into (6.1) we obtain the claimed bound. □

We now prove an upper bound on $\mathcal{I}_1$ by choosing appropriate trial functions f and w. We also prove a lower bound on $\mathcal{I}_1$, which shows the limitation of the method.

LEMMA 6.2. *If $d = 1$, then*

$$\frac{2}{3} \leq \mathcal{I}_1 \leq 0.747112.$$

In particular,

$$\tilde{K}_1 \geq (1.456)^{-2}\,K_1^{\mathrm{cl}}. \tag{6.4}$$

PROOF. Let $f, w \geq 0$ with $\int_0^\infty f(s)^2\,ds = 1$ and denote $a = \int_0^\infty w(s)^2\,ds$. Then

$$g(t) = \int_0^\infty w(s)f(st)\,ds \leq \left(\int_0^\infty w(s)^2\,ds\right)^{1/2} \left(\int_0^\infty f(st)^2\,ds\right)^{1/2} = a^{1/2}t^{-1/2}$$

and therefore $|1-g(t)| \geq (1-a^{1/2}t^{-1/2})_+$ for all $t > 0$. Thus,

$$\int_0^\infty \frac{(1-g(t))^2}{t^{3/2}}\,dt \geq \int_0^\infty \frac{(1-a^{1/2}t^{-1/2})_+^2}{t^{3/2}}\,dt = \frac{2}{3}a^{-1/2},$$

which implies that $\mathcal{I}_1 \geq \frac{2}{3}$, as claimed.

In order to prove the upper bound on $\mathcal{I}_1$ we choose

$$f(s) = (1+\mu_0 s^{4.5})^{-0.25}, \qquad w(s) = c_0\frac{(1-s^{0.36})^{2.1}}{1+s}\chi_{[0,1]}(s),$$

where μ_0 and c_0 are determined by $\int_0^\infty f(s)^2\,ds = \int_0^\infty w(s)\,ds = 1$. A numerical computation leads to the claimed bound on $\mathcal{I}_1$. □

We now complete the proof of Theorem 1.5. In fact, we will argue by duality and prove (3.7). This will follow by dualizing Theorem 6.1 and applying the Laptev–Weidl method of lifting the dimension [**116**].

Let $\mathcal{G}$ be a separable Hilbert space. We first observe that the inequality in Theorem 6.1 remains valid, with the same constant $\tilde{K}_d$, for functions $u_1, \ldots, u_N \in H^1(\mathbb{R}^d, \mathcal{G})$ that are orthonormal in $L^2(\mathbb{R}^d, \mathcal{G})$. This follows by a simple approximation argument using a sequence of finite dimensional projections on $\mathcal{G}$ that converges strongly to the identity.

Next, by the same duality argument as in the proof of Theorem 3.1 we infer that

$$\sum_n |E_n(-\Delta + W)| \le \tilde{L}_d \int_{\mathbb{R}^d} \operatorname{Tr}_{\mathcal{G}} \left(W(x)_-^{1+\frac{d}{2}} \right) dx \tag{6.5}$$

for any measurable function W from $\mathbb{R}^d$ into the selfadjoint operators on $\mathcal{G}$ such that $W(x)_-^{1+\frac{d}{2}}$ is trace class for almost every x and such that the integral of its trace is finite. The operator $-\Delta + W$ on the left side of (6.5) acts in $L^2(\mathbb{R}^d, \mathcal{G})$. The constant $\tilde{L}_d$ is related to the constant $\tilde{K}_d$ in Theorem 6.1 by

$$\left((1+\tfrac{d}{2})\tilde{L}_d \right)^{1+\frac{2}{d}} \left((1+\tfrac{2}{d})\tilde{K}_d \right)^{1+\frac{d}{2}} = 1\,. \tag{6.6}$$

Now let $V \in L^{1+\frac{d}{2}}(\mathbb{R}^d)$ be real. We introduce coordinates $x = (x', x_d) \in \mathbb{R}^{d-1} \times \mathbb{R}$ in $\mathbb{R}^d$ and write

$$-\Delta + V = -\tfrac{d^2}{dx_d^2} + W \qquad \text{in } L^2(\mathbb{R}^d) = L^2(\mathbb{R}, \mathcal{G})$$

where W acts as 'multiplication' by

$$W(x_d) := -\Delta' + V(\cdot, x_d) \qquad \text{in } \mathcal{G} := L^2(\mathbb{R}^{d-1})\,.$$

Applying inequality (6.5) with $d = 1$, we obtain

$$\sum_n |E_n(-\Delta + V)| = \sum_n |E_n(-\tfrac{d^2}{d^2 x_d} + W)| \le \tilde{L}_1 \int_{\mathbb{R}} \operatorname{Tr}_{\mathcal{G}} \left(W(x_d)_-^{\frac{3}{2}} \right) dx_d\,.$$

Moreover, by the Lieb–Thirring inequality of Laptev and Weidl [**116**], for any $x_d \in \mathbb{R}$,

$$\operatorname{Tr}_{\mathcal{G}} \left(W(x_d)_-^{\frac{3}{2}} \right) = \sum_n |E_n(-\Delta' + V(\cdot, x_d))|^{3/2} \le L^{\mathrm{cl}}_{3/2,d-1} \int_{\mathbb{R}^{d-1}} V(x', x_d)_-^{\frac{3}{2}+\frac{d-1}{2}}\,dx'\,.$$

Inserting this into the above bound we obtain

$$\sum_n |E_n(-\Delta + V)| \le \tilde{L}_1\, L^{\mathrm{cl}}_{3/2,d-1} \int_{\mathbb{R}^d} V(x)_-^{\frac{3}{2}+\frac{d-1}{2}}\,dx\,. \tag{6.7}$$

It remains to bound the constant on the right side using the explicit bound on $\tilde{K}_1$ from Lemma 6.2. We first note that

$$L^{\mathrm{cl}}_{1,1} L^{\mathrm{cl}}_{3/2,d-1} = L^{\mathrm{cl}}_d\,.$$

This follows either using the explicit expression for $L^{\rm cl}_{\gamma,d}$ together with identities for gamma functions or, more conceptually, from

$$\begin{aligned} L_d^{\rm cl} &= \int_{\mathbb{R}^d} (|\xi|^2-1)_- \frac{d\xi}{(2\pi)^d} = \int_{\mathbb{R}^{d-1}} \int_{\mathbb{R}} ((\xi')^2+\xi_d^2-1)_- \frac{d\xi_d}{2\pi}\frac{d\xi'}{(2\pi)^{d-1}} \\ &= \int_{\mathbb{R}^{d-1}} ((\xi')^2-1)_-^{\frac32} \int_{\mathbb{R}} (\eta_d^2-1)_- \frac{d\eta_d}{2\pi}\frac{d\xi'}{(2\pi)^{d-1}} = L^{\rm cl}_{3/2,d-1} L^{\rm cl}_{1,1}\,, \end{aligned}$$

where we changed variables $\xi_d = ((\xi')^2-1)_-^{1/2}\eta_d$. Moreover, by (3.6) and (6.6),

$$\left(\tilde L_1/L_1^{\rm cl}\right)^3 \left(\tilde K_1/K_1^{\rm cl}\right)^{\frac32} = 1\,.$$

Thus, by (6.4),

$$\tilde L_1\, L^{\rm cl}_{3/2,d-1} = \frac{\tilde L_1}{L_1^{\rm cl}}\, L_d^{\rm cl} = \left(\frac{K_1^{\rm cl}}{\tilde K_1}\right)^{\frac12} L_d^{\rm cl} \le 1.456\, L_d^{\rm cl}\,.$$

Inserting this into (6.7) we obtain (3.7), as claimed.

6.2. Proof of (b) in Proposition 4.5. Here we focus on the assertions in part (b) of Proposition 4.5 concerning $\gamma>0$. Those concerning $\gamma=0$ have already been discussed in Subsection 4.6.

We follow [**56**]. Let us define $L^{(N)}_{\gamma,d}$ to be the best constant in the inequality

$$\sum_{n=1}^N |E_n(-\Delta+V)|^\gamma \le L^{(N)}_{\gamma,d} \int_{\mathbb{R}^d} V(x)_-^{\gamma+\frac d2}\,dx\,. \tag{6.8}$$

Then, clearly, $L^{(N)}_{\gamma,d} \le L^{(N+1)}_{\gamma,d}$ and $L_{\gamma,d} = \lim_{N\to\infty} L^{(N)}_{\gamma,d}$. The remaining part of assertion (b) in Proposition 4.5 is therefore a consequence of the following result.

PROPOSITION 6.3. *If $\gamma > \max\{2-d/2,0\}$, then $L^{(2)}_{\gamma,d} > L^{(1)}_{\gamma,d}$.*

Define p by $p' = \gamma + d/2$ and recall from Subsection 4.2 that inequality (4.3) has an optimizer Q. After a translation, a dilation and multiplication by a constant we can assume that Q is positive, centered at the origin and satisfies

$$-\Delta Q - Q^{2p-1} = -Q \qquad \text{in } \mathbb{R}^d\,.$$

We abbreviate $m := \int_{\mathbb{R}^d} Q^2\,dx$ and record two identities for the function Q, namely,

$$\int_{\mathbb{R}^d} |\nabla Q|^2\,dx - \int_{\mathbb{R}^d} Q^{2p}\,dx = -m \qquad \text{and} \tag{6.9}$$

$$\left(\frac d2 - 1\right)\int_{\mathbb{R}^d} |\nabla Q|^2\,dx - \frac{d}{2p}\int_{\mathbb{R}^d} Q^{2p}\,dx = -\frac d2 m\,. \tag{6.10}$$

They follow by multiplying the equation for Q by Q and by $x\cdot\nabla Q$, respectively.

Next, for a parameter $R>0$, let

$$Q_\pm(x) = Q(x \pm (\tfrac R2, 0))$$

and

$$V = -(Q_+^2 + Q_-^2)^{p-1}\,.$$

The main ingredient of the proof of Proposition 6.3 is the following bound.

LEMMA 6.4. *As $R \to \infty$,*

$$|E_1(-\Delta+V)|^\gamma + |E_2(-\Delta+V)|^\gamma \geq 2\left(1+\frac{\gamma}{m}A+o(A)\right)$$

where

$$A = \frac{1}{2}\int_{\mathbb{R}^d}\left((Q_+^2+Q_-^2)^p - Q_+^{2p} - Q_-^{2p}\right)dx \to 0\,.$$

Before proving this lemma, let us use it to deduce Proposition 6.3. We clearly have

$$\int_{\mathbb{R}^d} V_-^{\gamma+d/2}\,dx = 2\int_{\mathbb{R}^d} Q^{2p}\,dx + 2A\,,$$

so

$$\begin{aligned}\frac{|E_1(-\Delta+V)|^\gamma + |E_2(-\Delta+V)|^\gamma}{\int_{\mathbb{R}^d} V_-^{\gamma+d/2}\,dx} &\geq \frac{1}{\int_{\mathbb{R}^d} Q^{2p}\,dx}\,\frac{1+\frac{\gamma}{m}A+o(A)}{1+\left(\int_{\mathbb{R}^d} Q^{2p}\,dx\right)^{-1}A}\\ &= L_{\gamma,d}^{(1)}\left(1+\left(\gamma-\frac{m}{\int_{\mathbb{R}^d} Q^{2p}\,dx}\right)\frac{A}{m}+o(A)\right).\end{aligned}$$

Here we used the fact that

$$\frac{1}{\int_{\mathbb{R}^d} Q^{2p}\,dx} = L_{\gamma,d}^{(1)}\,,$$

which follows from (4.4), (6.9) and (6.10). Using the latter two identities again, we find

$$\gamma - \frac{m}{\int_{\mathbb{R}^d} Q^{2p}\,dx} = \frac{\gamma}{p}\,,$$

and therefore

$$L_{\gamma,d}^{(2)} \geq L_{\gamma,d}^{(1)}\left(1+\frac{\gamma}{p}\frac{A}{m}+o(A)\right),$$

which completes the proof of the proposition. Thus, it remains to prove the lemma.

PROOF. Clearly, $E := \int_{\mathbb{R}^d} Q_+Q_-\,dx \to 0$ as $R \to \infty$, and therefore in the following we may assume that $|E| < m$. Then the two functions $\psi^{(\pm)}$ defined by

$$\begin{pmatrix}\psi^{(+)}\\ \psi^{(-)}\end{pmatrix} := \begin{pmatrix} m & E\\ E & m\end{pmatrix}^{-1/2}\begin{pmatrix}Q_+\\ Q_-\end{pmatrix}$$

are orthonormal in $L^2(\mathbb{R}^d)$. Let

$$\mathcal{H} := \begin{pmatrix}\langle\psi^{(+)},(-\Delta+V)\psi^{(+)}\rangle & \langle\psi^{(+)},(-\Delta+V)\psi^{(-)}\rangle\\ \langle\psi^{(-)},(-\Delta+V)\psi^{(+)}\rangle & \langle\psi^{(-)},(-\Delta+V)\psi^{(-)}\rangle\end{pmatrix}.$$

By the variational principle, the two lowest eigenvalues of $-\Delta+V$ are not larger than the corresponding eigenvalues of $\mathcal{H}$ and therefore, in particular,

$$|E_1(-\Delta+V)|^\gamma + |E_2(-\Delta+V)|^\gamma \geq \operatorname{Tr}\mathcal{H}_-^\gamma\,.$$

We have

$$\mathcal{H} = h + \begin{pmatrix}0 & \delta\\ \delta & 0\end{pmatrix}$$

with

$$h = \langle\psi^{(+)},(-\Delta+V)\psi^{(+)}\rangle = \langle\psi^{(-)},(-\Delta+V)\psi^{(-)}\rangle$$

and

$$\delta = \langle\psi^{(+)},(-\Delta+V)\psi^{(-)}\rangle = \langle\psi^{(-)},(-\Delta+V)\psi^{(+)}\rangle\,.$$

It is easy to see that $h \to -1$ and $\delta \to 0$ as $R \to \infty$, and therefore

$$\operatorname{Tr} \mathcal{H}_-^\gamma = 2|h|^\gamma - \gamma |h|^{\gamma-1} \operatorname{Tr} \begin{pmatrix} 0 & \delta \\ \delta & 0 \end{pmatrix} + \mathcal{O}(\delta^2) = 2|h|^\gamma + \mathcal{O}(\delta^2).$$

Next, let us expand h. We have

$$|\nabla \psi^{(+)}|^2 + |\nabla \psi^{(-)}|^2 = \frac{m}{m^2 - E^2} \left(|\nabla Q_+|^2 + |\nabla Q_-|^2 \right) - \frac{2E}{M^2 - E^2} \nabla Q_+ \cdot \nabla Q_-$$

and therefore, using the equation for Q,

$$\begin{aligned} \int_{\mathbb{R}^d} \left(|\nabla \psi^{(+)}|^2 + |\nabla \psi^{(-)}|^2 \right) dx = -2 + \frac{2m}{m^2 - E^2} \int_{\mathbb{R}^d} Q^{2p} \, dx \\ - \frac{E}{m^2 - E^2} \int_{\mathbb{R}^d} \left(Q_+^{2p-2} + Q_-^{2p-2} \right) Q_+ Q_- \, dx . \end{aligned}$$

Similarly,

$$(\psi^{(+)})^2 + (\psi^{(-)})^2 = \frac{m}{m^2 - E^2} \left(Q_+^2 + Q_-^2 \right) - \frac{2E}{M^2 - E^2} Q_+ Q_-$$

and therefore

$$\begin{aligned} h &= \frac{1}{2} \left(\langle \psi^{(+)}, (-\Delta + V) \psi^{(+)} \rangle + \langle \psi^{(-)}, (-\Delta + V) \psi^{(-)} \rangle \right) \\ &= -1 - \frac{m}{m^2 - E^2} A + \frac{E}{m^2 - E^2} B , \end{aligned}$$

where A is as in the lemma and

$$B := \int_{\mathbb{R}^d} Q_+ Q_- \left((Q_+^2 + Q_-^2)^{p-1} - \frac{1}{2} \left(Q_+^{2p-2} + Q_-^{2p-2} \right) \right) dx .$$

Using $Q(x) \leq C(1+|x|)^{-(d-1)/2} e^{-|x|}$ we can bound

$$E = \mathcal{O}_\varepsilon(e^{-(1-\varepsilon)R}) \qquad \text{and} \qquad B = \mathcal{O}_\varepsilon(e^{-(1-\varepsilon)R})$$

and obtain

$$|h|^\gamma = (-h)^\gamma = (1 + m^{-1} A)^\gamma + \mathcal{O}_\varepsilon(e^{-(2-\varepsilon)R}) = 1 + \gamma m^{-1} A + \mathcal{O}(A^2) + \mathcal{O}_\varepsilon(e^{-(2-\varepsilon)R}) .$$

This gives the desired expansion of h expansion.

Next, we show $\delta = \mathcal{O}_\varepsilon(e^{-(2-\varepsilon)R})$. By a similar computation as before, we find that

$$\nabla \psi^{(+)} \cdot \nabla \psi^{(-)} = -\frac{E}{m^2 - E^2} \frac{1}{2} \left(|\nabla Q_+|^2 + |\nabla Q_-|^2 \right) + \frac{m}{m^2 - E^2} \nabla Q_+ \cdot \nabla Q_-$$

and therefore, by the equation,

$$\begin{aligned} \int_{\mathbb{R}^d} \nabla \psi^{(+)} \cdot \nabla \psi^{(-)} dx = -\frac{E}{m^2 - E^2} \int_{\mathbb{R}^d} Q^{2p} \, dx \\ + \frac{m}{m^2 - E^2} \frac{1}{2} \int_{\mathbb{R}^d} Q_+ Q_- (Q_+^{2p-2} + Q_-^{2p-2}) \, dx. \end{aligned}$$

Moreover,

$$\psi^{(+)} \psi^{(-)} = -\frac{E}{m^2 - E^2} \frac{1}{2} \left(Q_+^2 + Q_-^2 \right) + \frac{m}{m^2 - E^2} Q_+ Q_-$$

and therefore

$$\int_{\mathbb{R}^d} V\psi^{(+)}\psi^{(-)}\,dx = \frac{E}{m^2-E^2}\frac{1}{2}\int_{\mathbb{R}^d}(Q_+^2+Q_-^2)^p\,dx$$
$$-\frac{m}{m^2-E^2}\int_{\mathbb{R}^d} Q_+Q_-(Q_+^2+Q_-^2)^{p-1}\,dx.$$

Thus,

$$\begin{aligned}\delta &= \frac{1}{2}\left(\langle\psi^{(+)},(-\Delta+V)\psi^{(-)}\rangle+\langle\psi^{(-)},(-\Delta+V)\psi^{(+)}\rangle\right)\\ &= \frac{E}{m^2-E^2}A-\frac{m}{m^2-E^2}B\,.\end{aligned}$$

Together with the above bounds on E and B, this gives the claimed bound on δ.

To summarize, so far we have shown that

$$|E_1(-\Delta+V)|^\gamma+|E_2(-\Delta+V)|^\gamma \geq 2\left(1+\frac{\gamma}{m}A+\mathcal{O}_\varepsilon(e^{-(2-\varepsilon)R})+\mathcal{O}(A^2)\right).$$

To get a lower bound on A we use $Q(x)\geq c(1+|x|)^{-(d-1)/2}e^{-|x|}$. Therefore [**87**], the integrand in the definition of A is $\geq c_\varepsilon e^{-(p+\varepsilon)R}$ if $|x|\leq 1$, which gives

$$A\geq c'_\varepsilon e^{-(p+\varepsilon)R}\,.$$

This dominates the error term $\mathcal{O}_\varepsilon(e^{-(2-\varepsilon)R})$ if $p<2$ (that is, $\gamma+d/2>2$) and completes the proof of the lemma. □

In fact, a stronger conclusion than that in Proposition 6.3 can be shown [**58**]. Namely, if $\gamma>\max\{2-d/2,0\}$, then $L_{\gamma,d}^{(2N)}>L_{\gamma,d}^{(N)}$ for all $N\geq 1$. Therefore, in particular,

$$L_{\gamma,d}^{(N)}<L_{\gamma,d}\qquad\text{for all } N\geq 1 \text{ if }\begin{cases}\gamma>3/2 & \text{if } d=1\,,\\ \gamma>1 & \text{if } d=2\,,\\ \gamma>1/2 & \text{if } d=3\,,\\ \gamma>0 & \text{if } d\geq 4\,.\end{cases}$$

This shows that the best Lieb–Thirring constant cannot be attained for a potential having finitely many negative eigenvalues for the corresponding values of γ.

A proof of this stronger conclusion under the additional assumption $\gamma\geq 1$ appeared in [**56**]. It uses the following equivalence, which generalizes (4.4) to general N, provided $\gamma\geq 1$.

LEMMA 6.5. *Let $1\leq\gamma<\infty$ and $1<p\leq 1+\frac{2}{d}$ be related by $\gamma=p'-\frac{d}{2}$, and let $N\in\mathbb{N}$. Then inequality (6.8) is equivalent to the inequality*

$$\sum_{n=1}^N\|\nabla u_n\|^2\geq K_{p,d}^{(N)}\left(\int_{\mathbb{R}^d}\left(\sum_{n=1}^N|u_n|^2\right)^p dx\right)^{\frac{2}{d(p-1)}}\left(\sum_{n=1}^N\|u_n\|^{\frac{2(1-(1-2/d)p)}{1+2/d-p}}\right)^{-\frac{2}{d(p-1)}+1}$$

for all $(u_n)\subset H^1(\mathbb{R}^d)$ that are orthogonal in $L^2(\mathbb{R}^d)$, in the sense that the optimal constants satisfy

$$L_{p'-d/2,d}^{(N)}\left(K_{p,d}^{(N)}\right)^{\frac{d}{2}}=\left(\frac{d}{2p'}\right)^{\frac{d}{2}}\left(\frac{2p'-d}{2p'}\right)^{\frac{2p'-d}{2}}.$$

Note that the u_n are orthogonal and not necessarily orthonormal. Lemma 6.5 follows by an argument similarly as in the proof of Theorem 3.1, see [**56**]. The analogue corresponding to $N = \infty$ can be found in [**145**].

References

[1] A. A. Abramov, A. Aslanyan, and E. B. Davies, *Bounds on complex eigenvalues and resonances*, J. Phys. A **34** (2001), no. 1, 57–72, DOI 10.1088/0305-4470/34/1/304. MR1819914

[2] M. Aizenman and E. H. Lieb, *On semiclassical bounds for eigenvalues of Schrödinger operators*, Phys. Lett. A **66** (1978), no. 6, 427–429, DOI 10.1016/0375-9601(78)90385-7. MR598768

[3] T. Aubin, *Problèmes isopérimétriques et espaces de Sobolev* (French), J. Differential Geometry **11** (1976), no. 4, 573–598. MR448404

[4] V. Bach, W. de Siqueira Pedra, and S. N. Lakaev, *Bounds on the discrete spectrum of lattice Schrödinger operators*, J. Math. Phys. **59** (2018), no. 2, 022109, 25, DOI 10.1063/1.5006641. MR3766378

[5] J. R. Baxter, *Inequalities for potentials of particle systems*, Illinois J. Math. **24** (1980), no. 4, 645–652. MR586803

[6] R. Benguria and M. Loss, *A simple proof of a theorem of Laptev and Weidl*, Math. Res. Lett. **7** (2000), no. 2-3, 195–203, DOI 10.4310/MRL.2000.v7.n2.a5. MR1764316

[7] R. D. Benguria and M. Loss, *Connection between the Lieb-Thirring conjecture for Schrödinger operators and an isoperimetric problem for ovals on the plane*, Partial differential equations and inverse problems, Contemp. Math., vol. 362, Amer. Math. Soc., Providence, RI, 2004, pp. 53–61, DOI 10.1090/conm/362/06604. MR2091490

[8] F. A. Berezin, *Covariant and contravariant symbols of operators* (Russian), Izv. Akad. Nauk SSSR Ser. Mat. **36** (1972), 1134–1167. MR0350504

[9] J. Bernstein and T. Mettler, *One-dimensional projective structures, convex curves and the ovals of Benguria and Loss*, Comm. Math. Phys. **336** (2015), no. 2, 933–952, DOI 10.1007/s00220-014-2275-7. MR3322391

[10] N. Bez, Y. Hong, S. Lee, S. Nakamura, and Y. Sawano, *On the Strichartz estimates for orthonormal systems of initial data with regularity*, Adv. Math. **354** (2019), 106736, 37, DOI 10.1016/j.aim.2019.106736. MR3985036

[11] N. Bez, S. Lee, and S. Nakamura, *Maximal estimates for the Schrödinger equation with orthonormal initial data*, Selecta Math. (N.S.) **26** (2020), no. 4, Paper No. 52, 24, DOI 10.1007/s00029-020-00582-6. MR4125984

[12] N. Bez, S. Lee, and S. Nakamura, *Strichartz estimates for orthonormal families of initial data and weighted oscillatory integral estimates*, Forum Math. Sigma **9** (2021), Paper No. e1, 52, DOI 10.1017/fms.2020.64. MR4202486

[13] M. Š. Birman, *On the spectrum of singular boundary-value problems* (Russian), Mat. Sb. (N.S.) **55 (97)** (1961), 125–174. MR0142896

[14] M. Sh. Birman and A. Laptev, *The negative discrete spectrum of a two-dimensional Schrödinger operator*, Comm. Pure Appl. Math. **49** (1996), no. 9, 967–997, DOI 10.1002/(SICI)1097-0312(199609)49:9⟨967::AID-CPA3⟩3.3.CO;2-O. MR1399202

[15] M. Š. Birman and M. Z. Solomjak, *Quantitative analysis in Sobolev imbedding theorems and applications to spectral theory*, American Mathematical Society Translations, Series 2, vol. 114, American Mathematical Society, Providence, R.I., 1980. Translated from the Russian by F. A. Cezus. MR562305

[16] X. Blanc and M. Lewin, *Existence of the thermodynamic limit for disordered quantum Coulomb systems*, J. Math. Phys. **53** (2012), no. 9, 095209, 32, DOI 10.1063/1.4729052. MR2905791

[17] Ph. Blanchard and J. Stubbe, *Bound states for Schrödinger Hamiltonians: phase space methods and applications*, Rev. Math. Phys. **8** (1996), no. 4, 503–547, DOI 10.1142/S0129055X96000172. MR1405763

[18] G. A. Bley and S. Fournais, *Hardy-Lieb-Thirring inequalities for fractional Pauli operators*, Comm. Math. Phys. **365** (2019), no. 2, 651–683, DOI 10.1007/s00220-018-3204-y. MR3907954

[19] S. Bögli, *Schrödinger operator with non-zero accumulation points of complex eigenvalues*, Comm. Math. Phys. **352** (2017), no. 2, 629–639, DOI 10.1007/s00220-016-2806-5. MR3627408

[20] S. Bögli, F. Štampach, *On Lieb–Thirring inequalities for one-dimensional non-self-adjoint Jacobi and Schrödinger operators*. Preprint (2020), arXiv:2004.09794.

[21] A. Borichev, R. L. Frank, A. Volberg, *Counting eigenvalues of Schrödinger operator with complex fast decreasing potential*. Preprint (2018), arXiv:1811.05591.

[22] A. Borichev, L. Golinskii, and S. Kupin, *A Blaschke-type condition and its application to complex Jacobi matrices*, Bull. Lond. Math. Soc. **41** (2009), no. 1, 117–123, DOI 10.1112/blms/bdn109. MR2481997

[23] L. Bugliaro, C. Fefferman, J. Fröhlich, G. M. Graf, and J. Stubbe, *A Lieb-Thirring bound for a magnetic Pauli Hamiltonian*, Comm. Math. Phys. **187** (1997), no. 3, 567–582, DOI 10.1007/s002200050149. MR1468314

[24] A. Burchard and L. E. Thomas, *On an isoperimetric inequality for a Schrödinger operator depending on the curvature of a loop*, J. Geom. Anal. **15** (2005), no. 4, 543–563, DOI 10.1007/BF02922244. MR2203162

[25] E. A. Carlen, R. L. Frank, and E. H. Lieb, *Stability estimates for the lowest eigenvalue of a Schrödinger operator*, Geom. Funct. Anal. **24** (2014), no. 1, 63–84, DOI 10.1007/s00039-014-0253-z. MR3177378

[26] J. G. Conlon, *A new proof of the Cwikel-Lieb-Rosenbljum bound*, Rocky Mountain J. Math. **15** (1985), no. 1, 117–122, DOI 10.1216/RMJ-1985-15-1-117. MR779256

[27] P. Constantin, C. Foias, and R. Temam, *Attractors representing turbulent flows*, Mem. Amer. Math. Soc. **53** (1985), no. 314, vii+67, DOI 10.1090/memo/0314. MR776345

[28] J.-C. Cuenin, *Improved eigenvalue bounds for Schrödinger operators with slowly decaying potentials*, Comm. Math. Phys. **376** (2020), no. 3, 2147–2160, DOI 10.1007/s00220-019-03635-w. MR4104544

[29] M. Cwikel, *Weak type estimates for singular values and the number of bound states of Schrödinger operators*, Ann. of Math. (2) **106** (1977), no. 1, 93–100, DOI 10.2307/1971160. MR473576

[30] D. Damanik and C. Remling, *Schrödinger operators with many bound states*, Duke Math. J. **136** (2007), no. 1, 51–80, DOI 10.1215/S0012-7094-07-13612-3. MR2271295

[31] E. B. Davies and J. Nath, *Schrödinger operators with slowly decaying potentials*, J. Comput. Appl. Math. **148** (2002), no. 1, 1–28, DOI 10.1016/S0377-0427(02)00570-8. On the occasion of the 65th birthday of Professor Michael Eastham. MR1946184

[32] M. Demuth, M. Hansmann, and G. Katriel, *On the discrete spectrum of non-selfadjoint operators*, J. Funct. Anal. **257** (2009), no. 9, 2742–2759, DOI 10.1016/j.jfa.2009.07.018. MR2559715

[33] M. Demuth, M. Hansmann, and G. Katriel, *Eigenvalues of non-selfadjoint operators: a comparison of two approaches*, Mathematical physics, spectral theory and stochastic analysis, Oper. Theory Adv. Appl., vol. 232, Birkhäuser/Springer Basel AG, Basel, 2013, pp. 107–163, DOI 10.1007/978-3-0348-0591-9_2. MR3077277

[34] J. Denzler, *Existence and regularity for a curvature dependent variational problem*, Trans. Amer. Math. Soc. **367** (2015), no. 6, 3829–3845, DOI 10.1090/S0002-9947-2014-06188-6. MR3324911

[35] J. Dolbeault, P. Felmer, M. Loss, and E. Paturel, *Lieb-Thirring type inequalities and Gagliardo-Nirenberg inequalities for systems*, J. Funct. Anal. **238** (2006), no. 1, 193–220, DOI 10.1016/j.jfa.2005.11.008. MR2253013

[36] J. Dolbeault, A. Laptev, and M. Loss, *Lieb-Thirring inequalities with improved constants*, J. Eur. Math. Soc. (JEMS) **10** (2008), no. 4, 1121–1126, DOI 10.4171/JEMS/142. MR2443931

[37] F. J. Dyson and A. Lenard, *Stability of matter. I*, J. Math. Phys. **8** (1967), no. 3, 423–434, DOI 10.1063/1.1705209. MR2408896

[38] A. Lenard and F. J. Dyson, *Stability of matter. II*, J. Math. Phys. **9** (1968), no. 5, 698–711, DOI 10.1063/1.1664631. MR2408897

[39] A. Eden and C. Foias, *A simple proof of the generalized Lieb-Thirring inequalities in one-space dimension*, J. Math. Anal. Appl. **162** (1991), no. 1, 250–254, DOI 10.1016/0022-247X(91)90191-2. MR1135275

[40] T. Ekholm and R. L. Frank, *On Lieb-Thirring inequalities for Schrödinger operators with virtual level*, Comm. Math. Phys. **264** (2006), no. 3, 725–740, DOI 10.1007/s00220-006-1521-z. MR2217288

[41] L. Erdős, *Magnetic Lieb-Thirring inequalities*, Comm. Math. Phys. **170** (1995), no. 3, 629–668. MR1337136

[42] L. Erdős, M. Loss, and V. Vougalter, *Diamagnetic behavior of sums of Dirichlet eigenvalues* (English, with English and French summaries), Ann. Inst. Fourier (Grenoble) **50** (2000), no. 3, 891–907. MR1779898

[43] L. Erdős and J. P. Solovej, *Semiclassical eigenvalue estimates for the Pauli operator with strong nonhomogeneous magnetic fields. I. Nonasymptotic Lieb-Thirring-type estimate*, Duke Math. J. **96** (1999), no. 1, 127–173, DOI 10.1215/S0012-7094-99-09604-7. MR1663923

[44] L. Erdős and J. P. Solovej, *Uniform Lieb-Thirring inequality for the three-dimensional Pauli operator with a strong non-homogeneous magnetic field*, Ann. Henri Poincaré **5** (2004), no. 4, 671–741, DOI 10.1007/s00023-004-0180-x. MR2090449

[45] L. Erdős and J. P. Solovej, *Magnetic Lieb-Thirring inequalities with optimal dependence on the field strength*, J. Statist. Phys. **116** (2004), no. 1-4, 475–506, DOI 10.1023/B:JOSS.0000037216.45270.1d. MR2083151

[46] L. Fanelli, D. Krejčiřík, and L. Vega, *Spectral stability of Schrödinger operators with subordinated complex potentials*, J. Spectr. Theory **8** (2018), no. 2, 575–604, DOI 10.4171/JST/208. MR3812810

[47] C. L. Fefferman, *The uncertainty principle*, Bull. Amer. Math. Soc. (N.S.) **9** (1983), no. 2, 129–206, DOI 10.1090/S0273-0979-1983-15154-6. MR707957

[48] C. Förster and J. Östensson, *Lieb-Thirring inequalities for higher order differential operators*, Math. Nachr. **281** (2008), no. 2, 199–213, DOI 10.1002/mana.200510595. MR2387360

[49] R. L. Frank, *A simple proof of Hardy-Lieb-Thirring inequalities*, Comm. Math. Phys. **290** (2009), no. 2, 789–800, DOI 10.1007/s00220-009-0759-7. MR2525639

[50] R. L. Frank, *Remarks on eigenvalue estimates and semigroup domination*, Spectral and scattering theory for quantum magnetic systems, Contemp. Math., vol. 500, Amer. Math. Soc., Providence, RI, 2009, pp. 63–86, DOI 10.1090/conm/500/09821. MR2655143

[51] R. L. Frank, *Eigenvalue bounds for Schrödinger operators with complex potentials*, Bull. Lond. Math. Soc. **43** (2011), no. 4, 745–750, DOI 10.1112/blms/bdr008. MR2820160

[52] R. L. Frank, *Cwikel's theorem and the CLR inequality*, J. Spectr. Theory **4** (2014), no. 1, 1–21, DOI 10.4171/JST/59. MR3181383

[53] R. L. Frank, *Ground states of semi-linear PDE.* Lecture notes from the 'Summerschool on Current Topics in Mathematical Physics', CIRM Marseille, Sept. 2013

[54] R. L. Frank, *Eigenvalue bounds for Schrödinger operators with complex potentials. III*, Trans. Amer. Math. Soc. **370** (2018), no. 1, 219–240, DOI 10.1090/tran/6936. MR3717979

[55] R. L. Frank, *Eigenvalue bounds for the fractional Laplacian: a review*, Recent developments in nonlocal theory, De Gruyter, Berlin, 2018, pp. 210–235, DOI 10.1515/9783110571561-007. MR3824213

[56] R. L. Frank, D. Gontier, and M. Lewin, *The nonlinear Schrödinger equation for orthonormal functions II: Application to Lieb-Thirring inequalities*, Comm. Math. Phys. **384** (2021), no. 3, 1783–1828, DOI 10.1007/s00220-021-04039-5. MR4268833

[57] R. L. Frank, D. Gontier, M. Lewin, *The periodic Lieb–Thirring inequality.* In: Partial Differential Equations, Spectral Theory, and Mathematical Physics. The Ari Laptev Anniversary Volume, EMS Publishing House. EMS Series of Congress Reports Vol. 18, 2021. DOI 10.4171/ECR/18-1/8.

[58] R. L. Frank, D. Gontier, M. Lewin, *Optimizers for the finite-rank Lieb–Thirring inequality.* Preprint (2021).

[59] R. L. Frank, D. Hundertmark, M. Jex, and P. T. Nam, *The Lieb-Thirring inequality revisited*, J. Eur. Math. Soc. (JEMS) **23** (2021), no. 8, 2583–2600, DOI 10.4171/jems/1062. MR4269422

[60] R. L. Frank and A. Laptev, *Bound on the number of negative eigenvalues of two-dimensional Schrödinger operators on domains*, Algebra i Analiz **30** (2018), no. 3, 250–272, DOI 10.1090/spmj/1559; English transl., St. Petersburg Math. J. **30** (2019), no. 3, 573–589. MR3812007

[61] R. L. Frank, A. Laptev, E. H. Lieb, and R. Seiringer, *Lieb-Thirring inequalities for Schrödinger operators with complex-valued potentials*, Lett. Math. Phys. **77** (2006), no. 3, 309–316, DOI 10.1007/s11005-006-0095-1. MR2260376

[62] R. L. Frank, A. Laptev, and O. Safronov, *On the number of eigenvalues of Schrödinger operators with complex potentials*, J. Lond. Math. Soc. (2) **94** (2016), no. 2, 377–390, DOI 10.1112/jlms/jdw039. MR3556444

[63] R. L. Frank and S. Larson, *Two-term spectral asymptotics for the Dirichlet Laplacian in a Lipschitz domain*, J. Reine Angew. Math. **766** (2020), 195–228, DOI 10.1515/crelle-2019-0019. MR4145207

[64] R. L. Frank and S. Larson, *On the error in the two-term Weyl formula for the Dirichlet Laplacian*, J. Math. Phys. **61** (2020), no. 4, 043504, 14, DOI 10.1063/1.5145003. MR4086994

[65] R. L. Frank, M. Lewin, E. H. Lieb, R. Seiringer, *Energy cost to make a hole in the Fermi sea*. Phys. Rev. Lett. **106** (2011), 150402.

[66] R. L. Frank, M. Lewin, E. H. Lieb, and R. Seiringer, *A positive density analogue of the Lieb-Thirring inequality*, Duke Math. J. **162** (2013), no. 3, 435–495, DOI 10.1215/00127094-2019477. MR3024090

[67] R. L. Frank, M. Lewin, E. H. Lieb, and R. Seiringer, *Strichartz inequality for orthonormal functions*, J. Eur. Math. Soc. (JEMS) **16** (2014), no. 7, 1507–1526, DOI 10.4171/JEMS/467. MR3254332

[68] R. L. Frank, E. H. Lieb, and R. Seiringer, *Stability of relativistic matter with magnetic fields for nuclear charges up to the critical value*, Comm. Math. Phys. **275** (2007), no. 2, 479–489, DOI 10.1007/s00220-007-0307-2. MR2335782

[69] R. L. Frank, E. H. Lieb, and R. Seiringer, *Hardy-Lieb-Thirring inequalities for fractional Schrödinger operators*, J. Amer. Math. Soc. **21** (2008), no. 4, 925–950, DOI 10.1090/S0894-0347-07-00582-6. MR2425175

[70] R. L. Frank, E. H. Lieb, and R. Seiringer, *Equivalence of Sobolev inequalities and Lieb-Thirring inequalities*, XVIth International Congress on Mathematical Physics, World Sci. Publ., Hackensack, NJ, 2010, pp. 523–535, DOI 10.1142/9789814304634_0045. MR2730819

[71] R. L. Frank and M. Loss, *Hardy-Sobolev-Maz'ya inequalities for arbitrary domains* (English, with English and French summaries), J. Math. Pures Appl. (9) **97** (2012), no. 1, 39–54, DOI 10.1016/j.matpur.2011.04.004. MR2863763

[72] R. L. Frank, M. Loss, and T. Weidl, *Pólya's conjecture in the presence of a constant magnetic field*, J. Eur. Math. Soc. (JEMS) **11** (2009), no. 6, 1365–1383, DOI 10.4171/JEMS/184. MR2557138

[73] R. L. Frank and A. Pushnitski, *Trace class conditions for functions of Schrödinger operators*, Comm. Math. Phys. **335** (2015), no. 1, 477–496, DOI 10.1007/s00220-014-2205-8. MR3314510

[74] R. L. Frank and A. Pushnitski, *Kato smoothness and functions of perturbed self-adjoint operators*, Adv. Math. **351** (2019), 343–387, DOI 10.1016/j.aim.2019.05.002. MR3952573

[75] R. L. Frank and A. Pushnitski, *Schatten class conditions for functions of Schrödinger operators*, Ann. Henri Poincaré **20** (2019), no. 11, 3543–3562, DOI 10.1007/s00023-019-00838-8. MR4019196

[76] R. L. Frank and J. Sabin, *Restriction theorems for orthonormal functions, Strichartz inequalities, and uniform Sobolev estimates*, Amer. J. Math. **139** (2017), no. 6, 1649–1691, DOI 10.1353/ajm.2017.0041. MR3730931

[77] R. L. Frank and J. Sabin, *The Stein-Tomas inequality in trace ideals*, Séminaire Laurent Schwartz—Équations aux dérivées partielles et applications. Année 2015–2016, Ed. Éc. Polytech., Palaiseau, 2017, pp. Exp. No. XV, 12. MR3616317

[78] R. L. Frank and J. Sabin, *Spectral cluster bounds for orthonormal systems and oscillatory integral operators in Schatten spaces*, Adv. Math. **317** (2017), 157–192, DOI 10.1016/j.aim.2017.06.023. MR3682666

[79] R. L. Frank and R. Seiringer, *Lieb-Thirring inequality for a model of particles with point interactions*, J. Math. Phys. **53** (2012), no. 9, 095201, 11, DOI 10.1063/1.3697416. MR2905783

[80] R. L. Frank and B. Simon, *Eigenvalue bounds for Schrödinger operators with complex potentials. II*, J. Spectr. Theory **7** (2017), no. 3, 633–658, DOI 10.4171/JST/173. MR3713021

[81] J. K. Freericks, E. H. Lieb, and D. Ueltschi, *Segregation in the Falicov-Kimball model*, Comm. Math. Phys. **227** (2002), no. 2, 243–279, DOI 10.1007/s002200200632. MR1903646

[82] L. Geisinger, A. Laptev, and T. Weidl, *Geometrical versions of improved Berezin-Li-Yau inequalities*, J. Spectr. Theory **1** (2011), no. 1, 87–109, DOI 10.4171/JST/4. MR2820887

[83] C. S. Gardner, J. M. Greene, M. D. Kruskal, and R. M. Miura, *Korteweg-deVries equation and generalization. VI. Methods for exact solution*, Comm. Pure Appl. Math. **27** (1974), 97–133, DOI 10.1002/cpa.3160270108. MR336122

[84] B. Gidas, W. M. Ni, and L. Nirenberg, *Symmetry of positive solutions of nonlinear elliptic equations in* $\mathbf{R}^n$, Mathematical analysis and applications, Part A, Adv. in Math. Suppl. Stud., vol. 7, Academic Press, New York-London, 1981, pp. 369–402. MR634248

[85] V. Glaser, H. Grosse, and A. Martin, *Bounds on the number of eigenvalues of the Schrödinger operator*, Comm. Math. Phys. **59** (1978), no. 2, 197–212. MR491613

[86] S. Golden, *Lower bounds for the Helmholtz function*, Phys. Rev. (2) **137** (1965), B1127–B1128. MR189691

[87] D. Gontier, M. Lewin, and F. Q. Nazar, *The nonlinear Schrödinger equation for orthonormal functions: existence of ground states*, Arch. Ration. Mech. Anal. **240** (2021), no. 3, 1203–1254, DOI 10.1007/s00205-021-01634-7. MR4264945

[88] A. Grigor'yan and N. Nadirashvili, *Negative eigenvalues of two-dimensional Schrödinger operators*, Arch. Ration. Mech. Anal. **217** (2015), no. 3, 975–1028, DOI 10.1007/s00205-015-0848-z. MR3356993

[89] A. Grigor'yan, N. Nadirashvili, and Y. Sire, *A lower bound for the number of negative eigenvalues of Schrödinger operators*, J. Differential Geom. **102** (2016), no. 3, 395–408. MR3466803

[90] A. Grigor′yan, Y. Netrusov, and S.-T. Yau, *Eigenvalues of elliptic operators and geometric applications*, Surveys in differential geometry. Vol. IX, Surv. Differ. Geom., vol. 9, Int. Press, Somerville, MA, 2004, pp. 147–217, DOI 10.4310/SDG.2004.v9.n1.a5. MR2195408

[91] C. Hainzl, M. Lewin, and J. P. Solovej, *The thermodynamic limit of quantum Coulomb systems. II. Applications*, Adv. Math. **221** (2009), no. 2, 488–546, DOI 10.1016/j.aim.2008.12.011. MR2508929

[92] M. Hansmann, *An eigenvalue estimate and its application to non-selfadjoint Jacobi and Schrödinger operators*, Lett. Math. Phys. **98** (2011), no. 1, 79–95, DOI 10.1007/s11005-011-0494-9. MR2836430

[93] B. Helffer and D. Robert, *Riesz means of bounded states and semi-classical limit connected with a Lieb-Thirring conjecture. II* (English, with French summary), Ann. Inst. H. Poincaré Phys. Théor. **53** (1990), no. 2, 139–147. MR1079775

[94] Y. Hong, S. Kwon, and H. Yoon, *Global existence versus finite time blowup dichotomy for the system of nonlinear Schrödinger equations* (English, with English and French summaries), J. Math. Pures Appl. (9) **125** (2019), 283–320, DOI 10.1016/j.matpur.2018.12.003. MR3944206

[95] D. Hundertmark, *Some bound state problems in quantum mechanics*, Spectral theory and mathematical physics: a Festschrift in honor of Barry Simon's 60th birthday, Proc. Sympos. Pure Math., vol. 76, Amer. Math. Soc., Providence, RI, 2007, pp. 463–496, DOI 10.1090/pspum/076.1/2310215. MR2310215

[96] D. Hundertmark, P. Kunstmann, T. Ried, S. Vugalter, *Cwikel's bound reloaded.* Preprint (2019), `arXiv:1809.05069`.

[97] D. Hundertmark, A. Laptev, and T. Weidl, *New bounds on the Lieb-Thirring constants*, Invent. Math. **140** (2000), no. 3, 693–704, DOI 10.1007/s002220000077. MR1760755

[98] D. Hundertmark, E. H. Lieb, and L. E. Thomas, *A sharp bound for an eigenvalue moment of the one-dimensional Schrödinger operator*, Adv. Theor. Math. Phys. **2** (1998), no. 4, 719–731, DOI 10.4310/ATMP.1998.v2.n4.a2. MR1663336

[99] D. Hundertmark and B. Simon, *Lieb-Thirring inequalities for Jacobi matrices*, J. Approx. Theory **118** (2002), no. 1, 106–130, DOI 10.1006/jath.2002.3704. MR1928259

[100] V. Ja. Ivriĭ, *The second term of the spectral asymptotics for a Laplace-Beltrami operator on manifolds with boundary* (Russian), Funktsional. Anal. i Prilozhen. **14** (1980), no. 2, 25–34. MR575202

[101] N. Nadirashvili, *Isoperimetric inequality for the second eigenvalue of a sphere*, J. Differential Geom. **61** (2002), no. 2, 335–340. MR1972149

[102] J. B. Keller, *Lower bounds and isoperimetric inequalities for eigenvalues of the Schrödinger equation*, J. Mathematical Phys. **2** (1961), 262–266, DOI 10.1063/1.1703708. MR121101

[103] N. N. Khuri, A. Martin, T. T. Wu, *Bound states in* n *dimensions (especially* $n = 1$ *and* $n = 2$*).* Few-Body Systems **31** (2002), 83–89.

[104] R. Killip and B. Simon, *Sum rules for Jacobi matrices and their applications to spectral theory*, Ann. of Math. (2) **158** (2003), no. 1, 253–321, DOI 10.4007/annals.2003.158.253. MR1999923

[105] R. Killip and B. Simon, *Sum rules and spectral measures of Schrödinger operators with L^2 potentials*, Ann. of Math. (2) **170** (2009), no. 2, 739–782, DOI 10.4007/annals.2009.170.739. MR2552106

[106] N. Korevaar, *Upper bounds for eigenvalues of conformal metrics*, J. Differential Geom. **37** (1993), no. 1, 73–93. MR1198600

[107] H. Kovařík, S. Vugalter, and T. Weidl, *Spectral estimates for two-dimensional Schrödinger operators with application to quantum layers*, Comm. Math. Phys. **275** (2007), no. 3, 827–838, DOI 10.1007/s00220-007-0318-z. MR2336366

[108] H. Kovařík, S. Vugalter, and T. Weidl, *Two-dimensional Berezin-Li-Yau inequalities with a correction term*, Comm. Math. Phys. **287** (2009), no. 3, 959–981, DOI 10.1007/s00220-008-0692-1. MR2486669

[109] M. Kwaśnicki, R. S. Laugesen, and B. A. Siudeja, *Pólya's conjecture fails for the fractional Laplacian*, J. Spectr. Theory **9** (2019), no. 1, 127–135, DOI 10.4171/JST/242. MR3900781

[110] M. K. Kwong, *Uniqueness of positive solutions of $\Delta u - u + u^p = 0$ in $\mathbf{R}^n$*, Arch. Rational Mech. Anal. **105** (1989), no. 3, 243–266, DOI 10.1007/BF00251502. MR969899

[111] A. Laptev, *Dirichlet and Neumann eigenvalue problems on domains in Euclidean spaces*, J. Funct. Anal. **151** (1997), no. 2, 531–545, DOI 10.1006/jfan.1997.3155. MR1491551

[112] A. Laptev, *Spectral inequalities for partial differential equations and their applications*, Fifth International Congress of Chinese Mathematicians. Part 1, 2, AMS/IP Stud. Adv. Math., 51, pt. 1, vol. 2, Amer. Math. Soc., Providence, RI, 2012, pp. 629–643. MR2908096

[113] A. Laptev and O. Safronov, *Eigenvalue estimates for Schrödinger operators with complex potentials*, Comm. Math. Phys. **292** (2009), no. 1, 29–54, DOI 10.1007/s00220-009-0883-4. MR2540070

[114] A. Laptev and M. Solomyak, *On the negative spectrum of the two-dimensional Schrödinger operator with radial potential*, Comm. Math. Phys. **314** (2012), no. 1, 229–241, DOI 10.1007/s00220-012-1501-4. MR2954515

[115] A. Laptev and M. Solomyak, *On spectral estimates for two-dimensional Schrödinger operators*, J. Spectr. Theory **3** (2013), no. 4, 505–515, DOI 10.4171/JST/53. MR3122220

[116] A. Laptev and T. Weidl, *Sharp Lieb-Thirring inequalities in high dimensions*, Acta Math. **184** (2000), no. 1, 87–111, DOI 10.1007/BF02392782. MR1756570

[117] A. Laptev and T. Weidl, *Recent results on Lieb-Thirring inequalities*, Journées "Équations aux Dérivées Partielles" (La Chapelle sur Erdre, 2000), Univ. Nantes, Nantes, 2000, pp. Exp. No. XX, 14. MR1775696

[118] S. Larson, *Asymptotic shape optimization for Riesz means of the Dirichlet Laplacian over convex domains*, J. Spectr. Theory **9** (2019), no. 3, 857–895, DOI 10.4171/JST/265. MR4003544

[119] S. Larson, D. Lundholm, and P. T. Nam, *Lieb-Thirring inequalities for wave functions vanishing on the diagonal set* (English, with English and French summaries), Ann. H. Lebesgue **4** (2021), 251–282, DOI 10.5802/ahl.72. MR4213161

[120] J. L. Lebowitz, E. H. Lieb, *Existence of thermodynamics for real matter with Coulomb forces.* Phys. Rev. Lett. **22** (1969), 631–634.

[121] Y. Lee and I. Seo, *A note on eigenvalue bounds for Schrödinger operators*, J. Math. Anal. Appl. **470** (2019), no. 1, 340–347, DOI 10.1016/j.jmaa.2018.10.006. MR3865141

[122] D. Levin and M. Solomyak, *The Rozenblum-Lieb-Cwikel inequality for Markov generators*, J. Anal. Math. **71** (1997), 173–193, DOI 10.1007/BF02788029. MR1454250

[123] A. Levitt, *Best constants in Lieb-Thirring inequalities: a numerical investigation*, J. Spectr. Theory **4** (2014), no. 1, 153–175, DOI 10.4171/JST/65. MR3181389

[124] M. Lewin, E. H. Lieb, R. Seiringer, *Universal functionals in density functional theory.* Chapter in a book "Density Functional Theory" edited by Éric Cancès, Gero Friesecke & Lin Lin, to appear.

[125] M. Lewin and J. Sabin, *The Hartree equation for infinitely many particles, II: Dispersion and scattering in 2D*, Anal. PDE **7** (2014), no. 6, 1339–1363, DOI 10.2140/apde.2014.7.1339. MR3270166

[126] M. Lewin and J. Sabin, *The Hartree equation for infinitely many particles I. Well-posedness theory*, Comm. Math. Phys. **334** (2015), no. 1, 117–170, DOI 10.1007/s00220-014-2098-6. MR3304272

[127] M. Lewin and J. Sabin, *The Hartree and Vlasov equations at positive density*, Comm. Partial Differential Equations **45** (2020), no. 12, 1702–1754, DOI 10.1080/03605302.2020.1803355. MR4176914

[128] P. Li and S. T. Yau, *On the Schrödinger equation and the eigenvalue problem*, Comm. Math. Phys. **88** (1983), no. 3, 309–318. MR701919

[129] E. Lieb, *Bounds on the eigenvalues of the Laplace and Schroedinger operators*, Bull. Amer. Math. Soc. **82** (1976), no. 5, 751–753, DOI 10.1090/S0002-9904-1976-14149-3. MR407909

[130] E. H. Lieb, *The number of bound states of one-body Schroedinger operators and the Weyl problem*, Geometry of the Laplace operator (Proc. Sympos. Pure Math., Univ. Hawaii, Honolulu, Hawaii, 1979), Proc. Sympos. Pure Math., XXXVI, Amer. Math. Soc., Providence, R.I., 1980, pp. 241–252. MR573436

[131] E. H. Lieb, *Thomas-Fermi and related theories of atoms and molecules*, Rev. Modern Phys. **53** (1981), no. 4, 603–641, DOI 10.1103/RevModPhys.53.603. MR629207

[132] E. H. Lieb, *Sharp constants in the Hardy-Littlewood-Sobolev and related inequalities*, Ann. of Math. (2) **118** (1983), no. 2, 349–374, DOI 10.2307/2007032. MR717827

[133] E. H. Lieb, *An L^p bound for the Riesz and Bessel potentials of orthonormal functions*, J. Funct. Anal. **51** (1983), no. 2, 159–165, DOI 10.1016/0022-1236(83)90023-X. MR701053

[134] E. H. Lieb, *Density functionals for Coulomb systems.* Int. J. Quantum Chem. **24** (1983), 243–277.

[135] E. H. Lieb, *On characteristic exponents in turbulence*, Comm. Math. Phys. **92** (1984), no. 4, 473–480. MR736404

[136] E. H. Lieb, *Kinetic energy bounds and their application to the stability of matter*, Schrödinger operators (Sønderborg, 1988), Lecture Notes in Phys., vol. 345, Springer, Berlin, 1989, pp. 371–382, DOI 10.1007/3-540-51783-9_24. MR1037324

[137] E. H. Lieb and J. L. Lebowitz, *The constitution of matter: Existence of thermodynamics for systems composed of electrons and nuclei*, Advances in Math. **9** (1972), 316–398, DOI 10.1016/0001-8708(72)90023-0. MR339751

[138] E. H. Lieb and M. Loss, *Analysis*, 2nd ed., Graduate Studies in Mathematics, vol. 14, American Mathematical Society, Providence, RI, 2001, DOI 10.1090/gsm/014. MR1817225

[139] E. H. Lieb, M. Loss, and J. P. Solovej, *Stability of matter in magnetic fields*, Phys. Rev. Lett. **75** (1995), no. 6, 985–989, DOI 10.1103/PhysRevLett.75.985. MR1342707

[140] E. H. Lieb and R. Seiringer, *The stability of matter in quantum mechanics*, Cambridge University Press, Cambridge, 2010. MR2583992

[141] E. H. Lieb and B. Simon, *The Thomas-Fermi theory of atoms, molecules and solids*, Advances in Math. **23** (1977), no. 1, 22–116, DOI 10.1016/0001-8708(77)90108-6. MR428944

[142] E. H. Lieb, W. E. Thirring, *Bound on kinetic energy of fermions which proves stability of matter.* Phys. Rev. Lett. **35** (1975), 687–689.

[143] E. H. Lieb, W. E. Thirring, *Inequalities for the moments of the eigenvalues of the Schrödinger Hamiltonian and their relation to Sobolev inequalities.* In: Studies in Mathematical Physics, Princeton University Press, 1976, 269–303.

[144] H. Linde, *A lower bound for the ground state energy of a Schrödinger operator on a loop*, Proc. Amer. Math. Soc. **134** (2006), no. 12, 3629–3635, DOI 10.1090/S0002-9939-06-08483-8. MR2240676

[145] P.-L. Lions and T. Paul, *Sur les mesures de Wigner* (French, with English and French summaries), Rev. Mat. Iberoamericana **9** (1993), no. 3, 553–618, DOI 10.4171/RMI/143. MR1251718

[146] P.-L. Lions and B. Perthame, *Propagation of moments and regularity for the 3-dimensional Vlasov-Poisson system* (English, with French summary), Invent. Math. **105** (1991), no. 2, 415–430, DOI 10.1007/BF01232273. MR1115549

[147] D. Lundholm, P. T. Nam, and F. Portmann, *Fractional Hardy-Lieb-Thirring and related inequalities for interacting systems*, Arch. Ration. Mech. Anal. **219** (2016), no. 3, 1343–1382, DOI 10.1007/s00205-015-0923-5. MR3448930

[148] D. Lundholm, F. Portmann, and J. P. Solovej, *Lieb-Thirring bounds for interacting Bose gases*, Comm. Math. Phys. **335** (2015), no. 2, 1019–1056, DOI 10.1007/s00220-014-2278-4. MR3316649

[149] D. Lundholm and R. Seiringer, *Fermionic behavior of ideal anyons*, Lett. Math. Phys. **108** (2018), no. 11, 2523–2541, DOI 10.1007/s11005-018-1091-y. MR3861388

[150] D. Lundholm and J. P. Solovej, *Hardy and Lieb-Thirring inequalities for anyons*, Comm. Math. Phys. **322** (2013), no. 3, 883–908, DOI 10.1007/s00220-013-1748-4. MR3079335

[151] D. Lundholm and J. P. Solovej, *Local exclusion and Lieb-Thirring inequalities for intermediate and fractional statistics*, Ann. Henri Poincaré **15** (2014), no. 6, 1061–1107, DOI 10.1007/s00023-013-0273-5. MR3205745

[152] K. McLeod and J. Serrin, *Uniqueness of positive radial solutions of* $\Delta u + f(u) = 0$ *in* $\mathbf{R}^n$, Arch. Rational Mech. Anal. **99** (1987), no. 2, 115–145, DOI 10.1007/BF00275874. MR886933

[153] S. Molchanov and B. Vainberg, *Bargmann type estimates of the counting function for general Schrödinger operators*, J. Math. Sci. (N.Y.) **184** (2012), no. 4, 457–508, DOI 10.1007/s10958-012-0877-1. Problems in mathematical analysis. No. 65. MR2962816

[154] C. Morpurgo, *Sharp inequalities for functional integrals and traces of conformally invariant operators*, Duke Math. J. **114** (2002), no. 3, 477–553, DOI 10.1215/S0012-7094-02-11433-1. MR1924571

[155] B. v. Sz. Nagy, *Über Integralungleichungen zwischen einer Funktion und ihrer Ableitung* (German), Acta Univ. Szeged. Sect. Sci. Math. **10** (1941), 64–74. MR4277

[156] S. Nakamura, *The orthonormal Strichartz inequality on torus*, Trans. Amer. Math. Soc. **373** (2020), no. 2, 1455–1476, DOI 10.1090/tran/7982. MR4068269

[157] P. T. Nam, *Lieb-Thirring inequality with semiclassical constant and gradient error term*, J. Funct. Anal. **274** (2018), no. 6, 1739–1746, DOI 10.1016/j.jfa.2017.08.007. MR3758547

[158] Y. Netrusov and T. Weidl, *On Lieb-Thirring inequalities for higher order operators with critical and subcritical powers*, Comm. Math. Phys. **182** (1996), no. 2, 355–370. MR1447297

[159] B. S. Pavlov, *On a non-selfadjoint Schrödinger operator* (Russian), Probl. Math. Phys., No. I, Spectral Theory and Wave Processes (Russian), Izdat. Leningrad. Univ., Leningrad, 1966, pp. 102–132. MR0203530

[160] B. S. Pavlov, *On a non-selfadjoint Schrödinger operator. II* (Russian), Problems of Mathematical Physics, No. 2, Spectral Theory, Diffraction Problems (Russian), Izdat. Leningrad. Univ., Leningrad, 1967, pp. 133–157. MR0234319

[161] G. Pólya, *On the eigenvalues of vibrating membranes*, Proc. London Math. Soc. (3) **11** (1961), 419–433, DOI 10.1112/plms/s3-11.1.419. MR129219

[162] A. Pushnitski, *Estimates for the spectral shift function of the polyharmonic operator*, J. Math. Phys. **40** (1999), no. 11, 5578–5592, DOI 10.1063/1.533047. MR1722328

[163] X. Ren and J. Wei, *On a two-dimensional elliptic problem with large exponent in nonlinearity*, Trans. Amer. Math. Soc. **343** (1994), no. 2, 749–763, DOI 10.2307/2154740. MR1232190

[164] E. Rodemich, *The Sobolev inequality with best possible constant.* Analysis Seminar Caltech, Spring 1966.

[165] G. Rosen, *Minimum value for c in the Sobolev inequality* $\varphi^3\| \leq c\nabla\varphi\|^3$, SIAM J. Appl. Math. **21** (1971), 30–32, DOI 10.1137/0121004. MR289739

[166] G. V. Rozenbljum, *Distribution of the discrete spectrum of singular differential operators* (Russian), Dokl. Akad. Nauk SSSR **202** (1972), 1012–1015. MR0295148

[167] G. V. Rozenbljum, *Distribution of the discrete spectrum of singular differential operators* (Russian), Izv. Vysš. Učebn. Zaved. Matematika **1(164)** (1976), 75–86. MR0430557

[168] G. Rozenblum and M. Solomyak, *On the spectral estimates for the Schrödinger operator on* $\mathbb{Z}^d$, $d \geq 3$, J. Math. Sci. (N.Y.) **159** (2009), no. 2, 241–263, DOI 10.1007/s10958-009-9436-9. Problems in mathematical analysis. No. 41. MR2544038

[169] M. Rumin, *Spectral density and Sobolev inequalities for pure and mixed states*, Geom. Funct. Anal. **20** (2010), no. 3, 817–844, DOI 10.1007/s00039-010-0075-6. MR2720233

[170] M. Rumin, *Balanced distribution-energy inequalities and related entropy bounds*, Duke Math. J. **160** (2011), no. 3, 567–597, DOI 10.1215/00127094-1444305. MR2852369

[171] J. Sabin, *Littlewood-Paley decomposition of operator densities and application to a new proof of the Lieb-Thirring inequality*, Math. Phys. Anal. Geom. **19** (2016), no. 2, Art. 11, 11, DOI 10.1007/s11040-016-9215-z. MR3508209

[172] O. Safronov, *Estimates for eigenvalues of the Schrödinger operator with a complex potential*, Bull. Lond. Math. Soc. **42** (2010), no. 3, 452–456, DOI 10.1112/blms/bdq007. MR2651940

[173] A. Sahovic, *New constants in discrete Lieb-Thirring inequalities for Jacobi matrices*, J. Math. Sci. (N.Y.) **166** (2010), no. 3, 319–327, DOI 10.1007/s10958-010-9869-1. Problems in mathematical analysis. No. 45. MR2839035

[174] L. Schimmer, *Spectral inequalities for Jacobi operators and related sharp Lieb-Thirring inequalities on the continuum*, Comm. Math. Phys. **334** (2015), no. 1, 473–505, DOI 10.1007/s00220-014-2137-3. MR3304283

[175] U.-W. Schmincke, *On Schrödinger's factorization method for Sturm-Liouville operators*, Proc. Roy. Soc. Edinburgh Sect. A **80** (1978), no. 1-2, 67–84, DOI 10.1017/S0308210500010143. MR529570

[176] J. Schwinger, *On the bound states of a given potential*, Proc. Nat. Acad. Sci. U.S.A. **47** (1961), 122–129, DOI 10.1073/pnas.47.1.122. MR129798

[177] E. Shargorodsky, *An estimate for the Morse index of a Stokes wave*, Arch. Ration. Mech. Anal. **209** (2013), no. 1, 41–59, DOI 10.1007/s00205-013-0614-z. MR3054598

[178] E. Shargorodsky, *On negative eigenvalues of two-dimensional Schrödinger operators*, Proc. Lond. Math. Soc. (3) **108** (2014), no. 2, 441–483, DOI 10.1112/plms/pdt036. MR3166359

[179] B. Simon, *The bound state of weakly coupled Schrödinger operators in one and two dimensions*, Ann. Physics **97** (1976), no. 2, 279–288, DOI 10.1016/0003-4916(76)90038-5. MR404846

[180] B. Simon, *Analysis with weak trace ideals and the number of bound states of Schrödinger operators*, Trans. Amer. Math. Soc. **224** (1976), no. 2, 367–380, DOI 10.2307/1997482. MR423128

[181] B. Simon, *Functional integration and quantum physics*, 2nd ed., AMS Chelsea Publishing, Providence, RI, 2005, DOI 10.1090/chel/351. MR2105995

[182] A. V. Sobolev, *Efficient bounds for the spectral shift function* (English, with English and French summaries), Ann. Inst. H. Poincaré Phys. Théor. **58** (1993), no. 1, 55–83. MR1208792

[183] A. V. Sobolev, *On the Lieb-Thirring estimates for the Pauli operator*, Duke Math. J. **82** (1996), no. 3, 607–635, DOI 10.1215/S0012-7094-96-08225-3. MR1387686

[184] M. Solomyak, *Piecewise-polynomial approximation of functions from $H^l((0,1)^d)$, $2l = d$, and applications to the spectral theory of the Schrödinger operator*, Israel J. Math. **86** (1994), no. 1-3, 253–275, DOI 10.1007/BF02773681. MR1276138

[185] J.-P. Françoise, G. L. Naber, and T. S. Tsun (eds.), *Encyclopedia of mathematical physics. Vol. 5*, Academic Press/Elsevier Science, Oxford, 2006. MR2238867

[186] M. Stoiciu, *An estimate for the number of bound states of the Schrödinger operator in two dimensions*, Proc. Amer. Math. Soc. **132** (2004), no. 4, 1143–1151, DOI 10.1090/S0002-9939-03-07257-5. MR2045431

[187] J. Stubbe, *Universal monotonicity of eigenvalue moments and sharp Lieb-Thirring inequalities*, J. Eur. Math. Soc. (JEMS) **12** (2010), no. 6, 1347–1353, DOI 10.4171/JEMS/233. MR2734344

[188] K. Symanzik, *Proof and refinement of an inequality of Feynman*. J. Math. Phys. **6** (1965), 1155–1156

[189] G. Talenti, *Best constant in Sobolev inequality*, Ann. Mat. Pura Appl. (4) **110** (1976), 353–372, DOI 10.1007/BF02418013. MR463908

[190] R. Temam, *Infinite-dimensional dynamical systems in mechanics and physics*, 2nd ed., Applied Mathematical Sciences, vol. 68, Springer-Verlag, New York, 1997, DOI 10.1007/978-1-4612-0645-3. MR1441312

[191] C. J. Thompson, *Inequality with applications in statistical mechanics*, J. Mathematical Phys. **6** (1965), 1812–1813, DOI 10.1063/1.1704727. MR189688

[192] T. Weidl, *On the Lieb-Thirring constants $L_{\gamma,1}$ for $\gamma \geq 1/2$*, Comm. Math. Phys. **178** (1996), no. 1, 135–146. MR1387945

[193] T. Weidl, *Another look at Cwikel's inequality*, Differential operators and spectral theory, Amer. Math. Soc. Transl. Ser. 2, vol. 189, Amer. Math. Soc., Providence, RI, 1999, pp. 247–254, DOI 10.1090/trans2/189/19. MR1730517

[194] T. Weidl, *Improved Berezin-Li-Yau inequalities with a remainder term*, Spectral theory of differential operators, Amer. Math. Soc. Transl. Ser. 2, vol. 225, Amer. Math. Soc., Providence, RI, 2008, pp. 253–263, DOI 10.1090/trans2/225/17. MR2509788

[195] M. I. Weinstein, *Nonlinear Schrödinger equations and sharp interpolation estimates*, Comm. Math. Phys. **87** (1982/83), no. 4, 567–576. MR691044

Mathematics 253-37, Caltech, Pasadena, California 91125; Mathematisches Institut, Ludwig-Maximilans Universität München, Theresienstr. 39, 80333 München, Germany; and Munich Center for Quantum Science and Technology (MCQST), Schellingstr. 4, 80799 München, Germany

Email address: rlfrank@caltech.edu, r.frank@lmu.de

Proceedings of Symposia in Pure Mathematics
Volume **104**, 2021
https://doi.org/10.1090/pspum/104/01878

Some topological properties of surface bundles

Ursula Hamenstädt

ABSTRACT. We describe the second integral cohomology group of a surface bundle as the group of Chern classes of fiberwise holomorphic complex line bundles and use this to obtain some new information on this group.

1. Introduction

A *surface bundle over a surface* is a smooth closed 4-manifold E which fibers over a closed oriented surface B of genus $h \geq 0$, with fiber a closed oriented surface S_g of genus $g \geq 0$.

Assume that $g, h \geq 2$. Then the monodromy of the bundle determines a homomorphism ρ of the fundamental group $\pi_1(B)$ of B into the *mapping class group* Γ_g of S_g, that is, the group of isotopy classes of orientation preserving diffeomorphisms of S_g. Thus the geometry and topology of surface bundles over surfaces is intimately related to properties of the mapping class group.

Natural topological invariants of such surface bundles E are the *Euler characteristic* $\chi(E)$ and the *signature* $\sigma(E)$. For the Euler characteristic we have

$$\chi(E) = \chi(B)\chi(S_g) = (2h-2)(2g-2).$$

The signature can be computed as follows.

The tangent bundle ν of the fibers of the surface bundle, called the *vertical tangent bundle* in the sequel, is a real two-dimensional oriented smooth subbundle of the tangent bundle TE of E. Hence it can be equipped with the structure of a complex line bundle.

Choose a smooth Riemannian metric on TE and let $\nu^\perp$ be the orthogonal complement of ν in TE for this metric. Then the differential of the projection $\Pi : E \to B$ maps each fiber of $\nu^\perp$ isomorphically onto a fiber of TB and hence as a smooth vector bundle, $\nu^\perp$ is isomorphic to the bundle Π^*TB. Now TB can be equipped with the structure of a complex line bundle as well, and

$$TE = \nu \oplus \Pi^*TB$$

is the direct sum of two complex line bundles (this is a decomposition of TE as a smooth vector bundle). In particular, the first and second Chern class of TE are defined, and they are independent of the choices made.

2020 *Mathematics Subject Classification.* Primary 57R22, 57R20.

Partially supported by the Hausdorff Center Bonn. The completion of this work was also supported by the National Science Foundation under Grant no. 144140.

By Hirzebruch's signature theorem, the signature $\sigma(E)$ of E equals

$$\sigma(E) = \frac{1}{3}p_1(E)$$

where $p_1(E)$ is the first Pontryagin number of E. We then have

$$\sigma(E) = \frac{1}{3}(c_1(E)^2 - 2c_2(E)) = \frac{1}{3}(c_1(E)^2 - 2\chi(E)) \tag{1}$$

where as is customary, $c_1(E)^2$ and $c_2(E)$ denote Chern numbers of E.

As $TE = \nu \oplus \Pi^*TB$, the total Chern class of TE equals

$$\begin{aligned} c(TE) &= (1 + c_1(\nu)) \cup (1 + c_1(\Pi^*TB)) \\ &= 1 + c_1(\nu) + c_1(\Pi^*TB) + c_1(\nu) \cup c_1(\Pi^*TB) \end{aligned}$$

and hence since $c_1(\Pi^*TB) \cup c_1(\Pi^*TB)(E) = 0$ we have

$$3\sigma(E) = c_1(\nu) \cup c_1(\nu)(E). \tag{2}$$

Let $\mathcal{M}_g$ be the *moduli space of complex curves of genus* g, that, is the moduli space of complex structures on S_g up to biholomorphic equivalence, and let $\mathcal{U} \to \mathcal{M}_g$ be the *universal curve* whose fiber over a point $X \in \mathcal{M}_g$ is just the complex curve X. Let us now assume that E is a *Kodaira fibration*, that is, E is a complex manifold, and the complex structures of the fibers vary nontrivially. This is equivalent to stating that there is a complex structure on the base B, and there is a nonconstant holomorphic map $\varphi : B \to \mathcal{M}_g$ such that $E = \varphi^*\mathcal{U}$. By the classification of complex surfaces, a Kodaira fibration is of general type and hence projective and Kähler.

The Miaoka inequality for complex surfaces Y of general type states that $c_1^2(Y) \le 3c_2(Y)$. Therefore by equation (1) which is valid for all closed oriented 4-manifolds, we have

$$3|\sigma(Y)| \le |\chi(Y)|.$$

Equality holds if and only if Y is a quotient of the ball. On the other hand, Kapovich [**Ka98**] showed that no surface bundle over a surface is a quotient of the ball and hence we always have $3|\sigma(E)| < |\chi(E)|$ for all Kodaira fibrations E. It is also known that the signature of a Kodaira fibration does not vanish.

For surface bundles over surfaces which do not admit a complex structure, much less is known about the relation between signature and Euler characteristic. The most general result to date seems to be a theorem of Kotschick [**K98**]. Using Seibert Witten invariants, he showed

$$2|\sigma(E)| \le |\chi(E)| \tag{3}$$

for all surface bundles over surfaces. If E admits an Einstein metric, then the stronger inequality

$$3|\sigma(E)| < |\chi(E)|$$

holds true, generalizing the Miaoka inequality for Kodaira fibrations. The following conjecture was formulated among others in [**K98**].

CONJECTURE. $3|\sigma(E)| \le |\chi(E)|$ *holds true for all surface bundles over surfaces.*

Perhaps the motivation for this conjecture stems from the general belief that aspherical smooth closed 4-manifolds should admit an Einstein metric.

The conjecture can be viewed as a twisted higher dimensional version of the Milnor-Wood inequality for the Euler number of a flat circle bundle over a closed oriented surface. Namely, call a circle bundle $H \to M$ over a manifold M (or any CW-complex) *flat* [**M58, W71**] if the following holds true. Let $\mathrm{Top}^+(S^1)$ be the group of orientation preserving homeomorphisms of the circle S^1. We require that there is a homomorphism $\eta : \pi_1(M) \to \mathrm{Top}^+(S^1)$ such that

$$H = \tilde{M} \times S^1 / \pi_1(M)$$

where $\tilde{M}$ is the universal covering of M and $\pi_1(M)$ acts on $\tilde{M} \times S^1$ via

$$(x, t)g = (xg, \eta(g)^{-1}(t)).$$

The same definition applies if M is a good orbifold, that is, M is the quotient of a smooth simply connected manifold $\tilde{M}$ by the action of a group of diffeomorphisms which acts properly discontinuously, but not necessarily freely.

The celebrated Milnor Wood inequality bounds the absolute value of the Euler number (or first Chern class) of a complex line bundle with flat circle subbundle over a closed surface by the absolute value of the Euler characteristic of the surface [**M58, W71**].

Now as was pointed out by Morita [**Mo88**], the circle subbundle of the vertical tangent bundle of a surface bundle $\Pi : E \to B$ over an arbitrary smooth base B is flat. In this vein, the conjecture predicts a twisted higher dimensional analog of the Milnor Wood inequality.

The goal of this article is to provide a geometric perspective on the topology of surface bundles over a surface. We begin with summarizing some constructions of Kodaira fibrations in Section 2. In Section 3 we give a geometric proof of Morita's theorem (see also Chapter 5 of [**FM12**]).

A *section* of a surface bundle $\Pi : E \to B$ is a smooth map $\sigma : B \to E$ such that $\Pi \circ \sigma = \mathrm{Id}$. In Section 4 we study the self-intersection number of a section of a surface bundle over a surface and compute all such self-intersection numbers for the trivial bundle. We also point out that Morita's theorem yields an elementary and purely topological proof of the following extension of Proposition 1 of [**BKM13**] (see also [**Bow11**]) which was originally established using Seiberg Witten invariants.

PROPOSITION. *Let $E \to B$ be a surface bundle over a surface. Let Σ be closed surface and let $f : \Sigma \to E$ be a smooth map; then*

$$|c_1(f^*\nu)(\Sigma)| \leq |\chi(\Sigma)|.$$

In Section 5 we describe the second integral cohomology group of a surface bundle over a smooth base in an explicit way as the group of first Chern classes of complex line bundles. We apply this discussion to show an analog of a result of Morita who computed the cohomology of a surface bundle with rational coefficients (Proposition 3.1 of [**Mo87**]). The following is also related to the work of Harer [**H83**] who computed for $g \geq 5$ the second homology group of $\mathcal{M}_g$ with integral coefficients (see also the more recent and more complete account in [**KS03**]).

THEOREM. *Let $E \to B$ be a surface bundle with fiber genus g. If E admits a section then there exists an integral class $e \in H^2(E, \mathbb{Z})$ such that $(2g-2)e = c_1(\nu)$, and*

$$H^p(E, \mathbb{Z}) \equiv H^p(B, \mathbb{Z}) \oplus H^{p-1}(B; H^1(S_g, \mathbb{Z})) \oplus eH^{p-2}(B, \mathbb{Z})$$

In view of the result of Chen and Salter [**CS18**] that the pullback of the universal curve to any finite orbifold cover of $\mathcal{M}_g$ does not admit a section, it seems that most surface bundles do not admit sections.

2. Constructions

In this section we review some constructions of Kodaira fibrations from the literature.

The best known construction method of Kodaira fibrations goes back to an idea of Atiyah and Kodaira. Their examples are branched covers over a product of two complex curves, and they fiber in two different ways. A variation of this idea was used by Bryan and Donagi [**BD02**] to show that for any integers $h, n \geq 2$, there exists a connected algebraic surface $X_{h,n}$ of signature $\sigma(X_{h,n}) = \frac{4}{3}h(h-1)(n^2-1)n^{2h-3}$ that admits two smooth fibrations $\theta_1 : X_{h,n} \to C$ and $\theta_2 : X_{h,n} \to D$ with base and fiber genus (b_i, f_i) equal to

$$(b_1, f_1) = (h, h(hn-1)n^{2h-2}+1) \text{ and}$$
$$(b_2, f_2) = (h(h-1)n^{2h-2}+1, hn)$$

respectively. Note that the smallest fibre genus of the surfaces in the above family equals 4.

Taking $n = h = 2$, we conclude that there is a surface bundle with fiber genus 4 and base genus 9 with

$$\frac{\sigma(E)}{\chi(E)} = \frac{16}{96} = \frac{1}{6}.$$

According to my knowledge, this is the example with the largest known ratio between signature and Euler characteristic.

Complete intersections provide a more indirect way to construct Kodaira fibrations (see [**Ar17**] for a recent discussion). To be more specific, call a Kodaira fibration with fiber S_g of genus g *generic* if the fundamental group $\pi_1(B)$ of the base surjects onto a finite index subgroup of the mapping class group Γ_g by the monodromy homomorphism ρ. Generic Kodaira fibrations can be constructed as follows.

The action of a diffeomorphism of S_g on the first homology group $H_1(S_g, \mathbb{Z})$ preserves the intersection form and only depends on the isotopy class of the diffeomorphism. Thus there exists a surjective [**FM12**] homomorphism

$$\Psi : \Gamma_g \to \mathrm{Sp}(2g, \mathbb{Z}).$$

For every $n \geq 3$, the kernel of the induced homomorphism $\Gamma_g \to \mathrm{Sp}(2g, \mathbb{Z}/n\mathbb{Z})$ is torsion free and determines the *fine moduli space of genus g curves with level n structure* $\mathcal{M}_g[n]$, which is a complex manifold.

Let $\mathcal{M}_g[n]^*$ be the Satake compactification of $\mathcal{M}_g[n]$. If $g \geq 3$, then the boundary $\mathcal{M}_g[n]^* - \mathcal{M}_g[n]$ has complex codimension at least 2. Therefore a curve $C \subset \mathcal{M}_g[n]^*$ given as an intersection of general ample divisors lies entirely in $\mathcal{M}_g[n]$. By the weak Lefschetz theorem, the inclusion $C \to \mathcal{M}_g[n]$ induces a surjection of fundamental groups. As the fundamental group of $\mathcal{M}_g[n]$ is a finite index subgroup of the mapping class group, the restriction to C of the universal curve defines a generic Kodaira fibration.

The Atiyah Kodaira examples which are branched covers over the product of two complex curves are not generic. Namely, if $E \to B$ is a generic Kodaira fibration, then the image of the monodromy group under the homomorphism Ψ is a Zariski dense subgroup of $Sp(2g, \mathbb{R})$. However, if $E \to B$ is an Atiyah Kodaira example, then the group $\Psi(\rho(\pi_1(B)))$ fixes a symplectic plane in $H^1(B, \mathbb{R})$ and hence by duality, $\Psi(\rho(\pi_1(B)))$ is not Zariski dense in $Sp(2g, \mathbb{R})$. Interestingly, Flapan [**Fl17**] used complete intersections to construct Kodaira fibrations with fiber of genus 3 which also have this property. She also classified all $\mathbb{Q}$-algebraic subgroups of $Sp(6, \mathbb{R})$ which arise as the smallest algebraic group containing the image of the monodromy group of a Kodaira fibration with fiber of genus 3.

On the other hand, Bregman [**Br18**] established that variations of the Atiyah Kodaira construction may in some sense be universal for Kodaira fibrations whose monodromy fixes a symplectic plane in $H^1(B, \mathbb{R})$. He showed that if the dimension of the fixed point set of the mondromy of a Kodaira fibration E acting on the holomorphic one-forms of a fixed fiber equals d for some $1 \leq d \leq 2$, then there exists a genus d curve D and a ramified covering $F : E \to D \times B$ inducing an isomorphism on first cohomology with rational coefficients.

There are also explicit constructions of surface bundles over surfaces with non-trivial signature which do not admit a complex structure [**B12**]. The fiber genus of such a surface bundle is at least 4. The article [**EKKOS02**] constructs surface bundles over surfaces with positive signature for any fiber genus $g \geq 3$. In contrast, the signature of a surface bundle over a surface with fiber genus 2 always vanishes. This follows from the fact that the second cohomology group $H^2(\Gamma_g, \mathbb{Z})$ is isomorphic to $H_2(\Gamma_g, \mathbb{Z})/\text{torsion}$ (see p.158 of [**FM12**]), on the other hand we have $H_2(\Gamma_2, \mathbb{Z}) = \mathbb{Z}/2\mathbb{Z}$ [**KS03**].

3. Flat circle bundles

Consider the universal curve $\Pi : \mathcal{U} \to \mathcal{M}_g$ over the moduli space $\mathcal{M}_g$ of genus g curves. Its fiber over a point $X \in \mathcal{M}_g$ is just the complex curve X. The tangent bundle ν of the fibers of this bundle is a holomorphic complex line bundle on the complex orbifold $\mathcal{U}$.

The following observation (which is due to Morita [**Mo88**]) is based on some facts which were probably already known to Nielsen.

PROPOSITION 3.1. *The circle subbundle of the bundle $\nu \to \mathcal{U}$ is flat.*

PROOF. Let $\Gamma_{g,1}$ be the mapping class group of a surface of genus g with one marked point (puncture), and denote by $\Theta : \Gamma_{g,1} \to \Gamma_g$ the homomorphism induced by the puncture forgetful map. This homomorphism fits into the *Birman exact sequence* [**Bi74, FM12**]

$$1 \to \pi_1(S_g) \to \Gamma_{g,1} \xrightarrow{\Theta} \Gamma_g \to 1.$$

Via this sequence, the group $\Gamma_{g,1}$ is the orbifold fundamental group of the universal curve.

We claim that the group $\Gamma_{g,1}$ admits an action on the circle S^1 by orientation preserving homeomorphisms, where we view S^1 as the ideal boundary $\partial\mathbb{H}^2$ of the hyperbolic plane $\mathbb{H}^2$. Namely, let $x \in S_g$ be a fixed point. The group $\Gamma_{g,1}$ can be viewed as the group of isotopy classes of orientation preserving diffeomorphisms of the surface S_g preserving x. Isotopies are also required to fix x. Any orientation preserving diffeomorphism f of S_g which fixes x induces an automorphism $f_* \in \text{Aut}(\pi_1(S_g, x))$, and since the group of diffeomorphisms of S_g isotopic to the identity is contractible, the isotopy class of f is uniquely determined by the induced map f_*.

The group $PSL(2,\mathbb{R})$ is just the group of orientation preserving isometries of the hyperbolic plane $\mathbb{H}^2$, or, equivalently, it is the group of biholomorphic automorphisms of the unit disk $D \subset \mathbb{C}$. The choice of a hyperbolic structure on S_g then determines the conjugacy class of an embedding $\pi_1(S_g) \to PSL(2,\mathbb{R})$, with discrete cocompact image.

Since the group $PSL(2,\mathbb{R})$ acts simply transitively on the unit tangent bundle $T^1\mathbb{H}^2$ of the hyperbolic plane, we can choose an identification of $PSL(2,\mathbb{R})$ with $T^1\mathbb{H}^2$ which maps the identity to the point $0 \in D = \mathbb{H}^2$. We also may assume that 0 is a preimage of the point $x \in S_g$. With these identifications, the group $\pi_1(S_g, x)$ determines an embedding $\pi_1(S_g, x) \to PSL(2,\mathbb{R})$, unique up to conjugation with the central subgroup $SO(2) \subset PSL(2,\mathbb{R})$, that is, the stabilizer of the basepoint 0. Fix once and for all such an embedding.

A diffeomorphism f of S_g fixing x is a bilipschitz map for the hyperbolic structure of S_g. Thus f can be lifted to a $\pi_1(S_g, x)$-equivariant bilipschitz map $\tilde{f} : \mathbb{H}^2 = D \to D$ which fixes 0. This means that the map $\tilde{f}$ satisfies $\tilde{f}(\psi y) = f_*(\psi)(\tilde{f}(y))$ for all $y \in D$ and all $\psi \in \pi_1(S_g, x) \subset PSL(2,\mathbb{R})$. Equivariance and the requirement that $\tilde{f}(0) = 0$ determines the lift $\tilde{f}$ completely.

Now any bilipschitz map of the hyperbolic plane which fixes the point 0 maps geodesic rays beginning at 0 to uniform quasi-geodesic rays issuing from the same point. Such a uniform quasi-geodesic ray is at uniformly bounded distance from a geodesic ray, and this geodesic ray is unique if its starting point is required to be the fixed point 0. As the ideal boundary $\partial D = S^1$ of the hyperbolic plane is just the set of geodesic rays issuing from 0, this shows that the map $\tilde{f}$ induces a homeomorphism $\Upsilon(f) \in \mathrm{Top}^+(S^1)$. Here preservation of orientation of $\Upsilon(f)$ follows from preservation of orientation of f. The homeomorphism $\Upsilon(f)$ only depends on the isotopy class of f provided that such an isotopy fixes the point x.

By construction, if u is another orientation preserving bilipschitz homemorphism of S_g fixing x, then $\Upsilon(u \circ f) = \Upsilon(u) \circ \Upsilon(f)$. As a consequence, the assignment $f \to \Upsilon(f)$ which associates to a diffeomorphism f of S_g fixing x the homeomorphism $\Upsilon(f) \in \mathrm{Top}^+(S^1)$ descends to a homomorphism $\hat{\Upsilon} : \Gamma_{g,1} \to \mathrm{Top}^+(S^1)$, unique up to conjugation in $\mathrm{Top}^+(S^1)$. This homomorphism then defines a flat circle bundle $H \to \mathcal{U}$. We claim that this circle bundle is (up to equivalence) the circle subbundle of the vertical tangent bundle of the surface bundle $\mathcal{U}$.

Namely, consider the space $\mathcal{T}(S_g^1)$ of all discrete faithful orientation preserving homomorphisms $\rho : \pi_1(S_g, x) \to PSL(2,\mathbb{R})$. The group $PSL(2,\mathbb{R})$ acts on this space by conjugation. Each orbit of this action is a fiber of the bundle $\mathcal{T}(S_g^1) \to \mathcal{T}(S_g)$ where $\mathcal{T}(S_g)$ denotes the Teichmüller space of marked complex structures on the closed oriented surface S_g of genus g. The quotient of $\mathcal{T}(S_g^1)$ by the action of the central circle group $SO(2) = S^1$ is the Teichmüller space $\mathcal{T}(S_{g,1})$ of all marked complex structures on an oriented surface $S_{g,1}$ of genus g with one marked point.

The circle bundle $PSL(2,\mathbb{R}) = T^1\mathbb{H}^2 \to \mathbb{H}^2$ has a $PSL(2,\mathbb{R})$-equivariant identification with $\mathbb{H}^2 \times S^1$ where the action of $PSL(2,\mathbb{R})$ on $S^1 = \partial\mathbb{H}^2$ is described as follows.

For a unit tangent vector $u \in T^1\mathbb{H}^2$ let γ_u be the geodesic ray with initial velocity u. The projection of the identity in $PSL(2,\mathbb{R})$ is a fixed basepoint $0 \in D = \mathbb{H}^2$. Identify $\partial\mathbb{H}^2$ with the fiber of the unit tangent bundle over this basepoint by associating to a unit tangent vector $v \in T_0\mathbb{H}^2$ the endpoint $\gamma_v(\infty)$ of the geodesic ray γ_v. For each $\alpha \in PSL(2,\mathbb{R})$, the differential $d\alpha(0)$ of α at the basepoint 0 then induces the homeomorphism $\gamma_v(\infty) \to \gamma_{d\alpha(v)}(\infty)$ of $\partial\mathbb{H}^2 = S^1$. The thus defined map $PSL(2,\mathbb{R}) \to \mathrm{Top}^+(S^1)$ is equivariant with respect to the action of $PSL(2,\mathbb{R})$ on itself by conjugation.

Now let $f : (S_g, x) \to (S_g, x)$ be an arbitrary diffeomorphism which fixes the point x. Taking a quotient by the action of $(0,\infty)$ on the tangent bundle of S_g by scaling shows that its differential induces an isomorphism of the circle bundle T^1S_g covering the base map f. On the other hand, using the above construction, the map f induces a second isomorphism of T^1S_g as follows. Lift f to a diffeomorphism $\tilde{f}$ of $\mathbb{H}^2$ which fixes 0 and is equivariant with respect to the action of $\pi_1(S_g, x)$ and its image under the automorphism f_*. For any point $y \in \mathbb{H}^2$, map a unit tangent vector $v \in T^1_y\mathbb{H}^2$ to the unit tangent vector $w \in T_{\tilde{f}(y)}\mathbb{H}^2$ such that $\gamma_w(\infty) = \Upsilon(f)\gamma_v(\infty)$. As this construction depends continuously on v and is equivariant with respect to the action of $\pi_1(S_g, x)$, it descends to an isomorphism of T^1S_g covering f.

The circle subbundle of the vertical tangent bundle of the universal curve $\mathcal{U}$ is the quotient of the vertical tangent bundle of $\mathcal{T}(S_{g,1})$, that is of $\mathcal{T}(S^1_g)$, by the action of $\Gamma_{g,1}$ via the tangent map of isotopy classes of diffeomorphisms fixing the basepoint x. Thus to show that this circle bundle is indeed the flat bundle defined by the homomorphism $\hat{\Upsilon} : \Gamma_{g,1} \to \mathrm{Top}^+(S^1)$, it suffices to show that for any diffeomorphism $f : (S_g, x) \to (S_g, x)$, the isomorphism of T^1S_g induced by df is homotopic to the isomorphism induced by $\Upsilon(f)$.

To show that this is indeed the case lift as before the diffeomorphism f to an f_*-equivariant diffeomorphism $\tilde{f}$ of $\mathbb{H}^2$ fixing 0. We deform equivariantly the tangent map $d\tilde{f}$ of $\tilde{f}$ as follows.

For a number $r > 0$ and a point $y \in \mathbb{H}^2$, identify the fiber of the unit tangent bundle of $\mathbb{H}^2$ at y with the boundary $\partial B(y, r)$ of the ball of radius r about y using the exponential map of the hyperbolic plane. The image $\tilde{f}(\partial B(y, r))$ bounds a disk containing $\tilde{f}(y)$. Use the inverse of the exponential map at $\tilde{f}(y)$ to map this circle onto the fiber of the unit tangent bundle of $\mathbb{H}^2$ at $\tilde{f}(y)$. Doing this simultaneously for all $y \in \mathbb{H}^2$ defines a continuous map $\tilde{\zeta}_r : T^1\mathbb{H}^2 \to T^1\mathbb{H}^2$ which is equivariant with respect to the action of $\pi_1(S_g)$ and hence descends to a continuous map $\zeta_r : T^1S_g \to T^1S_g$ covering f. Clearly the maps ζ_r depend continuously on r, and as $r \to 0$, they converge to the map induced by df. Thus for all r, the map ζ_r is homotopic to the map induced by df.

Now by construction, as $r \to \infty$ the maps ζ_r converge to the map induced by $\Upsilon(f)$. As this construction is moreover equivariant with respect to isotopy, this shows that the action of $\Gamma_{g,1}$ on the vertical tangent bundle of the universal covering of the universal curve $\mathcal{U}$ coincides with the action defined by the homomorphism $\hat{\Upsilon}$. Hence the flat circle bundle on $\mathcal{U}$ constructed from the homomorphism $\hat{\Upsilon} : \Gamma_{g,1} \to \mathrm{Top}^+(S^1)$ indeed equals the circle subbundle of the vertical tangent bundle of $\mathcal{U}$. This is what we wanted to show. □

Let now $\Pi : E \to B$ be a surface bundle over an arbitrary smooth base B, with fibre S_g of genus $g \geq 2$. Any such surface bundle can be obtained as a pull-back of the universal curve $\mathcal{U} \to \mathcal{M}_g$ by a smooth (in the sense of orbifolds) map $f : B \to \mathcal{M}_g$. Thus we may assume that the fibres of E are equipped with a complex structure varying smoothly over the base. As a consequence, the vertical tangent bundle ν of E is a smooth complex line bundle over E.

Since $\mathcal{M}_g$ is a classifying space (in the orbifold sense) for its (orbifold) fundamental group, the homotopy class of a map $f : B \to \mathcal{M}_g$ is uniquely determined by the induced homomorphism $f_* = \rho : \pi_1(B) \to \Gamma_g$. Here as before, Γ_g is the mapping class group of S_g. Furthermore, homotopic maps give rise to homeomorphic surface bundles, so the bundle E is determined by ρ. We refer to [**Mo87**] for more details about these well known facts.

Let as before $\Theta : \Gamma_{g,1} \to \Gamma_g$ be the natural surjective homomorphism. Since E is the pull-back of the universal curve under the map f, there exists an exact diagram

$$\begin{array}{ccccccccc} 1 & \longrightarrow & \pi_1(S_g) & \longrightarrow & \pi_1(E) & \longrightarrow & \pi_1(B) & \longrightarrow & 1 \\ & & \downarrow & & \downarrow & & \downarrow & & \\ 1 & \longrightarrow & \pi_1(S_g) & \longrightarrow & \Theta^{-1}(\rho(\pi_1(B))) & \longrightarrow & \rho(\pi_1(B)) & \longrightarrow & 1. \end{array} \tag{4}$$

As a consequence, there exists a homomorphism $\pi_1(E) \to \Theta^{-1}(\rho(\pi_1(B))) \subset \Gamma_{g,1}$ whose restriction to the subgroup $\pi_1(S_g)$ is an isomorphism. By naturality under pull-back, in the case that B is a closed surface we obtain

COROLLARY 3.2. *Let $\Pi : E \to B$ be a surface bundle over a surface. Then $TE = \nu \oplus \Pi^* TB$ is a sum of two complex line bundles whose circle subbundles are flat.*

PROOF. We observed before that $TE = \nu \oplus \Pi^* TB$, and by Proposition 3.1, the circle subbundle of ν is a pull-back of a flat bundle and hence flat. On the other hand, as B is a surface of genus $h \geq 2$, the circle subbundle of the tangent bundle TB of B is flat as well and hence the same holds true for the circle subbundle of the pull-back $\Pi^* TB$. $\square$

4. Selfintersection numbers of sections

A *section* of a surface bundle $\Pi : E \to B$ is a smooth map $f : B \to E$ so that $\Pi \circ f = \mathrm{Id}$. The image $f(B)$ of a section f is a cycle in E which defines a homology class $[f(B)] \in H_k(E, \mathbb{Z})$ where $k = \dim(B)$. In the case that B is a surface, the self-intersection number $[f(B)]^2$ of this class is defined. Our next goal is to shed some light on this self-intersection number from a geometric point of view.

Let as before ν be the vertical line bundle of E, with first Chern class $c_1(\nu)$. Equivalently, $c_1(\nu)$ is the Euler class of the oriented 2-dimensional real oriented vector bundle ν. We note

LEMMA 4.1. *$[f(B)]^2 = c_1(\nu)(f(B))$ for any section $f : B \to E$.*

PROOF. Since $f(B)$ is a smoothly embedded surface in E, the self-intersection number of $f(B)$ equals the Euler number of the pull-back to B of the oriented normal bundle of $f(B)$ in E, that is, it equals the evaluation of the Euler class of this normal bundle on the homology class $[f(B)]$.

As f is a section, $f(B)$ is everywhere transverse to the fibers of $E \to B$. Thus this oriented normal bundle is isomorphic to the restriction of the vertical tangent bundle ν of E. $\square$

It was shown by Milnor [**M58**] and Wood [**W71**] that the Euler number $e(H)$ of a flat circle bundle $H \to B$ over a closed oriented surface B of genus $h \geq 2$ and the Euler characteristic $\chi(B)$ of B satisfy the inequality

$$|e(H)| \leq |\chi(B)|.$$

In view of this result, the conjecture stated in the introduction can be viewed as a higher dimensional analog of the Milnor Wood inequality.

By a result of Thom, for any compact CW-complex X, any homology class $\alpha \in H_2(X, \mathbb{Z})$ can be represented by a map from a closed surface into X, and if α is not a two-torsion class, then the surface can be chosen to be orientable. As a consequence of Proposition 3.1, we obtain

COROLLARY 4.2. *Let $\beta \in H_2(E, \mathbb{Z})$ be represented by a map $f : \Sigma \to E$ where Σ is a closed oriented surface. Then $|c_1(\nu)(\beta)| \leq |\chi(\Sigma)|$.*

PROOF. By Proposition 3.1, the pull-back by f of the circle subbundle of ν is a flat circle bundle over Σ. By naturality, we have

$$|c_1(\nu)(\beta)| = |f^*(c_1(\nu))(\Sigma)| \leq |\chi(\Sigma)|$$

by the Milnor Wood inequality. □

As an immediate consequence, we obtain the following result of Baykur, Korkmaz and Monden (Proposition 1 of [**BKM13**]) and Bowden [**Bow11**], bypassing the use of Seiberg-Witten invariants used to derive this statement in [**BKM13**] and [**Bow11**].

COROLLARY 4.3. *Let $f : B \to E$ be a section of a surface bundle $E \to B$; then $|[f(B)]^2| \leq |\chi(B)|$.*

PROOF. The section is defined by a smooth map $B \to E$ and hence the corollary follows from Lemma 4.1 and Corollary 4.2. □

Theorem 15 of [**BKM13**] shows that for every $g \geq 2, h \geq 1$ and every integer $k \in [-2h + 2, 2h - 2]$ there is a surface bundle with fibre S_g and base of genus h which admits a section of self-intersection number k. We complement this result by analyzing self-intersection numbers of sections of the trivial bundle.

PROPOSITION 4.4. *Let $E \to B$ be the trivial surface bundle with fibre genus $g \geq 2$ and base genus h. If $h < g$ then every section of E has self-intersection number zero. If $h \geq g$ then for each integer k with $h - 1 \geq |k|(g - 1)$ there is a section of self-intersection number $2k(g-1)$, and no other self-intersection numbers occur.*

PROOF. Let B be a surface of genus $h \geq 1$ and let $E = B \times S_g \to B$ be the trivial surface bundle. Then a section $f : B \to E$ is just a smooth map of the form $x \to (x, \varphi(x))$ where $\varphi : B \to S_g$ is smooth. Let $d \in \mathbb{Z}$ be the degree of φ. We claim that the self-intersection number of f equals $d(2 - 2g)$.

To see that this is the case, denoting as before by ν the vertical tangent bundle, we have $c_1(\nu)(f(B)) = \varphi^* c_1(TS_g)(B) = d(2 - 2g)$. By Lemma 4.1, the self-intersection number of the section f coincides with this Euler number.

This shows that the self-intersection number of a section of the trivial bundle is a multiple of $2g - 2$. The proposition now follows from the fact that for a surface B of genus h and every $k \in \mathbb{Z}$, there exists a smooth map $\varphi : B \to S_g$ of degree

k if and only if $|k| \leq \frac{h-1}{g-1}$. Then $x \in B \to (x, \varphi(x))$ is a section of $E \to B$ with self-intersection number $k(2-2g)$.

To show that the condition on k is sufficient for the existence of a map $B \to S_g$ of degree k, note that if $k \leq \frac{h-1}{g-1}$ is positive, then a map $B \to S_g$ of degree k can be constructed as follows. Let $\psi : \Sigma \to S_g$ be an unbranched cover of degree k. The Euler characteristic of Σ fulfills $|\chi(\Sigma)| = k|\chi(S_g)| \leq |\chi(B)|$. Thus there is a degree one map $\zeta : B \to \Sigma$ which pinches a subsurface of B of genus g' to a point, where $g' = h - 1 - k(g-1) \geq 0$ [**Ed79**]. The composition $\psi \circ \zeta : B \to S_g$ has degree k. A map of degree $-k$ can be taken as a composition of this map with an orientation reversing diffeomorphism of B.

On the other hand, by a result of Edwards [**Ed79**], any map $B \to S_g$ of degree $k \geq 1$ is homotopic to the composition of a pinch $B \to B'$ with a branched cover $B' \to S_g$. A nontrivial pinch collapses a subsurface of B bounded by an essential separating simple closed curve to a point and hence it strictly decreases the genus. Thus the genus q of B' is not bigger than the genus h of B.

Now if $B' \to S_g$ is a branched cover of degree k and if b is the total number of branch points, counted with multiplicity, then by the Hurwitz formula [**FK80**],

$$2q - 2 = b + k(2g-2).$$

This implies that $|k| \leq \frac{q-1}{g-1}$ and hence $|k| \leq \frac{h-1}{g-1}$. As any map $B \to S_g$ can be precomposed with an orientation reversing diffeomorphism of B to yield a map of nonnegative degree, this completes the proof of the proposition. □

5. Cohomology of surface bundles

This final section collects some results on the second cohomology group of a surface bundle over a base B which is an arbitrary smooth closed manifold. We also give some additional information in the case that B is a surface.

The cohomology with rational coefficients of a surface bundle over a smooth base was computed by Morita. The following is Proposition 3.1 of [**Mo87**].

PROPOSITION 5.1. *Let $\Pi : E \to B$ be a surface bundle over a smooth base B, with fiber S_g. Let $k = \mathbb{Q}$ or $\mathbb{Z}/p\mathbb{Z}$ where p is a prime not dividing $2g-2$. Then the homomorphism $\Pi^* : H^*(B,k) \to H^*(E,k)$ is injective, and for all q we have*

$$H^q(E,k) \cong H^q(B,k) \oplus H^{q-1}(B, H^1(S_g,k)) \oplus c_1(\nu)H^{q-2}(B,k).$$

In general, we can not hope that the proposition passes on to cohomology with integral coefficients. The reason is that for a surface bundle $E \to B$ with fiber of genus $g \geq 2$, the fiber inclusion $\iota : S_g \to E$ may not induce a surjection $\iota^* : H^2(E,\mathbb{Z}) \to H^2(S_g,\mathbb{Z}) = \mathbb{Z}$. Namely, the Euler class $e \in H^2(S_g,\mathbb{Z})$ of the tangent bundle of S_g has a $2g-2$-th root, but there may not exist such a root for the class $c_1(\nu)$ where as before, $c_1(\nu) \in H^2(E,\mathbb{Z})$ denotes the first Chern class of the vertical tangent bundle of E.

In the following observation, the component $H^1(B, H^1(S_g,\mathbb{Q})) \subset H^2(E,\mathbb{Q})$ is as in Proposition 5.1.

PROPOSITION 5.2. *Let $E \to B$ be a surface bundle over a surface. Then there exists an embedding $H^1(B, H^1(S_g,\mathbb{Z})) \to H^2(E,\mathbb{Z})$ which induces an isomorphism $H^1(B, H^1(S_g,\mathbb{Z})) \otimes \mathbb{Q} \to H^1(B, H^1(S_g,\mathbb{Q})) \subset H^2(E,\mathbb{Q})$.*

PROOF. The standard Leray spectral sequence for the fiber bundle $\Pi : E \to B$ starts with a finite good cover $\mathcal{U} = \{U_i \mid 1 \le i \le k\}$ of B consisting of open sets $U_i \in \mathcal{U}$ with the following properties.

(1) Each set $U_i \in \mathcal{U}$ is diffeomorphic to an open disk $D \subset \mathbb{R}^2$.
(2) Each intersection $U_i \cap U_j$ or $U_i \cap U_j \cap U_k$ is contractible or empty.
(3) The intersection of any four distinct of the sets U_i is empty.

Such a covering can be constructed from a triangulation T of B as follows.

For each vertex x of T, choose a disk neighborhood D_x of x so that the closures of these disks are pairwise disjoint. We also require that each edge e of T intersects a disk D_x if and only if the edge is incident on x, and in this case, the intersection of e with D_x is a connected subarc of e. Furthermore, we require that a two-simplex f intersects a disk D_x if and only if x is a vertex of f, and in this case, $D_x \cap f$ is a disk. Call these disks of vertex type.

For each edge e of T, choose a disk D_e containing a neighborhood of $e - \cup_x D_x$. This can be done in such a way that the disks D_e are pairwise disjoint, that for a vertex x of T, the intersection $D_e \cap D_x$ is empty or a disk, and that the later holds true if and only if e is incident on x. Call such a disk of edge type. The union of the disks of vertex and edge type covers a neighborhood of the one-skeleton of T, and the intersection of any two of these disks either is a disk or empty. The intersection of any three of the disks is empty.

Finally choose a disk for each two-simplex f of T which is contained in f and covers $f - \cup_e D_e - \cup_x D_x$. Call these disks of face type. They can be chosen in such a way that the resulting family of disks covers B and that furthermore, if the intersection of any three of the disks is non-empty, then these disks are of distinct type. The resulting cover $\mathcal{U}$ is called a *good cover* of B. The restriction of E to U_i is trivial for all $U_i \in \mathcal{U}$.

The good cover $\mathcal{U}$ determines a first quadrant double chain complex $K^{p,q}$ with $0 \le p \le 2$ which can be used to compute $H^2(E, \mathbb{Z})$ using the Leray spectral sequence for the sheaf $\mathcal{F}$ of locally constant $\mathbb{Z}$-valued functions on E. Leray's theorem (see Theorem 14.18 of [**BT82**]) shows that the E_2-term of the spectral sequence with coefficients $\mathbb{Z}$ has the form

$$E_2^{p,q} = H^p(\mathcal{U}, H^q(S_g, \mathbb{Z})).$$

This spectral sequence converges to $H^*(E, \mathbb{Z})$ by the generalized Mayer-Vietoris principle, because $\Pi^{-1}(\mathcal{U})$ is a cover of E, see p.169 and Theorem 15.11 of [**BT82**] for details on these facts.

Since $K^{p,q}$ is trivial for $p \ge 3$, for $r \ge 2$ and every $k \ge 2$ the differential $d_r : E_k^{1,1} \to E_k^{1+r,2-r}$ vanishes. Since the spectral sequence converges to $H^*(E, \mathbb{Z})$, this implies that indeed we have an embedding $H^1(\mathcal{U}, H^1(S_g, \mathbb{Z})) = H^1(B, H^1(S_g, \mathbb{Z})) \to H^2(E, \mathbb{Z})$.

We claim that the image group is precisely the subgroup of $H^2(E, \mathbb{Z})$ whose cup product with $C = \Pi^* H^2(B, \mathbb{Z}) \oplus c_1(\nu) H^0(B, \mathbb{Z})$ vanishes. Note to this end that C is a free subgroup of $H^2(E, \mathbb{Z})$ of rank two, and the restriction of the cup product to C is non-degenerate. Now the degree two part of the E_2-term of the spectral sequence decomposes as $E_2 = E_2^{2,0} \oplus E_2^{1,1} \oplus E_2^{0,2}$. The cup product defines a homomorphism

$$E_2^{2,0} \otimes E_2^{1,1} \to E_2^{3,1} = 0,$$

and similarly, the cup product defines a homomorphism

$$E_2^{1,1} \otimes E_2^{0,2} \to E_2^{1,3} = 0.$$

As this argument is also valid with coefficients in $\mathbb{Q}$, and cup product is natural with respect to taking tensor product with $\mathbb{Q}$, this completes the proof. □

REMARK 5.3. We do not know an example of a surface bundle for which the conclusion of Proposition 5.2 is violated. In particular, by [**H83**], it holds true for the universal curve, in fact we have $H^1(\Gamma_g, H^1(S_g,\mathbb{Z})) = 0$ for all g.

The final goal of this article is to give a geometric interpretation of the subgroup $H^1(B, H^1(S_g,\mathbb{Z}))$ of $H^2(E,\mathbb{Z})$ for a surface bundle over a surface $\Pi : E \to B$ and prove the theorem from the introduction. We begin with some results which hold true for an arbitrary surface bundle $E \to B$ over a smooth base. Assume as before that $E \to B$ is obtained by a smooth map $B \to \mathcal{M}_g$. This means that each of the fibers of E has a complex structure varying smoothly over the base.

Abel's theorem shows that the *Picard group* $\mathrm{Pic}(X, 2g-2)$ of all complex line bundles of degree $2g-2$ over a Riemann surface X can be identified with the *Jacobian* $\mathcal{J}(X)$ of X as follows [**FK80, GH78**].

Choose a geometric symplectic basis $a_1, b_1, \dots, a_g, b_g$ of $H_1(S_g,\mathbb{Z})$. This means that a_i, b_i are oriented non-separating simple closed curves in S_g so that a_i, b_i intersect in precisely one point, and $a_i \cap a_j = a_i \cap b_j = b_i \cap b_j = \emptyset$ for all $i \neq j$. This choice then determines a basis $\omega_1, \dots, \omega_g$ of the g-dimensional [**FK80**] complex vector space $H^{1,0}(X,\mathbb{C})$ of holomorphic one-forms on X so that $\omega_i(a_j) = \delta_{ij}$. The imaginary parts of ω_i are linearly independent over $\mathbb{R}$ and hence the one-forms $\omega_1, \dots, \omega_g$ determine a lattice $\Lambda(X)$ in $\mathbb{C}^g$, obtained by integration over the geometric symplectic basis $a_1, b_1, \dots, a_g, b_g$ of $H_1(S_g,\mathbb{Z})$. The quotient of $\mathbb{C}^g$ by this lattice then is the Jacobian $\mathcal{J}(X)$ of X. The fundamental group A of $\mathcal{J}(X)$ is isomorphic to the integral homology group $H_1(S_g,\mathbb{Z}) = \mathbb{Z}^{2g}$ of S_g, and hence using duality provided by the symplectic form, to the group $H^1(S_g,\mathbb{Z})$.

If X varies in a smooth family, then the holomorphic one-forms $\omega_i = \omega_i(X)$ on X defined by $\omega_i(X_i)(a_j) = \delta_{i,j}$ also vary smoothly. This means that there exists a smooth fiber bundle $\Theta : W \to B$ whose fiber over X is just the Jacobian $\mathcal{J}(X)$ of X.

The bundle W is naturally a quotient of the *Hodge bundle*, the complex vector bundle $Z \to B$ whose fiber at a point $X \in B$ equals the complex vector space of holomorphic one-forms on X. This bundle is in general not trivial as a complex vector bundle. However, it is flat as a real vector bundle with symplectic fiber. Namely, the action of the mapping class group of S_g on the first cohomology of S_g defines a homomorphism $\rho : \Gamma_g \to Sp(2g,\mathbb{Z})$, and the Hodge bundle is the bundle

$$Z = \tilde{B} \times H^1(S_g,\mathbb{R})/\pi_1(B)$$

where the action of $\pi_1(B)$ is defined by $(x, Y)g = (xg, \rho(g)^{-1}Y)$.

The right action of $H^1(S_g,\mathbb{R})$ by translation commutes with the action by ρ and hence there is a quotient bundle $W = Z/H^1(S_g,\mathbb{Z}) \to B$ whose fiber at X just equals the Jacobian $J(X)$ of X.

The Jacobian $\mathcal{J}(X)$ parameterizes divisors of degree 0 on X up to linear equivalence, that is, up to adding a divisor of a meromorphic function. Thus $\mathcal{J}(X)$ can be viewed as the subgroup of the Picard group of X parameterizing holomorphic line bundles of degree zero. The group structure is given by the tensor product, with the trivial line bundle as the neutral element.

A *section* of the bundle $\Theta : W \to B$ is a smooth map $\sigma : B \to W$ such that $\Theta \circ \sigma = \mathrm{Id}$. Such a section then determines a splitting of the extension

$$1 \to A \to \pi_1(W) \xrightarrow{\Theta_*} \pi_1(B) \to 1, \tag{5}$$

that is, for some $x \in B$ it defines a homomorphism $\sigma_* : \pi_1(B, x) \to \pi_1(W, \sigma(x))$ such that $\Theta_* \circ \sigma_* = \mathrm{Id}$.

The following is a topological analog of a well known statement on group extensions as discussed in Proposition IV.2.1 of [**Bro82**]. Namely, if G is a discrete group and if A is any G-module, then A-conjugacy classes of splittings of the split extension

$$1 \to A \to A \rtimes G \to G \to 1 \tag{6}$$

are in 1-1-correspondence with the elements of $H^1(G, A)$.

In the topological setting, a conjugacy class of an element in the fundamental group $\pi_1(Y, y)$ of a path connected topological space Y is just a free homotopy class of loops in Y. Being able to move the basepoint continuously is the main difference to the setting of discrete groups. With this in mind, the next observation gives a topological interpretation of the sequence (6) in our setting. Here the G-module A is just the integral cohomology group $H^1(S_g, \mathbb{Z})$ with the monodromy action of $\pi_1(B)$ defined by the representation ρ.

PROPOSITION 5.4. *Homotopy classes of sections $B \to W$ form a group which is isomorphic to $H^1(B, H^1(\pi_1(S_g), \mathbb{Z}))$.*

PROOF. Let $\sigma : B \to W$ be a section. Then for some basepoint $x \in B$, the induced homomorphism $\sigma_* : \pi_1(B, x) \to \pi_1(W, \sigma(x))$ defines a splitting of the extension (5).

Let as before $Z \to B$ be Hodge bundle with fiber $H^1(S_g, \mathbb{R})$, viewed as an abelian group. Recall that we have $W = Z/H^1(S_g, \mathbb{Z})$. For each $x \in B$, there is a natural action of $H^1(S_g, \mathbb{R})$ on the fiber W_x of W over x. We claim that if η is another section of W then σ and η are homotopic if and only if there exists a section ρ of the bundle Z so that $\eta = \rho(\sigma)$, where the action of ρ is fiber preserving.

Namely, if ρ is any section of Z, then using the fiberwise group structure (or, alternatively, the fact that the fiber of Z is contractible), there is a smooth fiber preserving homotopy h_t of $\rho = h_1$ to the section h_0 of Z which associates to $x \in B$ the neutral element in $H^1(S_g, \mathbb{R}) = Z_x$. Then $s \to h_s\sigma$ is a fiber preserving homotopy between σ and $\rho\sigma$.

On the other hand, let us assume that η is homotopic to σ. Let h_t be a fiber preserving homotopy connecting $h_0 = \sigma$ to $h_1 = \eta$. Choose a point $x \in B$ and a preimage $q \in Z_x$ of $\sigma(x)$ in the fiber Z_x of Z at x. The path $t \to h(t, x) = h_t(x)$ lifts to a path $\tilde{h}(t, x)$ in Z_x beginning at q. We can write $\tilde{h}(1, x) = \beta(x) + q$ for some $\beta(x) \in H^1(S_g, \mathbb{R})$ (here we write the group multiplication additively). Now if $u \in Z_x$ is another preimage of $\sigma(x)$ in Z_x then $u = m + q$ for some $m \in A$, identified with the lattice in $H^1(S_g, \mathbb{R})$ defined by the complex structure on the fiber E_x of E. The path $t \to \tilde{h}(t, x) + m$ is the lift of $t \to h(t, x)$ through u. Thus the difference of the endpoints $\beta(x) = \tilde{h}(1, x) - \tilde{h}(0, x) \in H^1(S_g, \mathbb{R})$ does not depend on the choice of the preimage q of $\sigma(x)$ and hence only depends on h. Furthermore, the map

$x \to \beta(x)$ is continuous and hence defines a section of Z with $\eta = \beta(\sigma)$. This is what we wanted to show.

Let now $C^\infty(W)$ and $C^\infty(Z)$ be the sheaf of smooth sections of W and Z, respectively. Since W has a fiberwise structure of an abelian group, these are sheaves of abelian groups. Similarly we define the sheaf $C^\infty(H^1(S_g,\mathbb{Z}))$ of smooth sections of the fiber bundle with fiber the group $A = H^1(S_g,\mathbb{Z})$ (this is meant to be the twisted bundle). We then obtain a short exact sequence of sheaves

$$0 \to C^\infty(H^1(S_g,\mathbb{Z})) \to C^\infty(Z) \to C^\infty(W) \to 0.$$

It follows from the above discussion that $H^0(C^\infty(W))/\xi H^0(C^\infty(Z))$ can naturally be identified with the group of homotopy classes of sections of W where ξ is the projection map $H^0(C^\infty(Z)) \to H^0(C^\infty(W))$. On the other hand, the short exact sequence of sheaves induces a long exact cohomology sequence

$$\cdots \to H^0(C^\infty(Z)) \to H^0(C^\infty(W)) \to H^1(C^\infty(H^1(S_g,\mathbb{Z}))) \to H^1(C^\infty(Z)) \to \cdots.$$

Since the sheaf $C^\infty(Z)$ is the sheaf of smooth sections of a flat vector bundle $Z \to B$, it is fine and hence acyclic. To see this note that a morphism of $C^\infty(Z)$ is a smooth section of the bundle $Z^* \otimes Z$ over B whose fiber over x equals the vector space of endomorphisms of Z_x, that is, it equals the vector space $Z_x^* \otimes Z_x$. This vector space has a distinguished real one-dimensional subspace consisting of constant multiples of the identity, and these one-dimensional subspaces define a trivial one-dimensional real subbundle L of $Z^* \otimes Z$. A smooth section of L can be identified with a smooth function on B. The identity morphism corresponds to the real number 1 in this interpretation.

Now if $\mathcal{U} = \{U_i\}$ is a locally finite covering of B, then there is a subordinate partition of unity, and using the identification of the fiber of the bundle L with $\mathbb{R}$, this partition of unity defines a partition of unity for the sheaf $C^\infty(Z)$, showing that this sheaf is fine and hence acyclic. Thus we obtain the short exact sequence

$$H^0(C^\infty(Z)) \to H^0(C^\infty(W)) \to H^1(C^\infty(H^1(S_g,\mathbb{Z}))) \to 0.$$

But $H^1(C^\infty(H^1(S_g,\mathbb{Z}))) = H^1(B, H^1(S_g,\mathbb{Z}))$ by de Rham's theorem which completes the proof of the proposition. □

Let again $\sigma : B \to W$ be a smooth section. Then by Abel's theorem, for every $x \in B$, the value $\sigma(x)$ of σ at x can be thought of as a holomorphic line bundle of degree 0 on the fiber E_x of E at x depending smoothly on x. Thus σ defines a fiberwise holomorphic line bundle $L(\sigma) \to E$.

For our next observation, let us denote by $\mathcal{F}$ the sheaf of smooth functions on E whose restriction to a fiber is holomorphic, and let $\mathcal{F}^*$ be the subsheaf of functions which vanish nowhere. These are sheaves of abelian groups.

LEMMA 5.5. *The cohomology group $H^1(E,\mathcal{F}^*)$ parameterizes smooth complex line bundles on E whose restrictions to a fiber are holomorphic.*

PROOF. A smooth complex line bundle L on E whose restriction to a fiber is holomorphic is defined by some good cover $\mathcal{U} = \{U_i \mid i\}$ of E with the property that for each i, the intersection of U_i with a fiber is a disk or empty, and smooth trivializations of L on each of the open sets $U_i \in \mathcal{U}$ whose restrictions to the intersections of U_i with a fiber are holomorphic.

Then transition functions for L on $U_i \cap U_j$ are smooth $\mathbb{C}^*$-valued functions on $U_i \cap U_j$ whose restrictions to a fiber are holomorphic. Thus these functions define a one-cocycle for $\mathcal{U}$ with values in $\mathcal{F}^*$, and then they define a class in $H^1(E, \mathcal{F}^*)$.

Vice versa, each one-cocycle for $\mathcal{U}$ with values in $\mathcal{F}^*$ defines a smooth fiberwise holomorphic line bundle on E by gluing the trivial bundle over the sets U_i with the transition functions on $U_i \cap U_j$ defined by the cocycle. Passing to cohomology yields the lemma. □

Smooth line bundles on E are defined by classes in the cohomology group $H^1(E, (C^\infty)^*)$ where C^∞ is the sheaf of smooth functions on E and $(C^\infty)^*$ is the sheaf of smooth functions vanishing nowhere. The short exact sequence

$$0 \longrightarrow \mathbb{Z} \longrightarrow C^\infty \xrightarrow{\exp} (C^\infty)^* \longrightarrow 0$$

then induces a long exact sequence in cohomology

$$\cdots \longrightarrow H^1(E, C^\infty) \longrightarrow H^1(E, (C^\infty)^*) \xrightarrow{\delta} H^2(E, \mathbb{Z}) \longrightarrow \cdots .$$

Since the sheaf C^∞ is fine, this sequence describes explicitly the parameterization of the group of isomorphism classes of smooth line bundles on E by their Chern classes, that is, by the group $H^2(E, \mathbb{Z})$.

Our next goal is to verify that homotopic sections of the bundle W define smoothly equivalent line bundles, or, equivalently, line bundles with the same Chern class, and that this Chern class is just the cohomology class in $H^1(B, H^1(S_g, \mathbb{Z}))$ corresponding to this homotopy class by Proposition 5.4.

To this end note that since the second cohomology of E is representable, each class $\alpha \in H^2(E, \mathbb{Z})$ is the Chern class of a smooth complex line bundle, obtained as the pull-back of the tautological line bundle under a smooth map $f : E \to \mathbb{C}P^N$ for some sufficiently large N which defines α. Homotopic maps define isomorphic line bundles.

Now if $L \to E$ is a smooth complex line bundle, then the *degree* of L can be defined as the evaluation of its Chern class $c_1(L)$ on one (and hence on any) fiber. Denote as before by $c_1(\nu)$ the Chern class of the vertical cotangent bundle. By Proposition 5.2, the subgroup $H^1(B, H^1(S_g, \mathbb{Z}))$ of $H^2(E, \mathbb{Z})$ is contained in the kernel of the homomorphism $\alpha \to \alpha \cup c_1(\nu)$. The restriction of this homomorphism to $\Pi^* H^2(B, \mathbb{Z})$ is injective. The next proposition provides the connection between the constructions in this section.

PROPOSITION 5.6. *Let $E \to B$ be a surface bundle over a surface. Then the cohomology group $H^1(B, H^1(S_g, \mathbb{Z})) \subset H^2(E, \mathbb{Z})$ parameterizes isomorphism classes of fiberwise holomorphic line bundles L on E of degree* 0 *whose Chern class $c_1(L)$ satisfies $c_1(L) \cup c_1(\nu) = 0$.*

PROOF. By the above discussion, a smooth section σ of the bundle W defines on the one hand an equivalence class of a complex line bundle $L(\sigma)$ on E whose restrictions to a fiber is holomorphic of degree 0. On the other hand, by Proposition 5.4 and Proposition 5.2, it defines a cohomology class in $H^1(B, H^1(S_g, \mathbb{Z})) \subset H^2(E, \mathbb{Z})$. We have to show that this cohomology class is just the first Chern class of $L(\sigma)$. As smooth line bundles on E with the same Chern class are equivalent, this implies that homotopic sections of W define smoothly equivalent line bundles on E, a fact which can also be verified directly.

Consider again the sheaf $\mathcal{F}$ of smooth functions on E which are fiberwise holomorphic, the subsheaf $\mathcal{F}^*$ of functions in $\mathcal{F}$ which vanish nowhere, and the sheaf $\mathbb{Z}$ of locally constant $\mathbb{Z}$-valued functions.

The short exact sequence

$$0 \longrightarrow \mathbb{Z} \longrightarrow \mathcal{F} \xrightarrow{\exp} \mathcal{F}^* \longrightarrow 0 \tag{7}$$

induces a long exact sequence in cohomology. Since the sheaf C^∞ of smooth functions on E is fine, the inclusions $\mathcal{F} \to C^\infty$ and $\mathcal{F}^* \to (C^\infty)^*$ then determine an exact commutative diagram

$$\begin{array}{ccccccccc} \cdots & \xrightarrow{\exp} & H^1(E,\mathcal{F}^*) & \longrightarrow & H^2(E,\mathbb{Z}) & \longrightarrow & H^2(E,\mathcal{F}) & \longrightarrow & \cdots \\ & & \downarrow{\scriptstyle\eta} & & \downarrow{\scriptstyle \mathrm{Id}} & & \downarrow & & \\ \cdots & \xrightarrow{\exp} & H^1(E,(C^\infty)^*) & \xrightarrow{\delta} & H^2(E,\mathbb{Z}) & \longrightarrow & 0 & \longrightarrow & \cdots \end{array} \tag{8}$$

We claim that the homomorphism η is surjective. By exactness and since the diagram commutes, this follows if we can show that $H^2(E,\mathcal{F}) = 0$. However, if E_x is any fiber of E then we have $H^2(E_x,\mathcal{O}) = 0$ where as usual, $\mathcal{O}$ is the sheaf of holomorphic functions on E_x. Namely, by Serre duality, this cohomology group can be identified with the space of holomorphic two-forms on E_x, and this space vanishes since the complex dimension of E_x equals one. On the other hand, the sheaf of smooth functions on B is fine and hence the Leray spectral sequence shows that indeed, $H^2(E,\mathcal{F}) = H^0(B,H^2(E_x,\mathcal{O})) = 0$.

Since $H^1(E,C^\infty) = 0$, the homomorphism δ is an isomorphism. This yields that every smooth line bundle on E is smoothly equivalent to a line bundle whose restriction to a fiber is holomorphic. It also follows that the homomorphism η maps $H^1(E,\mathcal{F}^*)/\exp(H^1(E,\mathcal{F}))$ isomorphically onto $H^1(E,(C^\infty)^*)$. Hence associating to an element in this group its Chern class is an isomorphism.

On the other hand, for each $x \in B$ the vector space $H^1(E_x,\mathcal{O})$ is just the space of holomorphic one-forms on the fiber E_x of E over x by Serre duality. That is, $H^1(E_x,\mathcal{F})$ is the fiber at x of the Hodge bundle $Z \to B$. Thus using the fact that the sheaf of smooth sections of Z is fine (see the discussion in the proof of Proposition 5.4 for details), the Leray spectral sequence shows that $H^1(E,\mathcal{F}) = H^0(Z)$, the vector space of smooth sections of Z.

Now consider the part

$$\begin{array}{ccccccccc} \cdots & \longrightarrow & H^1(E,\mathbb{Z}) & \xrightarrow{exp} & H^1(E,\mathcal{F}) & \longrightarrow & H^1(E,\mathcal{F}^*) & \longrightarrow & \cdots \\ & & \downarrow{\scriptstyle \mathrm{Id}} & & \downarrow & & \downarrow{\scriptstyle\eta} & & \\ \cdots & \longrightarrow & H^1(E,\mathbb{Z}) & \longrightarrow & 0 & \longrightarrow & H^1(E,(C^\infty)^*) & \longrightarrow & \cdots \end{array} \tag{9}$$

of the above exact diagram. It shows that the kernel of η can be identified with $H^0(Z)/\exp H^1(E,\mathbb{Z})$. By Proposition 5.4 and its proof, this subgroup is precisely the group of sections of the bundle W which are homotopic to the trivial section. As a consequence, homotopic sections of W define line bundles with the same Chern class and hence line bundles which are smoothly equivalent. Furthermore, associating to a homotopy class of a section of W the Chern class of the line bundle it defines is an isomorphism of the space of all homotopy classes of sections onto $H^1(B,H^1(S_g,\mathbb{Z})) \subset H^2(E,\mathbb{Z})$. $\square$

REMARK 5.7. The assumption that $E \to B$ is a surface bundle over a surface was only used through the conclusion of Proposition 5.2.

REMARK 5.8. The proof of Proposition 5.6 also shows that any smooth complex line bundle on E is smoothly equivalent to a line bundle whose restriction to each fiber is holomorphic.

We use similar ideas to show

PROPOSITION 5.9. *Let $E \to B$ be a surface bundle which admits a section. Then there exists a cohomology class $e \in H^2(E, \mathbb{Z})$ with $(2g-2)e = c_1(\nu)$, and for all q we have*

$$H^q(E, \mathbb{Z}) = H^q(B, \mathbb{Z}) \oplus H^{q-1}(B, H^1(S_g, \mathbb{Z})) \oplus eH^{q-2}(B, \mathbb{Z}).$$

PROOF. Let $f : B \to E$ be a section of the surface bundle $\Pi : E \to B$. Assume as before that each fiber E_x of E is equipped with a complex structure varying smoothly with x. Then for each $x \in B$, the point $f(x) \in E_x$ can be thought of as a divisor in E_x defining a complex line bundle L_x of degree 1 on E_x. As these line bundles depend smoothly on x, they fit together to a fiberwise holomorphic line bundle L of fiberwise degree one.

Let $c_1(L) \in H^2(E, \mathbb{Z})$ be the Chern class of L. Consider the inclusion $\iota : E_x \to E$. As the fiberwise degree of L equals one, we know that $\iota^* c_1(L)$ is a generator of $H^2(E_x, \mathbb{Z})$. Thus the spectral sequence argument in the proof of Proposition 3.1 of [**Mo87**] applies to compute the cohomology of E with coefficients in $\mathbb{Z}$ (this argument only uses surjectivity of ι^* for the coefficient ring under consideration), yielding the formula in Proposition 5.1 but with coefficients $\mathbb{Z}$. □

REMARK 5.10. Although the existence of a section for a surface bundle $E \to B$ is simply equivalent to stating that the induced homomorphism $\pi_1(B) \to \Gamma_g$ lifts to a homomorphism $\pi_1(B) \to \Gamma_{g,1}$, we do not know how to characterize this property in purely topological terms of the surface bundle. In fact, if $E \to B$ is a surface bundle over a surface, then E is bordant to a surface bundle over a surface which admits a section, see [**H83**].

Proposition 5.9 describes a correspondence between line bundles on a surface bundle over a surface $\Pi : E \to B$, their Chern classes and their Poincaré dual. This can be extended as follows. Namely, a section $f : B \to E$ can be thought of as a section in the bundle over B whose fiber consists of all effective divisors of degree 1 on the fiber of E. This viewpoint generalizes as follows.

An effective divisor on a Riemann surface X of degree $k \geq 1$ is just a weighted collection of points on X with positive weights which sum up to k. Thus there is a natural topology on the total space $\mathcal{D}^k$ of all effective divisors of degree k on the fibers of E defined as follows. Let $V \to B$ be the fiber product of k copies of the fiber of E. There is a natural smooth fiber preserving free action of the symmetric group in k variables on V. Then $\mathcal{D}^k$ can be identified with the quotient of this action and hence it inherits from V the quotient topology. By abuse of notation, we denote again by Π the projection $\mathcal{D}^k \to B$.

Let us assume that there exists a section $\psi : B \to \mathcal{D}^k$. Associate to this section the fiberwise holomorphic line bundle $L(\psi)$ whose restriction of a fiber E_x is dual to the divisor $\psi(x)$, and associate to $L(\psi)$ its Chern class $c_1(L(\psi)) \in H^2(E, \mathbb{Z})$.

Now the section ψ of $\mathcal{D}^k$ defines a cycle in E which can be seen as follows. The projection of the fat diagonal of V is a submanifold N of $\mathcal{D}^k$ of fiberwise positive real codimension 2. Thus by transversality, we may assume that ψ is transverse to this submanifold. Then there are (at most) finitely many points $x_1, \ldots, x_m$ such that $\psi(x_i) \in N$, and for each i, the image of x_i in E_{x_i} consists of $m-1$ distinct points, with precisely one point of multiplicity 2.

Choose a triangulation of B containing the points x_i as vertices. For any point $x \notin \{x_1, \ldots, x_m\}$, the preimage of x in E_x defined by ψ (that is, the union of all points of $\psi(x)$) consists of precisely m points moving smoothly with the base. Thus each two-simplex of the triangulation has precisely k preimages in E, and the same holds true for all one-simplices. The preimages of the points x_i consist of only $m-1$ points. It follows from this construction that the union of these triangles is a surface $\Sigma \subset E$. The orientation of B induces an orientation of Σ. The restriction of the projection Π to Σ is a branched cover, ramified precisely at the points x_i. Thus Σ defines a homology class $\beta(\psi) \in H_2(E, \mathbb{Z})$. We have

PROPOSITION 5.11. *The class $\beta(\psi)$ is Poincaré dual to $c_1(L(\psi))$.*

PROOF. Let us recall how to construct from the embedded surface Σ which is transverse to the fibers of E a line bundle whose restriction to a fiber is holomorphic. Namely, for a point $x \in \Sigma$, choose a neighborhood U of x in E so that $U \cap \Sigma$ is a smooth disk. There exists a smooth $\mathbb{C}$-valued function f on U whose restriction to a fiber is holomorphic and with nowhere vanishing derivative, so that $U \cap \Sigma$ is the level set of level zero. Choose a covering of Σ by such sets, with corresponding functions. On the intersections of these sets, the quotients of these functions do not vanish. Thus these functions define a cocycle whose cohomology class defines a line bundle. This line bundle has a smooth section which is fiberwise holomorphic and vanishes precisely on Σ. In particular, this line bundle coincides with the line bundle $L(\psi)$.

Now if the section ψ intersects the fat diagonal N of $\mathcal{D}^k$, then at the finitely many intersection points with N, choose the function so that it has a double zero at that point and proceed as before.

Since every class in $H_2(E, \mathbb{Z})$ can be represented by a smooth map $f : M \to E$ where M is a closed oriented surface of some genus $h \geq 0$, for the proof of the proposition it now suffices to show the following.

Assume without loss of generality that $f(M)$ intersects Σ transversely in finitely many points $y_1, \ldots, y_p \in E - \cup_j \Pi^{-1}(x_j)$, with intersection index $\sigma(y_i) \in \pm 1$. We have to show that $c_1(L(\psi))(f(M)) = \sum_i \sigma(y_i)$.

As the line bundle $L(\psi)$ is trivial on $E - \Sigma$, the pull-back of $L(\psi)$ under f is a complex line bundle on M with a section which vanishes precisely to first order at the points y_i, and the index of this zero is ± 1 depending on whether the intersection is positive or negative. On the other hand, $c_1(f^*L(\psi))(M)$ equals the number of zeros of a section of $f^*L(\psi)$, counted with sign and multiplicities, provided that this section is transverse to the zero section. Together this means that $c_1(f^*L(\psi))) = \sum_i \sigma(y_i)$. Since f was an aribtrary map of a closed oriented surface M into E, we conclude that indeed, for any second homology class $\alpha \in H_2(E, \mathbb{Z})$ we have $\alpha \cdot \Sigma = c_1(L(\psi))(\alpha)$ as claimed. □

References

[Ar17] Donu Arapura, *Toward the structure of fibered fundamental groups of projective varieties* (English, with English and French summaries), J. Éc. polytech. Math. **4** (2017), 595–611, DOI 10.5802/jep.52. MR3665609

[B12] R. İnanç Baykur, *Non-holomorphic surface bundles and Lefschetz fibrations*, Math. Res. Lett. **19** (2012), no. 3, 567–574, DOI 10.4310/MRL.2012.v19.n3.a5. MR2998140

[BKM13] R. İnanç Baykur, Mustafa Korkmaz, and Naoyuki Monden, *Sections of surface bundles and Lefschetz fibrations*, Trans. Amer. Math. Soc. **365** (2013), no. 11, 5999–6016, DOI 10.1090/S0002-9947-2013-05840-0. MR3091273

[Bi74] Joan S. Birman, *Braids, links, and mapping class groups*, Princeton University Press, Princeton, N.J.; University of Tokyo Press, Tokyo, 1974. Annals of Mathematics Studies, No. 82. MR0375281

[BT82] Raoul Bott and Loring W. Tu, *Differential forms in algebraic topology*, Graduate Texts in Mathematics, vol. 82, Springer-Verlag, New York-Berlin, 1982. MR658304

[Bow11] Jonathan Bowden, *On closed leaves of foliations, multisections and stable commutator lengths*, J. Topol. Anal. **3** (2011), no. 4, 491–509, DOI 10.1142/S1793525311000696. MR2887673

[Br18] C. Bregman, *On Kodaira fibrations with invariant cohomology*, arXiv:1811.00584, to appear in Geometry & Topology.

[Bro82] Kenneth S. Brown, *Cohomology of groups*, Graduate Texts in Mathematics, vol. 87, Springer-Verlag, New York-Berlin, 1982. MR672956

[BD02] Jim Bryan and Ron Donagi, *Surface bundles over surfaces of small genus*, Geom. Topol. **6** (2002), 59–67, DOI 10.2140/gt.2002.6.59. MR1885589

[CS18] L. Chen, N. Salter, *The Birman exact sequence does not virtually split*, arXiv:1804.11235, to appear in Math. Res. Lett.

[Ed79] Allan L. Edmonds, *Deformation of maps to branched coverings in dimension two*, Ann. of Math. (2) **110** (1979), no. 1, 113–125, DOI 10.2307/1971246. MR541331

[EKKOS02] H. Endo, M. Korkmaz, D. Kotschick, B. Ozbagci, and A. Stipsicz, *Commutators, Lefschetz fibrations and the signatures of surface bundles*, Topology **41** (2002), no. 5, 961–977, DOI 10.1016/S0040-9383(01)00011-8. MR1923994

[FM12] Benson Farb and Dan Margalit, *A primer on mapping class groups*, Princeton Mathematical Series, vol. 49, Princeton University Press, Princeton, NJ, 2012. MR2850125

[FK80] H. M. Farkas and I. Kra, *Riemann surfaces*, 2nd ed., Graduate Texts in Mathematics, vol. 71, Springer-Verlag, New York, 1992, DOI 10.1007/978-1-4612-2034-3. MR1139765

[Fl17] L. Flapan, *Monodromy of Kodaira fibrations of genus 3*, arXiv:1709.03164, to appear in Math. Nachrichten.

[GH78] Phillip Griffiths and Joseph Harris, *Principles of algebraic geometry*, Wiley-Interscience [John Wiley & Sons], New York, 1978. Pure and Applied Mathematics. MR507725

[H83] John Harer, *The second homology group of the mapping class group of an orientable surface*, Invent. Math. **72** (1983), no. 2, 221–239, DOI 10.1007/BF01389321. MR700769

[Ka98] M. Kapovich, *On normal subgroups in the fundamental group of complex surfaces*, arXiv:math/9808085.

[KS03] Mustafa Korkmaz and András I. Stipsicz, *The second homology groups of mapping class groups of oriented surfaces*, Math. Proc. Cambridge Philos. Soc. **134** (2003), no. 3, 479–489, DOI 10.1017/S0305004102006461. MR1981213

[K98] D. Kotschick, *Signatures, monopoles and mapping class groups*, Math. Res. Lett. **5** (1998), no. 1-2, 227–234, DOI 10.4310/MRL.1998.v5.n2.a9. MR1617905

[M58] John Milnor, *On the existence of a connection with curvature zero*, Comment. Math. Helv. **32** (1958), 215–223, DOI 10.1007/BF02564579. MR95518

[Mo87] Shigeyuki Morita, *Characteristic classes of surface bundles*, Invent. Math. **90** (1987), no. 3, 551–577, DOI 10.1007/BF01389178. MR914849

[Mo88] S. Morita, *Characteristic classes of surface bundles and bounded cohomology*, A fête of topology, Academic Press, Boston, MA, 1988, pp. 233–257. MR928403

[W71] John W. Wood, *Bundles with totally disconnected structure group*, Comment. Math. Helv. **46** (1971), 257–273, DOI 10.1007/BF02566843. MR293655

Mathematisches Institut der Universität Bonn, Endenicher Allee 60, 53115 Bonn, Germany

Email address: `ursula@math.uni-bonn.de`

Proceedings of Symposia in Pure Mathematics
Volume **104**, 2021
https://doi.org/10.1090/pspum/104/01873

Some recents advances on Duke's equidistribution theorems

Philippe Michel

ABSTRACT. Duke's theorems are three equidistribution statements for sets of representations of large integers by ternary quadratic forms. In this paper we survey some recent development of these beautiful theorems.

CONTENTS

1. Introduction

In the *Disquisitiones Arithmeticae*, C.-F. Gauss made fundamental contributions to the theory of integral quadratic forms, especially in the binary and ternary case. In the ternary case he considered in particular the *euclidean* and the *discriminant* quadratic forms

$$\mathcal{E}_3(a,b,c) = a^2 + b^2 + c^2,\ \ \Delta(a,b,c) = b^2 - 4ac.$$

For q either of these forms and $d \in \mathbb{Z} - \{0\}$, we denote by

$$\mathcal{R}_q(d) := \{(a,b,c) \in \mathbb{Z}^3, q(a,b,c) = d,\ \ (a,b,c) = 1\}$$

the set of primitive representations of d by q.

Gauss gave necessary and sufficient conditions for $\mathcal{R}_q(d)$ to be non-empty (these are special cases of the Hasse principle):

- For $\mathcal{E}_3$, the condition is that d is positive and not of the shape $4^k(8l + 7)$, $k, l \in \mathbb{N}$: this is the *Three squares* or *Gauss-Legendre Theorem*[1].
- For Δ, the condition is that $d \equiv 0, 1 \pmod 4$.

2020 *Mathematics Subject Classification.* Primary 11E20, 11F41, 11F66.

Partially supported by a DFG-SNF lead agency program grant (grant 200021L_175755).

[1]Gauss was the first to provide an unconditional proof

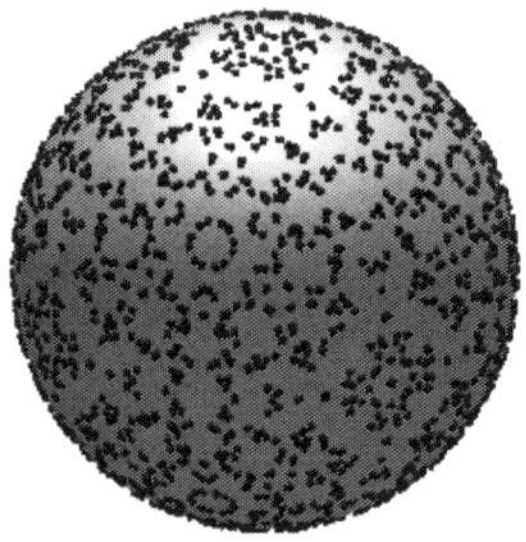

FIGURE 1.1. Equidistribution on the sphere ($q = \mathcal{E}_3$) for $d = 40001$.

In addition to existence, Gauss investigated the structure of the quotient space

$$[\mathcal{R}_q(d)] := \mathrm{SO}_q(\mathbb{Z})\backslash\mathcal{R}_q(d),$$

relative to the obvious action $\mathrm{SO}_q(\mathbb{Z}) \curvearrowright \mathcal{R}_q(d)$ and discovered that this set is in fact a *torsor* under the action of a finite abelian group[2], the *ideal class group* of a suitable order in the quadratic field $\mathbb{Q}(\sqrt{-d})$ for $q = \mathcal{E}_3$ or $\mathbb{Q}(\sqrt{d})$ for $q = \Delta$. Later, Dirichlet gave an analytic formula for the size of this group in terms of values of quadratic character L-functions (the *class number formula*).

1.1. Duke's equidistribution Theorems. Duke's Theorems [**Duk88**] (first conjectured by Linnik) are equidistribution results for the sets $\mathcal{R}_q(d)$ when d becomes large and can be seen as far reaching refinements of the work of Gauss:

Given $\lambda \in \mathbb{R}^\times$, we denote by

$$V_{q,\lambda}(\mathbb{R}) := \{(x, y, z) \in \mathbb{R}^3,\ q(x, y, z) = \lambda\}$$

the λ-level set of q : depending on the sign of λ, this level is, for $q = \mathcal{E}_3$, either a sphere or the empty set and, for $q = \Delta$, either a one or two-sheeted hyperboloid. Taking $\lambda = \mathrm{sign}(d) = \pm 1$, one has the obvious inclusion:

$$|d|^{-1/2}\mathcal{R}_q(d) \subset V_{q,\pm 1}(\mathbb{R}).$$

Here is a simplified version of Duke's theorems:

THEOREM (Duke's equidistribution theorems, 1st version). *As $d \to \infty$ along squarefree integers satisfying*

- $d > 0,\ d \not\equiv 7 \pmod 8$, *if* $q = \mathcal{E}_3$;
- $d \equiv 1, 2 \pmod 4$, *if* $q = \Delta$,

the set $|d|^{-1/2}.\mathcal{R}_q(d)$ become equidistributed on $V_{q,\pm 1}(\mathbb{R})$ ($\pm 1 = \mathrm{sign}(d)$) with respect to the unique (up to scalars) $\mathrm{SO}_q(\mathbb{R})$-invariant measure $\mu_{q,\pm 1}$.

REMARK 1.1. By "equidistribution" we mean that for any continuous compactly supported functions $\varphi_0, \varphi \in \mathcal{C}_c^0(V_{q,\pm 1}(\mathbb{R}))$ such that $\mu_{q,\pm 1}(\varphi_0) \neq 0$, one has

$$\sum_{(a,b,c)\in\mathcal{R}_q(d)} \varphi_0(|d|^{-1/2}(a, b, d)) \neq 0$$

[2] In fact, for $q = \Delta$ Gauss defined a group structure directly on $[\mathcal{R}_\Delta(d)]$: the law of composition of binary quadratic forms.

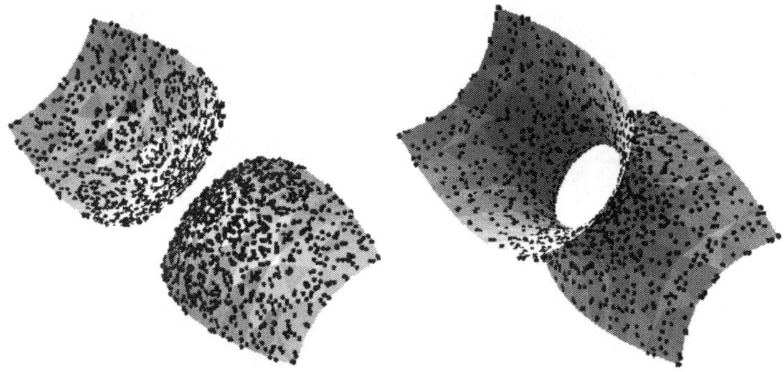

FIGURE 1.2. Equidistribution on hyperboloids ($q = \Delta$) for $d = -110003$ and $d = 10001$.

for d large enough (and satisfying the above congruences) and

$$\frac{\sum_{(a,b,c)\in\mathcal{R}_q(d)} \varphi(|d|^{-1/2}(a,b,d))}{\sum_{(a,b,c)\in\mathcal{R}_q(d)} \varphi_0(|d|^{-1/2}(a,b,d))} \to \frac{\mu_{q,\pm1}(\varphi)}{\mu_{q,\pm1}(\varphi_0)},\ d\to\infty.$$

REMARK 1.2. We have restricted to squarefree integers d's to simplify the statements but Duke's theorems holds for general d and for general ternary quadratic forms (see Remark 1.4 below).

REMARK 1.3. For the euclidean form $\mathcal{E}_3$, this result was also obtained independently by Fomenko and Golubeva at about the same time [**GF87**].

1.2. Dual formulation. Duke's equidistribution Theorems are proven using a far reaching generalisation of Weyl's equidistribution criterion [**Wey16**], that is by checking the convergence of a sequence of measures against test functions taken from a suitable generating set of functions on the relavant space. However in the present case, the spaces considered are not the obvious ones but suitable *duals* spaces. In this section we review the dual formulation of Duke's theorems.

There is not much to do for $q = \mathcal{E}_3$ (because $\mathcal{E}_3$ is definite) but for $q = \Delta$, this dual formulation takes quite different shapes depending on the sign of d.

Let

$$Y_0(1) = \mathrm{SL}_2(\mathbb{Z})\backslash\mathbb{H}.$$

denote the modular curve.

Given $d \equiv 1 \,(\mathrm{mod}\, 4)$ a squarefree integer; to $(a,b,c) \in \mathcal{R}_\Delta(d)$ a representation one associates

- If $d < 0$, the point

$$z_{[a,b,c]} := \mathrm{SL}_2(\mathbb{Z}).z_{(a,b,c)} \in Y_0(1) \text{ where } z_{(a,b,c)} := \frac{-b + i|d|^{1/2}}{2a}.$$

- If $d > 0$, the closed geodesic

$$\gamma_{[a,b,c]} = \mathrm{SL}_2(\mathbb{Z})\gamma_{(a,b,c)} \subset Y_0(1)$$

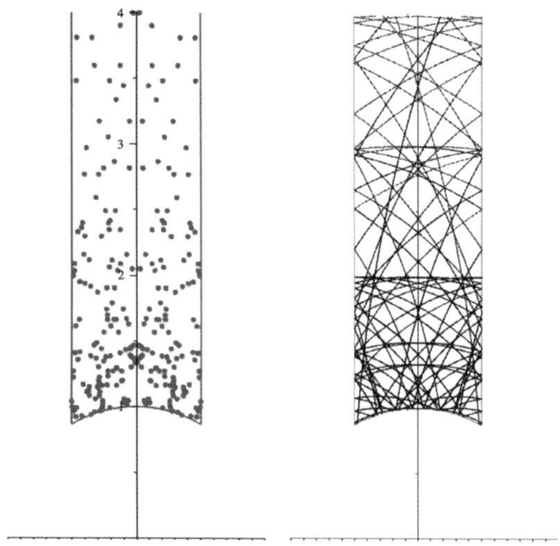

FIGURE 1.3. Equidistribution on the modular Riemann surface ($q = \Delta$) for $d = -418916$, and $d = 577$.

where $\gamma_{(a,b,c)} \subset \mathbb{H}$ is the semi-circle whose end-points are the real numbers

$$z^{\pm}_{(a,b,c)} = \frac{-b \pm |d|^{1/2}}{2a} \in \mathbb{R}.$$

In fact the quantities $z_{[a,b,c]}$ and $\gamma_{[a,b,c]}$ depend only on the *class* of the representation (a,b,c),

$$[a,b,c] := \mathrm{SO}_q(\mathbb{Z})(a,b,c) \in [\mathcal{R}_q(d)].$$

In particular, by the work of Gauss, the sets of all such points (called *Heegner points*) or geodesic curves are finite and have cardinality the class number $|[\mathcal{R}_q(d)]|$. We denote these sets by

$$\mathcal{H}_d := \{z_{[a,b,c]},\ (a,b,c) \in \mathcal{R}_\Delta(d)\},\ d < 0$$
$$\Gamma_d := \{\gamma_{[a,b,c]},\ (a,b,c) \in \mathcal{R}_\Delta(d)\},\ d > 0.$$

THEOREM (Duke's equidistribution theorem for Δ, 2nd version). *As $d \to \infty$ among squarefree integers $\equiv 1 \,(\mathrm{mod}\, 4)$ the sets of Heegner $\mathcal{H}_d$ and closed geodesics Γ_d are equidistributed on $Y_0(1)$ with respect to the hyperbolic probability measure $\mu_h = \frac{3}{\pi}\frac{dxdy}{y^2}$: for any $\varphi \in \mathcal{C}_c^0(Y_0(1))$, one has as $d \to \infty$*

$$\frac{1}{|\mathcal{H}_d|}\sum_{[a,b,c]} \varphi(z_{[a,b,c]}) \to \int_{Y_0(1)} \varphi(z)d\mu_h(z),\ \textit{if } d < 0$$

$$\frac{1}{|\Gamma_d|}\sum_{[a,b,c]} \frac{1}{\mathrm{length}(\gamma_{[a,b,c]})}\int_{\gamma_{[a,b,c]}} \varphi(t)dt \to \int_{Y_0(1)} \varphi(z)d\mu_h(z),\ \textit{if } d > 0.$$

1.3. The work of Duke and Schulze-Pillot on Hilbert's 11th problem. Shortly after [**Duk88**], Duke and Schulze-Pillot [**DSP90**] extended Duke's theorem for the Euclidean form to a general definite ternary quadratic form q. The chief difference is that as long as the discriminant of q is large enough, the set of genus classes is greater than one. A simplified version of [**DSP90**, Corollary] is the following:

THEOREM (Duke–Schulze-Pillot). *Let q be an integral valued definite ternary quadratic form. As $d \to \infty$, along the square-free integers that are locally representable by q (such that the equation $q(a,b,c) = d$ has a solution in $\mathbb{R}^3$ and in $\mathbb{Z}_p^3$ for every p)*

$$\mathcal{R}_q(d) \to \infty$$

(in particular is non-empty) and the set $|d|^{-1/2}\mathcal{R}_q(d)$ becomes equidistributed on $V_{q,\pm1}(\mathbb{R})$ for the unique $\mathrm{SO}_q(\mathbb{R})$-invariant probability measure.

REMARK 1.4. For general d, one can show that the same result holds as long as the valuation of d at any prime where q is anisotropic is bounded by a fixed number and the square-free part of d does not belong to a finite set of exceptional integers(depending on q). See [**SP04**, Cor 5.6] and the references herein.

Up to minor additional technicalities, this concludes the resolution of Hilbert's 11th problem (on representing integers by quadratic forms over a number field) for the field of rationals $\mathbb{Q}$.

For more than 30 years, Duke's equidistribution theorems have inspired many people from different areas of mathematics and have witnessed striking developments and extensions. In this survey, we will present some of these developments.

1.4. Plan of the paper.

- In the next Section 2 we discuss Duke's original proof which is via theta functions and has its origins in a breakthrough of Iwaniec.
- Shortly after, another proof emerged based on L-functions and the *subconvexity problem.* The two approaches are connected by two formulae due Waldspurger. We will discuss this in Section 3 and will highlight further extensions of Duke's theorems allowed by this approach (notably the extension to general number fields).
- As pointed out, Duke's theorems were first conjectured by Linnik in the 50's. These conjectures were supported by Linnik's proof of these theorems for d's restricted along subsequences of integers satisfying an additional congruence condition. Linnik's approach (*the ergodic method*) is not based on harmonic analysis Duke's original proof or the L-function approach are but on ideas methods from the theory of dynamical systems; the ergodic method used in a systematic way the torsor structure on $[\mathcal{R}_q(d)]$ discovered by Gauss. We will discuss Linnik's ergodic method and some of its recent developments in Section 4.

2. Duke's Equidistribution Theorems: the original proof

2.1. The dual formulation. All proofs of Duke's theorems are based on the dual formulation which we briefly explain here. We denote the special orthogonal group of the form q by

$$\mathrm{G} = \mathrm{SO}_q$$

(viewed as a $\mathbb{Q}$-algebraic group ; we denote by $\mathrm{G}(\mathbb{Z})$ the stwabilizer of the lattice $\mathbb{Z}^3 \subset \mathbb{R}^3$. Let

$$\mathbf{x} \in V_{q,\pm1}(\mathbb{R})$$

be a fixed base point. By Witt's theorem, the action $\mathrm{G}(\mathbb{R}) \curvearrowright V_{q,\pm1}(\mathbb{R})$ is transitive and we have an homeomorphism

$$V_{q,\pm1}(\mathbb{R}) \simeq \mathrm{G}(\mathbb{R})/\mathrm{G}_{\mathbf{x}}(\mathbb{R}) \tag{2.1}$$

(here $\mathrm{G}_{\mathbf{x}}(\mathbb{R})$ denote the stabilizer of $\mathbf{x}$); from this, the existence of a left $\mathrm{G}(\mathbb{R})$-invariant measure $\mu_{q,\pm1}$ follows as quotient of a Haar measure μ_{G} for $\mathrm{G}(\mathbb{R})$ by a Haar measure $\mu_{\mathrm{G}_{\mathbf{x}}}$ on $\mathrm{G}_{\mathbf{x}}(\mathbb{R})$; that measure is unique up to multiplication by a scalar.

Under the identification (2.1) each primitive representation of d by q is identified with a *right* $\mathrm{G}_{\mathbf{x}}(\mathbb{R})$-orbit:

$$|d|^{-1/2}(a,b,c) \simeq g_{(a,b,c)}.\mathrm{G}_{\mathbf{x}}(\mathbb{R}).$$

The left action $\mathrm{G}(\mathbb{Z}) \curvearrowright \mathcal{R}_q(d)$ partitions this set of orbits into a disjoint union of *left* orbits classes and this union is finite by the work of Gauss.

To resume, to any representation $(,a,b,c)$ we have

$$[a,b,c] := \mathrm{G}(\mathbb{Z}).(a,b,c) \simeq \mathrm{G}(\mathbb{Z}).g_{(a,b,c)}.\mathrm{G}_{\mathbf{x}}(\mathbb{R})/\mathrm{G}_{\mathbf{x}}(\mathbb{R}) \subset \mathrm{G}(\mathbb{R})/\mathrm{G}_{\mathbf{x}}(\mathbb{R}).$$

We then have the (almost) tautological:

Duality Principle. *To prove the equidistribution (as $d \to \infty$) of the finite set of left* $\mathrm{G}(\mathbb{Z})$*-orbits*

$$\mathcal{R}_q(d) = \bigsqcup_{[a,b,c]} \mathrm{G}(\mathbb{Z}).(a,b,c) \simeq \bigsqcup_{[a,b,c]} \mathrm{G}(\mathbb{Z}).g_{(a,b,c)}\mathrm{G}_{\mathbf{x}}(\mathbb{R})/\mathrm{G}_{\mathbf{x}}(\mathbb{R})$$

on the space

$$V_{q,\pm1}(\mathbb{R}) \simeq \mathrm{G}(\mathbb{R})/\mathrm{G}_{\mathbf{x}}(\mathbb{R})$$

equipped with the measure $\mu_{q,\pm1}$ it is sufficient to prove

– *the equidistribution of the finite set of (right)* $\mathrm{G}_{\mathbf{x}}(\mathbb{R})$*-orbits*

$$\{g_{[a,b,c]} := \mathrm{G}(\mathbb{Z})\backslash\mathrm{G}(\mathbb{Z}).g_{(a,b,c)}.\mathrm{G}_{\mathbf{x}}(\mathbb{R}),\ [a,b,c] \in [\mathcal{R}_q(d)]\}$$

equipped with the measure (induced by) $\mu_{\mathrm{G}_{\mathbf{x}}}$ on the quotient

$$[\mathrm{G}(\mathbb{R})] := \mathrm{G}(\mathbb{Z})\backslash\mathrm{G}(\mathbb{R}),$$

the later equipped with the probability measure $\mu_{[\mathbf{G}]}$ (say), quotient of the (suitably chosen) Haar measure μ_{G} by the counting measure on the discrete subgroup $\mathrm{G}(\mathbb{Z})$.

Let us relate the duality principle with the second formulation of Duke's theorems when $q = \Delta$: since q is isotropic, we have $\mathrm{SO}_q \simeq \mathrm{PGL}_2$ and the quotient

$$[\mathrm{G}(\mathbb{R})] \simeq \mathrm{PGL}_2(\mathbb{Z})\backslash \mathrm{PGL}_2(\mathbb{R})$$

is naturally identified with the *unit tangent bundle* of the modular curve

$$Y_0(1) = \mathrm{PGL}_2^+(\mathbb{Z})\backslash\mathbb{H} \simeq \mathrm{PGL}_2(\mathbb{Z})\backslash \mathrm{PGL}_2(\mathbb{R})/\mathrm{PSO}_2(\mathbb{R})$$

where $\mathrm{PSO}_2(\mathbb{R})$ is the stabilizer of $\imath \in \mathbb{H}$.

In fact, the second formulation of Duke's theorem is equivalent to this second equidistribution statement of the Duality principle as one *restricts* the computation of the limiting measure to test functions that are *right* $\mathrm{PSO}_2(\mathbb{R})$*-invariant.* For $d < 0$, this is equivalent to the first formulation of Duke's theorem, but for $d > 0$, this is strictly weaker.

2.2. Proof of Duke's theorems via theta series. By the duality principle, it is sufficient to show that for any $\varphi \in \mathcal{C}_c^0(\mathrm{G}(\mathbb{Z})\backslash\mathrm{G}(\mathbb{R}))$, that the corresponding *Weyl sum* has the correct limit, that is

$$W_\varphi(d) = \frac{1}{\mathrm{vol}(d)} \sum_{[a,b,c]} \int_{\mathrm{G}_{\mathbf{x}}(\mathbb{R})} \varphi(g_{(a,b,c)}.t) d\mu_{\mathrm{G}_{\mathbf{x}}}(t) \to \mu_{[\mathrm{G}]}(\varphi),\ d \to \infty, \tag{2.2}$$

where

$$\mathrm{vol}(d) := \sum_{[a,b,c]} \int_{\mathrm{G}_{\mathbf{x}}(\mathbb{R})} d\mu_{\mathrm{G}_{\mathbf{x}}}(t). \tag{2.3}$$

Note that by the class number formula and Siegel's theorem, one has

$$\mathrm{vol}(d) = |d|^{1/2+o(1)}$$

By an approximation argument (Weyl's equidistribution criterion), we may assume that φ is an *automorphic form* on the quotient $\mathrm{G}(\mathbb{Z})\backslash\mathrm{G}(\mathbb{R})$:

- For $q = \mathcal{E}_3$, we may in particular assume that φ is a non-constant harmonic homogeneous polynomial restricted to the sphere (because it is sufficient to consider $\mathrm{SO}_2(\mathbb{R})$-invariant functions). Let ℓ be the degree, then the Weyl sum $W_\varphi(d)$ equals the $|d|$-th Fourier coefficient, $a_\psi(|d|)$ of an holomorphic modular form ψ of level 4 and weight $k = 3/2 + \ell$ up a factor of size $|d|^{\ell/2+1/2+o(1)}$ by (2.3) [**Iwa97**, Chap. 10]. In [**Iwa87**], Iwaniec obtained the first not trivial bound for such coefficients (for weights $k \geq 5/2$ and square-free d), namely

$$a_\psi(|d|) \ll_\varphi |d|^{(k-1)/2+1/4-1/28+o(1)}, \tag{2.4}$$

  which yields

$$W_\varphi(d) \ll_\varphi |d|^{-1/28+o(1)} \tag{2.5}$$

  from which equidistribution follows in this case; in [**Duk88**], Duke extended Iwaniec's bound to all weights $k \geq 3/2$ (note for this specific equidistribution problem, when $\ell = 1$ the Weyl sums $W_\varphi(d)$ are zero due to parity reasons).
- For $q = \Delta$, we may assume that φ is a modular form (either an Eisenstein series or a non-holomorphic cusp form). Duke then uses a correspondence due to Maass to relate the Weyl sums $W_\varphi(d)$ to the d-th Fourier coefficients of non-holomorphic 1/2-integral weight forms; he then extends Iwaniec's bound to the non-holomorphic setting and proves (2.5). See [**Duk88**] for complete details.

3. L-functions and Waldspurger's formula

3.1. Duke's theorems via Subconvexity. In [**Iwa87**], Iwaniec hinted at an alternative approach to (2.4): if ψ is a Hecke eigenform of odd level, by a formula of Waldspurger [**Wal91**] (see also [**KS93**] for a classical formulation in the case of Maass forms) the square $|a_\psi(|d|)|^2$ equals, up to a factor of size $|d|^{k-1+o(1)}$, the central value of a twisted L-function

$$L(\widetilde{\psi}.\chi, 1/2)$$

where $\widetilde{\psi}$ is an holomorphic modular form of integral weight (a Shimura lift of ψ) $\chi = \chi_{-d}$ and is the Legendre symbol for the quadratic field $K = \mathbb{Q}(\sqrt{-d})$. This together with Iwaniec's bound yields the following *subconvexity* bound

$$L(\widetilde{\psi} \otimes \chi, 1/2) \ll_{\psi} |d|^{1/2-1/14+o(1)}. \tag{3.1}$$

Turning the tables, Iwaniec's observation establishes a direct link between Duke's equidistribution theorems and the *subconvexity problem* which is a major problem in the analytic theory of L-functions.

We refer to [**Mic07**] for an introduction to the subconvexity problem. For the purpose this paper, we will briefly recall what this is about: let $L(\pi, s)$ be an L-function attached to some automorphic datum π, normalized so that it satisfies a functional equation relating

$$L(\pi, s) \text{ to } \overline{L(\pi, 1-\overline{s})}.$$

In the subconvexity problem one is interested in the size of the central value $L(\pi, 1/2)$. There is a natural scale for measuring its size: to the L-function $L(\pi, s)$ is attached a positive real number, the *analytic conductor* $C(\pi)$ which is computed from the shape of the functional equation satisfied by $L(\pi, s)$. By general methods from complex analysis (the Phragmen-Lindeloef principle which is a consequence of the maximum principle) it is often not too difficult to obtain the following *convexity bound*

$$L(\pi, 1/2) \ll C(\pi)^{1/4+o(1)}.$$

The subconvexity problem aims at improving this general bound by replacing the exponent $1/4$ by any one strictly smaller.

What make the subconvexity problem particularly appealing and important is that several outstanding problems in number theory can be directly related to solving an instance the subconvexity problem and Duke's equidistribution theorems are major examples.

Indeed the analytic conductor of the twisted L-function $L(\widetilde{\psi} \otimes \chi, s)$ satisfies

$$C(\widetilde{\psi} \otimes \chi) \asymp_{\widetilde{\psi}} |d|^2$$

and therefore (3.1) improves the convexity bound by an exponent $1/14$.

The interesting point is that in [**DFI93**], Duke-Friedlander-Iwaniec obtained a subconvex bound as in (3.1) (with $1/14$ replaced by the weaker but still positive exponent $1/22$) by a method different entierely from [**Iwa87**]. This alternative proof is at the origin of many developments around Duke's theorems.

Quantitiative refinements. The most immediate improvement come from the quality of the subconvex exponent (3.1). Indeed it directly reflect on the speed of convergence of the sequence of measures constructed out of $\mathcal{R}_q(d)$.

At present the best known exponent is due to Conrey-Iwaniec, [**CI00**], with $1/14$ replaced by $1/6$ (see also [**PY20**] for a very recent extension of this result by Petrow and Young). An additional refinement consists in obtaining an as good as possible dependency in the ϕ variable (hybrid subconvexity): this allows to obtain equidistribution statements within domains which are shrinking with d at polynomial rate: see [**LMY13**] as well as [**You17, PY19**] for results in that direction. Another recent striking result is the work of Humphries and Radziwill which analyses precisely from deterministic and probabilistic viewpoints the discrepancy and the optimal scale in such equidistribution problems [**HM19**].

3.2. Waldspurger's formula. In this section we present Waldspurger formula which was mentioned above in its most general form. For this, it is convenient to first *adelize* the problem.

Given $(\mathbb{Z}^3, q)$ a ternary, integral valued, non-degenerate quadratic form, the following hold.

- There exists a quaternion algebra B with reduced norm Nr_{B}, such that the quadratic space $(\mathbb{Q}^3, q)$ is isometric to $(\mathrm{B}^0, \lambda\,\mathrm{Nr}_{\mathrm{B}})$ $\lambda \in \mathbb{Q}^\times$ is a constant, B^0 denote the space of trace 0 quaternions; therefore without loss of generality we may replace $\mathbb{Z}^3$ by a $\mathbb{Z}$-lattice $L \subset \mathrm{B}^0$ and write q for $\lambda\,\mathrm{Nr}_{\mathrm{B}}$.
- Under this isometry, we have an isomorphism of algebraic group
$$\mathrm{SO}_q \simeq \mathrm{SO}_{\mathrm{Nr}_{\mathrm{B}}\,|\mathrm{B}^0}\,\mathrm{PB} \simeq Z_{\mathrm{B}}^\times \backslash \mathrm{B}^\times$$
the later acting on B^0 by conjugation.
- Let $\mathrm{Genus}(q, L)$ be the (finite) set of genus classes of the quadratic lattice (q, L) and let (q_j, L_j), $j \in I$ be set of representatives of the genus classes; we denote their orthogonal groups by G_j. Let $K_f \subset \mathrm{G}(\mathbb{A}_f)$ be the stabilizer of the lattice $\widehat{L} = L \otimes_{\mathbb{Z}} \widehat{\mathbb{Z}}$. The genus $\mathrm{Genus}(q, L)$ is in bijection with the adelic quotient
$$\mathrm{G}(\mathbb{Q}) \backslash \mathrm{G}(\mathbb{A}_f)/K_f.$$
Moreover we have an identification
$$\bigsqcup_j \mathrm{G}_j(\mathbb{Z}) \backslash \mathrm{G}_j(\mathbb{R}) \simeq \mathrm{G}(\mathbb{Q}) \backslash \mathrm{G}(\mathbb{A})/K_f. \tag{3.2}$$
Therefore the natural space for general versions of Duke's equidistribution theorems to take place is the adelic quotient $[\mathrm{G}]/K_f$ where
$$[\mathrm{G}] := \mathrm{G}(\mathbb{Q}) \backslash \mathrm{G}(\mathbb{A}).$$
- Let (a, b, c) be a representation[3] of d by q and let
$$\mathrm{T} = \mathrm{G}_{(a,b,c)}$$
be its stabilizer. The group T is a maximal torus of G: more precisely, we have an isomorphism of algebraic groups
$$\mathrm{T} \simeq \mathrm{res}_{K/\mathbb{Q}} \mathbb{G}_m / \mathbb{G}_m, \ K = \mathbb{Q}(\sqrt{-d/\lambda}).$$
- Under the identification (3.2), the set of orbits associated to the representations of d by the various representatives of the genus of q,
$$\bigsqcup_j \bigsqcup_{[a,b,c] \in [\mathcal{R}_{q_j}(d)]} \mathrm{G}_j(\mathbb{Z}) \backslash \mathrm{G}_j(\mathbb{Z}).g_{(a,b,c)} \mathrm{G}_{j,\mathbf{x}_j}(\mathbb{R})$$
is covered by a finite union of (projections to $[\mathrm{G}]/K_f$ of) adelic T-orbits of the shape
$$[\mathrm{T}.g] = \mathrm{T}(\mathbb{Q}) \backslash \mathrm{T}(\mathbb{A}).g \subset [\mathrm{G}], \ \mathrm{g} = \mathrm{g}_\infty.\mathrm{g}_{\mathrm{f}} \in \mathrm{G}(\mathbb{A}).$$

[3] if it exists; if we don't know a priori one exists it is sufficient to replace q by another form in the genus which does represent d: there exists one if d is locally representable

It follows that, in order to prove Duke's equidistribution theorems it is sufficient to prove the equidistribution of the torus orbit

$$[\mathrm{T}.g].K_f/K_f \subset [\mathrm{G}]/K_f$$

and to evaluate the corresponding Weyl sums

$$\int_{[\mathrm{T}(\mathbb{Q})\backslash \mathrm{T}(\mathbb{A})]} \varphi(t.g)dt \tag{3.3}$$

for φ a right K_f-invariant function on [G]. In addition, by an approximation argument (see [**CU05**]) one may assume that φ is an automorphic form of the group G.

In his beautiful paper [**Wal85**], Walspurger gave an alternative expression for the norm square of the "periods" (3.3) in considerable generality. We will need the following notations:

- F is an arbitrary number field and v denote the generic letter for its various places.
- B/F is a quaternion algebra (possibly the split one), $\mathrm{G} = \mathrm{PB}$ and $[\mathrm{G}] := \mathrm{G}(F)\backslash \mathrm{G}(\mathbb{A}_F)$. We denote by $\mathcal{A}(\mathrm{G})$ the set of automorphic representations of G
- $\mathrm{T} \subset \mathrm{G}$ is a maximal non-split torus and $[\mathrm{T}] := \mathrm{T}(F)\backslash \mathrm{T}(\mathbb{A}_F)$. The group T is the stabilizer in G of some $\mathbf{x} \in \mathrm{B}^0$ such that $\mathrm{Nr}_{\mathrm{B}}(\mathbf{x}) = d = x^2$ is not a square in F. We denote by $K = F(\mathbf{x}) \simeq F(\sqrt{d})$ the quadratic field generated by $\mathbf{x}$ and denote by
$$D_K := \mathrm{Nr}_{F/\mathbb{Q}}(\mathrm{disc}(K)) = \prod_v D_{K_v}$$
the norm of the discriminant ideal; finally we recall the isomorphism of algebraic groups

$$\mathrm{T} \simeq \mathrm{res}_{K/F}\mathbb{G}_m/\mathbb{G}_m. \tag{3.4}$$

THEOREM (Waldspurger's formula). *Let $\varphi : [\mathrm{G}] \to \mathbb{C}$ be an automorphic form belonging to an automorphic representation $\pi \in \mathcal{A}(\mathrm{G})$ which is infinite dimensional. We assume that φ is factorizable: under any isomorphism $\pi \simeq \bigotimes_v' \pi_v$ the vector φ corresponds to a pure tensor $\otimes_v \varphi_v$. Let $\pi^{JL} \in \mathcal{A}(\mathrm{PGL}_{2,F})$ be the Jacquet-Langlands correspondant and π_K^{JL} its base change to $\mathrm{PGL}_{2,K}$. Let $\chi : [\mathrm{T}] \to \mathbb{C}^\times$ be a character of [T]. One has*

$$\frac{|\int_{[\mathrm{T}]} \varphi(tg)\chi(t)dt|^2}{\langle \varphi, \varphi \rangle} = c.L(\pi_K^{JL} \otimes \chi, 1/2) \prod_v \int_{\mathrm{T}_v} \frac{\langle g_v.\varphi_v, t_v.g_v.\varphi_v \rangle}{\langle \varphi_v, \varphi_v \rangle} \chi_v(t_v)dt_v. \tag{3.5}$$

In this formula, $L(\pi_K^{JL} \otimes \chi, 1/2)$ denote the Hecke L-function of π_K^{JL} twisted by the character χ (which we view as a character of $K^\times\backslash \mathbb{A}_K^\times$ via the isomorphism (3.4)) and $c = c(\pi, K) > 0$ is an explicit constant depending on π, K and the Haar measure normalizations.

REMARK 3.1. In this formula, the integration along the torus quotient $\mathrm{T}(\mathbb{Q})\backslash \mathrm{T}(\mathbb{A})$ is the adelic incarnation of the torsor structure on the set of representation classes $[\mathcal{R}_q(d)]$ discovered by Gauss: indeed the quotient

$$\mathrm{T}(\mathbb{Q})\backslash \mathrm{T}(\mathbb{A})/\mathrm{T}(\mathbb{R}).\mathrm{T}(\mathbb{A}_f) \cap g_f.K_f g_f^{-1}$$

is isomorphic to the class group $\mathrm{Pic}(\mathcal{O}_d)$ of an order $\mathcal{O}_d \subset K$.

REMARK 3.2. In the above formula, if we chose the measures on [T] and [G] to be the probability Haar measures and the local measure on $\mathrm{T}_v = \mathrm{T}(F_v)$ to be the Tate normalized (relative to a fixed non trivial additive character of $F\backslash\mathbb{A}_F$), it follows from Siegel's Theorem and the work of Goldfeld-Hoffstein-Lockhart that

$$c = (C(\pi^{JL})D_K)^{o_F(1)}$$

where $C(\pi^{JL})$ is the analytic conductor of π^{JL}

REMARK 3.3. Waldspurger established his formula for "cuspidal" representations. For infinite dimensional non-cuspidal representations, ie. the Eisenstein series, which occurs only in the split case $\mathrm{G} = \mathrm{PGL}_{2,F}$, there is an analogous formula due to Hecke (see [**Wie85**]).

From Waldspurger's formula one deduce the following bound for the Weyl sum, cf. [**ELMV11**, Thm 4.6]:

THEOREM 3.4. *Notations as above. There exists an absolute constant $\eta > 0$ (independent of F) such that*

$$\frac{\int_{[\mathrm{T}]} \varphi(tg)dt}{\langle\varphi,\varphi\rangle^{1/2}} \ll_\varphi D_{\mathrm{T}.g}^{-\eta}. \tag{3.6}$$

*Here $D_{[\mathrm{T}.g]} = \prod_v D_{\mathrm{T}_v.g_v}$ is the discriminant attached to the torus orbit $[\mathrm{T}.g]$ (see [**ELMV11**, §4.2]).*

PROOF. (Sketch)

By Waldspurger's formula applied to $\chi \equiv 1$, it is sufficient to bound all the factors on the righthand side of (3.5) (all but finitely many of which are equal to 1)

The local integrals are bounded using the *decay of matrix coefficients* (as t_v varies along T_v) (cf. [**MV10**, Lem. 4.4.2] and [**ELMV11**, Lem. 9.14]) and yield the bound

$$\int_{\mathrm{T}_v} \frac{\langle g_v.\varphi_v, t_v.g_v.\varphi_v\rangle}{\langle\varphi_v,\varphi_v\rangle} dt_v \ll (D_{\mathrm{T}_v.g_v}/D_{K_v})^{-\eta} D_{K_v}^{-1/2}$$

for some absolute constant $\eta > 0$ (the factor $D_{K_v}^{-1/2}$ comes from Tate's normalization of the measures).

For the L-value $L(\pi_K^{JL}, 1/2)$, we have the factorisation

$$L(\pi_K^{JL}, 1/2) = L(\pi^{JL}, 1/2)L(\pi^{JL} \otimes \chi_K, 1/2)$$

where χ_K is the quadratic character (the Legendre symbol) attached the the quadratic extension K/F. The second right factor is then bounded by

$$L(\pi^{JL} \otimes \chi_K, 1/2) \ll_\pi D_K^{1/2-\eta} \tag{3.7}$$

as a consequence of the resolution of the subconvexity problem in that case by Duke-Friedlander-Iwaniec for $F = \mathbb{Q}$ and Venkatesh for general number fields [**DFI93**, **Ven10**]. We have therefore

$$L(\pi_K^{JL}, 1/2) \ll D_K^{1/2-\eta}.$$

Combining all these bounds together, on obtains (3.6) follows. □

3.2.1. *The case of characters.* To conclude equidistribution, it remains to deal with the case where φ is a character, that is

$$\varphi = \chi_{\mathrm{B}} : g \in \mathrm{B}^\times(\mathbb{A}) \mapsto \chi(\mathrm{Nr}_{\mathrm{B}}(g)),\ \chi^2 = 1,\ \chi \neq 1.$$

In that case, the period is elementary to compute: we denote by K_χ/F the quadratic field whose "Legendre symbol" is χ and given a quadratic field K/F we denote by χ_K the corresponding "Legendre symbol". We have

$$\int_{[\mathrm{T}]} \chi(\mathrm{Nr}_{\mathrm{B}}(tg))dt = \chi(\mathrm{Nr}_{\mathrm{B}}(g)) \int_{[\mathrm{T}]} \chi(\mathrm{Nr}_{\mathrm{B}}(t))dt = \chi(\mathrm{Nr}_{\mathrm{B}}(g)) \int_{[\mathrm{T}]} \chi(\mathrm{Nr}_{K/\mathbb{Q}}(t))dt$$

$$(3.8) \qquad = \begin{cases} \chi(\mathrm{Nr}_{\mathrm{B}}(g)) & \text{if } K = K_\chi \\ 0 & \text{if } K \neq K_\chi \end{cases};$$

indeed for a quadratic field $K \subset \mathrm{B}$, one has for any $z \in K$

$$\mathrm{Nr}_{\mathrm{B}}(z) = \mathrm{Nr}_{K/F}(t)$$

and $\chi \circ \mathrm{Nr}_{K/F}$ is trivial on $[\mathrm{T}]$ if and only if $\chi = 1$ or χ_K.

We will say that a quadratic field of the shape K_χ as above is *K_f-exceptional.* For a given open compact subgroup K_f there are only finitely many K_f-exceptional quadratic fields.

From this we deduce the following general version of Duke's theorems over general number fields.

THEOREM (Duke's theorem over number fields). *Notations being as above, let $K_f \subset \mathrm{G}(\mathbb{A}_{F,f})$ be a fixed open compact subgroup and $([\mathrm{T}_d.g_d])_d$ be a sequence of torus orbits in $[\mathrm{G}]$. We assume that*

- $D_{\mathrm{T}_d.g_d} \to \infty,\ d \to \infty,$
- *for every d the quadratic fields K_d attached to T_d is not K_f-exceptional.*

As $d \to \infty$ the sequence of projected orbits

$$\mathrm{T}_d(F)\backslash \mathrm{T}_d(\mathbb{A}_F).g_d.K_f/K_f \subset \mathrm{G}(F)\backslash \mathrm{G}(\mathbb{A}_F)/K_f$$

becomes equidistributed relative to the natural (Haar) probability measure.

REMARK 3.5. The idea of using Waldspurger's formula in its adelic form to study equidistributions problems goes back to Clozel and Ullmo who used it for PGL_2 [**CU05**].

Reversing the discussion at the beginning of Section 3.2, one deduce from this, general equidistribution results for representations by ternary quadratic forms over number fields. A striking example is the work of Cogdell–Piatetsky-Shapiro–Sarnak [**Cog03**] who treated the case of totally definite ternary quadratic forms over a totally real field F. Indeed, this case was the last remaining case towards a complete resolution of Hilbert's 11-th problem. Using different methods than those of [**Ven10**] (very much inspired by the methods existing for $F = \mathbb{Q}$) they obtained the subconvex bound (3.7) when π^{JL} is associated with a (holomorphic) Hilbert modular form; from this they concluded as in [**DSP90**]that the Hasse principle for the equation $q(a,b,c) = d$ holds for sufficiently d outside of finitely many square classes (related to the exceptional fields discussed above).

3.3. Restriction along the torus. Besides the increased generality and uniformity in treatment, Waldspurger's formula offers other structural advantages.

Waldspurgers formula more generaly computed toric integrals with a twists by a (not necessarily trivial) character of the compact abelian group $[\mathrm{T}] = \mathrm{T}(F)\mathrm{T}(\mathbb{A}_F)$. This makes it possible to study more general integrals, of the shape

$$\int_{[\mathrm{T}]} \varphi(t.g)\Xi(t)dt$$

where $\Xi : [\mathrm{T}] \mapsto \mathbb{C}$ is a (sufficiently regular) function along $[\mathrm{T}]$ (for instance the characteristic function of some sufficiently big and regular subset of $[\mathrm{T}]$). Indeed, by Fourier analysis, one has the decomposition

$$\Xi(t) = \sum_{\chi \in \widehat{[\mathrm{T}]}} \widehat{\Xi}(\chi).\chi(t)$$

so that

$$\int_{[\mathrm{T}]} \varphi(t.g)\Xi(t)dt = \sum_{\chi \in \widehat{[\mathrm{T}]}} \widehat{\Xi}(\chi) \int_{[\mathrm{T}]} \varphi(t.g)\chi(t)dt.$$

and one is reduced to bounding the twisted integrals

$$\int_{[\mathrm{T}]} \varphi(t.g)\chi(t)dt.$$

It is possible to obtain bounds analogous to (3.6) for such integrals; the main ingredient for these is a subconvex bound for the L-value

$$L(\pi_K^{JL} \otimes \chi, 1/2) \ll_\pi C(\pi_K^{JL} \otimes \chi)^{1/4-\eta},\ \eta > 0.$$

Such bounds are known as a consequence of the resolution of the subconvexity problem for *Rankin-Selberg* L-functions (see [**Mic04, HM06, MV10**]). Indeed the twisted L-function $L(\pi_K^{JL} \otimes \chi, s)$ is the Rankin-Selberg L-function $L(\pi^{JL} \otimes \mathrm{Ind}(\chi), s)$ where $\mathrm{Ind}(\chi)$ is the automorphic induction of χ from $\mathrm{GL}_{1,K}$ to $\mathrm{GL}_{2,F}$.

4. Ergodic methods

4.1. Back to the future: Linnik's ergodic method. Amazingly, the importance to equidistribution problems of the torsor structure existing on $[\mathcal{R}_q(d)]$ had been recognized already by Linnik in the 50/60's.

By his so-called *ergodic method*, Linnik was able to conjecture Duke's equidistribution theorems and to prove them under an additional *congruence condition* on d [**Lin68**]:

THEOREM (Linnik, Skubenko). *Let* $\mathrm{B}/\mathbb{Q}$ *be a quaternion algebra and* $L \subset B^0$ *a lattice or trace* 0 *quaternions such that* $\mathrm{Nr}_{\mathrm{B}}(L) \subset \mathbb{Z}$. *Let* p *be a fixed prime at which* $\mathrm{B}(\mathbb{Q}_p)$ *is split.*

As $d \to \infty$ *along squarefree integers which are locally representable by* Nr_B, *and*

- *such that* p *split in* $K_d = \mathbb{Q}(\sqrt{d})$,

the set $\mathcal{R}_q(d)$ *becomes equidistributed on* $V_{\mathrm{Nr}_B,\pm 1}(\mathbb{R})$.

PROOF. (Rough sketch) A modern reformulation of Linnik's method was given in [**ELMV12, EMV13**] (for Δ and $\mathcal{E}_3$ but these are the main cases). We indicate the main steps and refer to the references above and to [**Lin68**] for complete details.

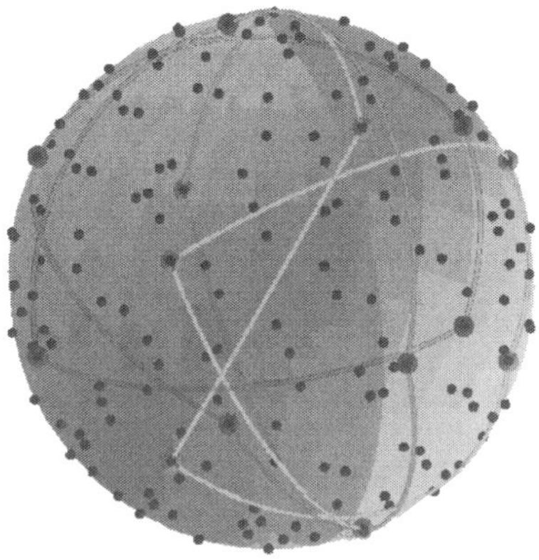

FIGURE 4.1. $q = \mathcal{E}_3$, $d =$: some orbits for the t_5-flow

(1) The splitting of $\mathbb{Q}(\sqrt{d})$ at p implies that for any d, $\mathrm{T}_d(\mathbb{Q}_p) \subset \mathrm{G}(\mathbb{Q}_p)$ contains an element conjugate to the matrix (recall that $\mathrm{B}(\mathbb{Q}_p) \simeq M_2(\mathbb{Q}_p)$)

$$t_p := \begin{pmatrix} p & 0 \\ 0 & 1 \end{pmatrix}.$$

(2) The sequence of torus orbits $[\mathrm{T}_d.g_d]$ furnishes a sequence of probability measures on $[\mathrm{G}]$ invariant under the action of the group $t_p^{\mathbb{Z}}$. In particular, any weak-$\star$ limit of such sequence has the same invariance.
(3) The *basic lemma* (of Linnik) (which is a soft form of a case of the Siegel mass formula) implies that the entropy of the t_p-action for any weak limits is *maximal.*
(4) Maximality of entropy implies that any weak-$\star$ limit is given by the Haar measure.

4.2. Fast forward: Duke's theorems for products. The torsor structure allows for further striking extensions of Duke's theorems: the joint equidistribution on products.

- One considers a tuple of (distinct) quaternion algebras $\mathrm{B}_i/\mathbb{Q}$, $i = 1, \cdots, s$ with associated projective groups G_i; a tuple of open-compact subgroups $K_{f,i} \subset \mathrm{G}_i(\mathbb{A}_f)$. One sets

$$\mathbf{G} = \prod_{i=1}^{s} \mathrm{G}_i, \ \mathbf{K}_f = \prod_{i=1}^{s} K_{f,i} \subset \mathbf{G}(\mathbb{A}_f)$$

- Given a torus $\mathrm{T} \simeq \mathrm{res}_{K/\mathbb{Q}}\mathbb{G}_m/\mathbb{G}_m$, $K = \mathbb{Q}(\sqrt{d})$ with a tuple of embeddings

$$\iota_i : \mathrm{T} \hookrightarrow \mathrm{G}_i$$

(take $z_i \in \mathrm{B}_i^0$ s.t. $z_i^2 = d$), one obtains a diagonal embedding

$$\iota : t \in \mathrm{T} \hookrightarrow \iota(t) = (\iota_i(t))_{i \leq s} \in \mathbf{T} \subset \mathbf{G}.$$

- The data of a tuple $\mathbf{g} = (g_i)_{i \leq s} \in \mathbf{G}(\mathbb{A})$ define an adelic torus orbit $[\mathbf{T}.\mathbf{g}] \subset [\mathbf{G}]$ whose discriminant is defined as $D_{\mathbf{T}.\mathbf{g}} := \min_{i \leq s} D_{\mathrm{T}_i.g_i}$.

With these data at hand, one has the following conjecture:

CONJECTURE 1. *Let* $([\mathbf{T}_d.\mathbf{g}_d])_d$ *be a sequence of tuple of adelic torus orbits as above satisfying* $\lim_{d\to\infty} D_{\mathbf{T}_d.\mathbf{g}_d} = \infty$.

As $d \to \infty$, *the torus orbits* $[\mathbf{T}_d.\mathbf{g}_d]$ *becomes equidistributed on* $[\mathbf{G}]/\mathbf{K}_f$ *for the subspace of continuous functions orthogonal to the* $\mathbf{K}_f$*-invariant characters.*

REMARK 4.1. We have restricted the conjecture to functions orthogonal to the $\mathbf{K}_f$-invariant characters to avoid imposing additional conditions on $\mathbf{K}_f$ to handle the case of characters. However, as in §3.2.1, the case of characters is elementary. See for instance [**ALMW20**, §10].

By restricting to a subsequence of tori defined by *two* congruence conditions, Einsiedler-Lindenstrauss [**EL19**, Thm. 1.8] proved an unconditional version of this conjecture:

THEOREM 4.2 (Joinings Theorem). *Let* v_1, v_2 *be two places of* $\mathbb{Q}$ *as which all the quaternions algebras* B_i *are split. Let* $([\mathbf{T}_d.\mathbf{g}_d])_d$ *be a sequence of tuple of adelic torus orbits as above satisfying* $D_{\mathbf{T}_d.\mathbf{g}_d} \to \infty$ *as* $d \to \infty$ *and*

- *for every* d, *the quadratic field* K_d *splits at* v_1, v_2.

Then we have, as $d \to \infty$, *the torus orbits* $[\mathbf{T}_d.\mathbf{g}_d]$ *becomes equidistributed on* $[\mathbf{G}]/\mathbf{K}_f$ *for the subspace of continuous functions orthogonal to the* $\mathbf{K}_f$*-invariant characters.*

REMARK 4.3. The condition that K splits at v_1, v_2 is similar to Linnik's condition in the ergodic method but here two places are needed.

PROOF. (Very rough sketch) Theorem 4.2 a consequence of a much more general result ([**EL19**, Thm. 1.4]) classifying *joinings* in products of locally homogeneous spaces

$$\mathrm{G}_1 \times \cdots \times \mathrm{G}_s(\mathbb{Q})\backslash \mathrm{G}_1 \times \cdots \times \mathrm{G}_s(\mathbb{A})/K_{f,1} \times \cdots \times K_{f,s},\ s \geq 2,$$

(a joinings is a weak-$\star$ limits of the sequence of measures projecting to the image of the Haar measure on each factor) for $\mathbb{Q}$-almost simple algebraic groups G_i. The key assumption making this classification possible is that the measures are invariant under *two* commuting actions (here at distinct places).

The conclusion is that any ergodic components of any such joinings is *algebraic* which means (very schematically) that such measure is supported along the orbit of an algebraic subgroup $\mathbf{L} \subset \mathrm{G}_1 \times \cdots \times \mathrm{G}_s$.

In the context of Theorem 4.2, the fact that we have a joinings follows from applying Duke's theorems on each factor (either via Waldspurger's formula and subconvex bounds or via Linnik's ergodic method using one of the two splitting conditions that one needs to assume anyway). The algebraicity of the joinings along with the fact that the groups G_i are pairwise non-isogenous (because the quaternions algebras B_i are distincts) readily implies that $\mathbf{L} = \mathbf{G}$. □

REMARK 4.4. To be a bit more correct, it would be necessary to replace the groups $\mathrm{G}_i, \mathbf{G}$ by their simply connected covers, which amounts to consider the groups $\mathrm{B}_i^{(1)}$ of quaternions of norm one instead of the projective group PB_i. This is also related with the issue of characters.

4.2.1. *An L-function approach to the joinings theorem.* The proof of Theorem 4.2 is ergodic theoretic and one may wonder whether a direct approach by bounding the Weyl sums for this problem is possible: that is given a tuple of automorphic forms $(\varphi_i)_{i\leq s}$, with φ_i automorphic for G_i, the corresponding Weyl sums are the adelic integral

$$\int_{[\mathrm{T}]}\prod_i \varphi_i(tg_i)dt. \tag{4.1}$$

Unfortunately for $s>1$, there does not seem to be a nice formula "à la " Waldspurger relating this integral to an L-function.

Surprisingly, Blomer and Brumley [**BB20**] have recently obtained a condtional proof the conjecture for *two* factors ($s=2$) using L-functions and *without any splitting condition* but by assuming instead the *General Riemann Hypothesis* for various L-functions.

The starting principle is to decompose the Weyl sum (4.1), using Plancherel:

$$\int_{[\mathrm{T}]}\varphi_1(tg_1)\varphi_2(tg_1)dt=\sum_\chi\int_{[\mathrm{T}]}\varphi_1(tg_1)\chi(t)dt\int_{[\mathrm{T}]}\varphi_2(t'g_1)\overline{\chi}(t')dt'.$$

This last expression is bounded by

$$\sum_\chi|\int_{[\mathrm{T}]}\varphi_1(tg_1)\chi(t)dt||\int_{[\mathrm{T}]}\varphi_2(t'g_1)\overline{\chi}(t')dt'|.$$

Applying Waldspurger's formula in its most general form leads to a sum, roughly of the shape

$$\sum_\chi h(\chi)L(\pi_1^{JL}\otimes\chi,1/2)^{1/2}L(\pi_2^{JL}\otimes\chi,1/2)^{1/2}$$

for $h(\chi)$ a rapidly decreasing weight function depending on g_i,φ_i. The expectation is that, since π_1 and π_2 are automorphic representations for two different groups, the squareroot L-values $L(\pi_i^{JL}\otimes\chi,1/2)^{1/2}$, $i=1,2$ should not correlate *in size.* Blomer and Brumley establish this under GRH and even save a positive power of $\log|d|$ over the trivial bound for the Weyl sum; their proof is inspired by the recent work of Lester and Radziwill on the Quantum Unique Ergocity conjecture for half-integral weight forms [**LR20**].

4.3. Sums of three squares and Heegner points. The first application of Theorem 4.2 was given by Aka-Einsiedler-Shapira [**AES16**]. We present here a simplified version.

Given $d\geq 3$ square-free and $\not\equiv 7\,(\mathrm{mod}\,8)$, let $(a,b,c)\in\mathcal{R}_{\mathcal{E}_3}(d)$ be representation; the intersection

$$(a,b,c)^\perp\cap\mathbb{Z}^3$$

is a lattice in the $\mathbb{R}$-plane $(a,b,c)^\perp$ of covolume $|d|^{1/2}$. To this lattice, one can associate a (well defined) $\mathbb{C}^\times$-homothety class of lattices $[\Lambda(a,b,c)]=[\mathbb{Z}+\mathbb{Z}.z(a,b,c)]$ in $\mathbb{C}\simeq\mathbb{R}^2$ (with $z(a,b,c)\in\mathbb{H}$) and therefore a class on the modular surface

$$z[a,b,c]=\mathrm{SL}_2(\mathbb{Z})z(a,b,c)\in Y_0(1),$$

which depends only on the class $[a,b,c]=\mathrm{SO}_3(\mathbb{Z}).(a,b,c)$. More precisely one has (cf. [**AES16**, §4.1.2])

$$z[a,b,c]=z_{[a',b',c']},\text{ for some explicit }[a',b',c']\in[\mathcal{R}_\Delta(d')];$$

here

$$d' = \begin{cases} -d & \text{if } d \equiv 3 \,(\mathrm{mod}\, 4) \\ -4d & \text{if } -d \equiv 1, 2 \,(\mathrm{mod}\, 4). \end{cases}$$

We have therefore a map

$$(a,b,c) \in \mathcal{R}_{\mathcal{E}_3}(d) \mapsto (\frac{(a,b,c)}{|d|^{1/2}}, z[a,b,c]) \in S^2 \times Y_0(1).$$

THEOREM 4.5. *Let q_1, q_2 be two odd primes. For $d \to \infty$, squarefree and such that*

- $d \not\equiv 7(8)$
- *q_1, q_2 split in $\mathbb{Q}(\sqrt{-d})$,*

the set of pairs

$$(\frac{(a,b,c)}{|d|^{1/2}}, z[a,b,c]), \ (a,b,c) \in \mathcal{R}_{\mathcal{E}_3}(d)$$

is equidistributed on $S^2 \times Y_0(1)$ with respect to the product of the natural probability measures.

The proof is an application of the Joinings Theorem for $s = 2$ and

$$\mathrm{B}_1 = \mathrm{B}_{2,\infty}, \ \mathrm{B}_2 = \mathrm{M}_2(\mathbb{Q})$$

where B_1 is the algebra of Hamilton quaternions and

$$K_{f,1} = \mathrm{P}\widehat{\mathcal{O}}_{2,\infty}^{\times}, \ K_{f,2} = \mathrm{PGL}_2(\widehat{\mathbb{Z}})$$

where $\mathcal{O}_{2,\infty}$ is the maximal (the Hurwitz quaternions) order of $\mathrm{B}_{2,\infty}$.

However there is an additional subtlety (see [**AES16**, Claim 3]) : the map from sums of three squares to Heegner points

$$[a,b,c] \in [\mathcal{R}_{\mathcal{E}_3}(d)] \mapsto z[a,b,c] \in \mathcal{H}_{d'}$$

is not always $1-1$: its image is an orbit of the subgroup of *square* classes in the ideal class group

$$\mathrm{Pic}(\mathcal{O}_{\mathbb{Q}(\sqrt{-d})})^2 \subset \mathrm{Pic}(\mathcal{O}_{\mathbb{Q}(\sqrt{-d})})$$

($\mathcal{O}_{\mathbb{Q}(\sqrt{-d})}$ the ring of integers of $\mathbb{Q}(\sqrt{-d})$) which is a strict subgroup (unless d is a prime $\equiv 3 \,(\mathrm{mod}\, 4)$). In concrete terms, chosing $(a,b,c) \in \mathcal{R}_{\mathcal{E}_3}(d)$ and letting $[a',b',c'] \in \mathcal{R}_{\Delta}(d')$ be the corresponding class of representation by Δ, one has to show that the following $\mathrm{Pic}(\mathcal{O}_{\mathbb{Q}(\sqrt{-d})})$-orbit

$$\{(d^{-1/2}[\mathfrak{a}] \star [a,b,c], [\mathfrak{a}]^2 \star z_{[a',b',c']}), \ [\mathfrak{a}] \in \mathrm{Pic}(\mathcal{O}_{\mathbb{Q}(\sqrt{-d})})\}$$

is equidistributed on the product $\mathrm{SO}_3(\mathbb{Z})\backslash S^2 \times Y_0(1)$. In adelic terms, this corresponds to an orbit of the shape

$$\{(t.g_1, t^2.g_2), \ t \in [\mathrm{T}]\} \subset [\mathrm{G}_1] \times [\mathrm{G}_2].$$

In order to apply the general result [**EL19**, Thm. 1.4], one has to verify that this orbit equidistributes on each factor. For the first, this is Duke's theorem for the sphere; for the second factor, this correspond to a restricted equidistribution problem for the orbit of a *finite index subgroup* of $[\mathrm{T}]$ (the index is $2^{\omega(d')-1}$ by Gauss's genus theory, with $\omega(d')$ the number of prime factors of d'). Such problem was discussed in §3.3; it amounts to proving subconvex bounds of the form

$$L(\pi_K^{JL} \otimes \chi, 1/2) \ll_\pi C(\pi_K^{JL} \otimes \chi)^{1/4-\eta}, \ \eta > 0$$

for χ a character of order ≤ 2; such bounds are known by [**MV10**]. Alternatively, this restricted equidistribution can also be obtained by using Linnik's ergodic method by taking advantage of one of the two fixed split places.

REMARK 4.6. In the case d squarefree which we discuss here, when χ has order ≤ 2, the L-function $L(\pi_K^{JL}\otimes\chi,s)$ admits a (Kronecker type) factorisation

$$L(\pi_K^{JL}\otimes\chi,s)L(\pi^{JL}\otimes\chi_1,s)L(\pi^{JL}\otimes\chi_2,s)$$

for χ_1,χ_2 two Dirichlet characters of order at most 2 and of coprime conductors d_1,d_2 satisfying $d_1d_2=|d'|$; thereofre the required subconvex bound follows already from [**DFI93**].

REMARK 4.7. The interested reader will find in [**AEW19**] another lovely extension of this result involving 6 factors and and orbits by the product of ideal class groups

$$\mathrm{Pic}(\mathcal{O}_{\mathbb{Q}(\sqrt{-d})})\times\mathrm{Pic}(\mathcal{O}_{\mathbb{Q}(\sqrt{-d})}).$$

4.4. Application to the reduction of CM elliptic curves. We describe yet another application from [**ALMW20**], of a more arithmetic geometric flavor.

We recall [**Sil09, Sil94**] that the modular surface $Y_0(1)=\mathrm{SL}_2(\mathbb{Z})\backslash\mathbb{H}$ parametrize elliptic curves over $\mathbb{C}$ up to isomorphism via the map

$$z\in\mathbb{H}\mapsto\Lambda_z=\mathbb{Z}+\mathbb{Z}.z\mapsto\mathbb{C}/\Lambda_z\mapsto E/\mathbb{C}.$$

We also recall that an elliptic curve E has *complex multiplications* if its ring of endomorphism $\mathrm{End}(E)$ is strictly larger than $\mathbb{Z}$; $\mathrm{End}(E)$ is then isomorphic to an order $\mathcal{O}\subset K\subset\mathbb{C}$ in an imaginary quadratic field: one then say that E has CM by $\mathcal{O}$. Equivalently $E\simeq\mathbb{C}/\Lambda$ has CM by $\mathcal{O}$ if an only if

$$\mathrm{End}(\Lambda):=\{z\in\mathbb{C},\ z\Lambda\subset\Lambda\}=\mathcal{O}.$$

We denote by

$$\mathcal{E}\ell\ell_{\mathcal{O}}=\{\text{Elliptic curves with CM by }\mathcal{O}\}/\text{Isomorphism}.$$

- Viewed as a subset of $Y_0(1)$, $\mathcal{E}\ell\ell_{\mathcal{O}}$ is precisely the set of Heegner points
$$\mathcal{H}_d=\{z_{[a,b,c]},\ (a,b,c)\in\mathcal{R}_\Delta(d)\}$$
where $d=\mathrm{disc}(\mathcal{O})$.
- In particular $\mathcal{E}\ell\ell_{\mathcal{O}}$ is a torsor under the action of $\mathrm{Pic}(\mathcal{O})$ given by:
$$\text{for } E\simeq\mathbb{C}/\Lambda,\ \mathfrak{a}\star E\simeq\mathbb{C}/\mathfrak{a}^{-1}\Lambda.$$

In particular, Duke's theorem (2nd version, $d<0$) can be interpreted as stating that, for $d\to\infty$, the set of elliptic curves with CM by $\mathcal{O}$ become equidistributed on the moduli space of all complex elliptic curves.

4.4.1. *Reduction of CM elliptic curves.* We also recall the following classical facts concerning CM elliptic curves due mainly to Deuring:

- A CM elliptic curve is defined of a number field (equivalently its j-invariant is algebraic): the ring class field of the order $\mathcal{O}$, $H_{\mathcal{O}}:=K(j(E))$ which is Galois over K with Galois group isomorphic to $\mathrm{Pic}(\mathcal{O}_K)$
- A CM elliptic curve has (potential) good reduction at every prime (Serre-Tate theory); equivalently its j-invariant is integral.

- If p is inert in K and $\mathfrak{p}$ is a place in $\overline{\mathbb{Q}}$ above p, its reduction $E \,(\mathrm{mod}\,\mathfrak{p})$ is a *supersingular* elliptic curve: the ring of absolute endomorphisms of $E \,(\mathrm{mod}\,\mathfrak{p})$ is (isomorphic to) a maximal order in the unique quaternion algebra ramified precisely at p and ∞:

$$\mathrm{End}(E \,(\mathrm{mod}\,\mathfrak{p})) \simeq \mathcal{O}_{p,\infty} \subset \mathrm{B}_{p,\infty}.$$

We now denote by

$$\mathcal{E}\ell\ell_p^{ss} = \{\text{Supersingular elliptic curves}/\overline{\mathbb{F}_p}\}/\text{Isomorphism}.$$

the set of supersingular elliptic curves over $\overline{\mathbb{F}_p}$. This is a finite set of cardinality

$$\frac{p-1}{12} + O(1)$$

and it is equipped with a natural probability measure μ_p defined for any $\overline{E} \in \mathcal{E}\ell\ell_p^{ss}$ by

$$\mu_p(\overline{E}) = \frac{1/|\mathrm{End}(\overline{E})^\times|}{\sum_{\overline{E}'} 1/|\mathrm{End}(\overline{E}')^\times|}.$$

We have therefore a *reduction modulo* p map

$$\mathrm{red}_p : E \in \mathcal{E}\ell\ell_{\mathcal{O}} \mapsto E \,(\mathrm{mod}\,\mathfrak{p}) \in \mathcal{E}\ell\ell_p^{ss}$$

which induces embeddings

$$\iota_p : \mathcal{O} \simeq \mathrm{End}(E) \hookrightarrow \mathrm{End}(E \,(\mathrm{mod}\,\mathfrak{p})) \simeq \mathcal{O}_{p,\infty}.$$

A natural question is:

how does the image of red_p *distributes on* $\mathcal{E}\ell\ell_p^{ss}$ *(as a multiset);*
in particular is red_p *surjective ?*

It was observed in [**Mic04**] (and independently [**EOY05**]) that the following holds as special case of Duke's equidistribution theorems:

THEOREM 4.8. *As $d \to \infty$ (such that $(d,p)=1$ and p is inert in $K = \mathbb{Q}(\sqrt{d})$) the map red_p is surjective and $\mathrm{red}_p(\mathcal{E}\ell\ell_{\mathcal{O}})$ (as a multiset) becomes equidistributed relative to the measure μ_p.*

PROOF. (Sketch) The finite space $\mathcal{E}\ell\ell_p^{ss}$ is identified with the adelic quotient

$$\mathcal{E}\ell\ell_p^{ss} \simeq \mathrm{G}(\mathbb{Q})\backslash\mathrm{G}(\mathbb{A})/\mathrm{G}(\mathbb{R})\mathrm{P}\widehat{\mathcal{O}}_{p,\infty}^\times,\ \mathrm{G} = \mathrm{PB}_{p,\infty}$$

and the embedding ι_p induces and embedding $\iota_p : \mathrm{T} \hookrightarrow \mathrm{G}$. Under the above identification the image $\mathrm{red}_p(\mathcal{E}\ell\ell_{\mathcal{O}})$ identifies (as a multiset) with the projection of the adelic orbit $[\iota_p(\mathrm{T})]$ whose discriminant is $\approx |d|$. Moreover since $\mathcal{O}_{p,\infty}$ is a maximal order there are no non-trivial characters to consider. □

4.4.2. *Simultaneous reduction of CM elliptic curves.* We can now fix a tuple of primes $\mathbf{p} = (p_1, \cdots, p_s)$ and a tuple of places in $\overline{\mathbb{Q}}$ above ($\mathfrak{P} = (\mathbf{p}_1, \cdots, \mathbf{p}_s)$) and for any quadratic order $\mathcal{O}$ such that all the p_i remain inert in K, we have a *simultaneous reduction map*

$$\mathrm{red}_{\mathbf{p}} : E \in \mathcal{E}\ell\ell_{\mathcal{O}} \mapsto (z_E, \mathrm{red}_{p_1}(E), \cdots, \mathrm{red}_{p_s}(E)) \in X_0(1) \times \prod_i \mathcal{E}\ell\ell_{p_i}^{ss}.$$

The following theorem is proven in [**ALMW20**]:

THEOREM 4.9. *Let q_1, q_2 be two primes $\neq p_i$. As $d \to \infty$ such that*

- *$(d, p_i) = 1$ and p_i is inert in K for every i,*
- *q_1, q_2 split in K*

the set $\mathrm{red}(\mathcal{E}\ell\ell_{\mathcal{O}})$ *becomes equidistributed with respect to the probability measure* $\mu_\infty \otimes \bigotimes_i \mu_{p_i}$.

REMARK 4.10. In particular, given a choice of supersingular elliptic curves $\overline{E}_i$ modulo p_i, $i = 1, \cdots, s$ for $\mathcal{O}$ any quadratic order of large enough discriminant and satisfying the conditions above there exists $\gg |\mathrm{Pic}(\mathcal{O})|$ elliptic curves with CM by $\mathcal{O}$ and reducing to $\overline{E}_i$ modulo $\mathfrak{p}_i$ for each $i = 1, \cdots, s$.

PROOF. (Sketch) The method is again an application of the Joinings Theorem for

$$\mathbf{G} = \mathrm{PGL}_2 \times \prod_i \mathrm{G}_{p_i},\ \mathbf{K}_f = \mathrm{PGL}_2(\widehat{\mathbb{Z}}) \times \prod_i \mathrm{P}\widehat{\mathcal{O}}^\times_{p_i,\infty}$$

with the diagonal embedding for T

$$\iota = (\mathrm{Id}, \iota_{p_1}, \cdots, \iota_{p_s}).$$

A non obvious point is to verify is that the image $\mathrm{red}_{\mathbf{p}}(\mathcal{E}\ell\ell_{\mathcal{O}})$ is indeed represented by some diagonal torus orbit $[\iota(\mathrm{T})\mathbf{g}]$. This is a consequence of a construction of Serre, the $\mathfrak{a}$-*transform*, cf. [**Ser67**], which is an algebraic version of the classical $\mathrm{Pic}(\mathcal{O})$-action for complex CM elliptic curves

$$\mathfrak{a} \star \mathbb{C}/\Lambda = \mathbb{C}/\mathfrak{a}^{-1}\Lambda;$$

The $\mathfrak{a}$-transform has the following property: for any ideal class $[\mathfrak{a}] \in \mathrm{Pic}(\mathcal{O})$, one has

$$([\mathfrak{a}] \star E) \,(\mathrm{mod}\, \mathfrak{p}) = [\iota_p(\mathfrak{a})] \star (E \,(\mathrm{mod}\, \mathfrak{p})).$$

Finally, since the open compact $\mathbf{K}_f$ is big enough there are no non-trivial quadratic characters to consider. We refer to [**ALMW20**] for complete details. □

We conclude this section with two remarks pointing towards further possible extensions.

REMARK 4.11. The question of the supersingular reduction of CM elliptic can be considerably refined.

Indeed, the set of supersingular elliptic curves $\mathcal{E}\ell\ell_p^{ss}$ is the indexing set of a disjoint union of quotients of p-adic unit disks: each quotient provides a parametrization of the space of deformations of the (canonical lift of the) formal group of each supersingular curve $\overline{E}$ (Woods Hole theory).

The formal group attached to a CM curve E reducing to $\overline{E}$ is such a deformation and defines a point on the corresponding quotient. It is then natural to ask how these points distribute as $\mathrm{disc}(\mathcal{O}) \to \infty$.

This question has been answered recently by Herrero, Menares and Rivera-Letellier [**HMRL21**]. After defining a canonical probability measure on this union of quotients, they proved that, as $\mathrm{disc}(\mathcal{O}) \to \infty$, the reduction of $\mathcal{E}\ell\ell_{\mathcal{O}}$ equidistribute relative to that measure. For this, they prove that the associated Weyl sums are related to representations of integers by the quaternionic norm $\mathrm{Nr}_{\mathrm{B}_{p,\infty}}$ in certain lattices of $\mathrm{B}^0_{p,\infty}$ (with index divisible by increasingly large powers of p) and therefore to Fourier coefficients of holomorphic 1/2-integral weight modular forms. The proof is therefore another instance of Duke's Theorems.

In a forthcoming joint work with Menares, we will revisit this question from the adelic and dynamical viewpoint which will make it possible to obtain a more precise versions of Theorem 4.9.

REMARK 4.12. On the occasion of his work on Mazur's conjectures, Cornut [**Cor02**], established a special case of Theorem 4.9 for quadratic orders highly ramified at a fixed place: for orders of the shape

$$\mathcal{O}_{q^k} = \mathbb{Z} + q^k\mathcal{O}_K$$

for K a *fixed* imaginary quadratic field, q a *fixed* prime and $k \to \infty$. Cornut's proof which was inspired by the earlier work of Vatsal [**Vat02**] did not use Theorem 4.2 (it did not exist at that time) but instead, Ratner's theory of joinings for unipotent actions [**Rat83**] which is sufficient due to the degenerating shape of the order.

Theorem 4.2 offer the possibility to obtain results analogous to those of Vatsal, Cornut and Cornut-Vatsal [**Cor02, Vat02, CV05**] for orders of the shape $\mathcal{O}_{q^k} = \mathbb{Z} + q^k\mathcal{O}_K$ where K is *fixed*, k is *fixed* and for a prime $q \to \infty$ and to derive arithmetic consequences along the lines of Cornut and Vatsal (work in progress with D. Ramakrishnan).

4.5. Application to the mixing conjecture. All the applications of [**EL19**, Thm. 1.8] discussed so far involve product of groups attached to *distinct* quaternions algebra and this assumption makes the determination of the group $\mathbf{L} \subset \mathbf{G}$ particularly easy.

The *mixing conjecture* which –for simplicity and concreteness– we present here in the context of Heegner points is an equidistribution conjecture involving products of the shape PB × PB.

Given $d < 0$ a negative squarefree (say) discriminant with associated imaginary quadratic field $K_d = \mathbb{Q}(\sqrt{d})$. Let

$$\mathcal{H}_d = \{z_{[a,b,c]},\ [a,b,c] \in [\mathcal{R}_\Delta(d)]\} \subset Y_0(1)$$

be the set of Heegner points.

Let

$$([\mathfrak{a}], [a,b,c]) \mapsto [\mathfrak{a}] \star [a,b,c]$$

denote the simple transitive action of the ideal class group $\mathrm{Pic}(\mathcal{O}_{K_d}) \curvearrowright [\mathcal{R}_\Delta(d)]$.

Given $[\mathfrak{a}] \in \mathrm{Pic}(\mathcal{O}_{K_d})$, we consider the question of the distribution of set of pairs of Heegner points shifted by the action of $[\mathfrak{a}]$:

$$\mathcal{H}_{d,\mathfrak{a}} = \{(z_{[a,b,c]}, z_{[\mathfrak{a}]\star[a,b,c]}),\ [a,b,c] \in [\mathcal{R}_\Delta(d)]\} \subset Y_0(1) \times Y_0(1)$$

as d and $[\mathfrak{a}]$ vary. For this, we denote by

$$\mathrm{Nr}([\mathfrak{a}]) := \min_{\mathfrak{a}' \subset [\mathfrak{a}]} \mathrm{Nr}_{K_d/\mathbb{Q}}(\mathfrak{a}') \geq 1$$

the minimal norm of an ideal in the class $[\mathfrak{a}]$.

– If $\mathrm{Nr}([\mathfrak{a}])$ remains constant as $d \to \infty$ (say equals $\mathrm{Nr}([\mathfrak{a}]) = N$) then all the pairs of Heegner points $(z_{[a,b,c]}, z_{[\mathfrak{a}]\star[a,b,c]})$ are located along the embedded modular curve

$$Y_0(N) \hookrightarrow Y_0(1) \times Y_0(1)$$

and by a variant of Duke's theorem for Heegner points, one can show that the pairs $\mathcal{H}_{d,\mathfrak{a}}$ become equidistributed along $Y_0(N)$ with respect to the hyperbolic probability measure.

– On the other hand, if $\mathrm{Nr}([\mathfrak{a}]) \to \infty$ with d, we expect the following:

CONJECTURE 2. *(Mixing conjecture) Let $([\mathfrak{a}]_d)_d$ be a sequence of ideal classes indexed by the squarefree negative discriminants. If $\mathrm{Nr}([\mathfrak{a}]_d) \to \infty$ as $d \to \infty$ the set $\mathcal{H}_{d,\mathfrak{a}}$ becomes equidistributed on $Y_0(1) \times Y_0(1)$ for the product of hyperbolic measures.*

REMARK 4.13. This conjecture was first proposed for representations as sums of three squares (on the product $S^2 \times S^2$) in the joint paper with Ellenberg and Venkatesh [**EMV13**].

There are several reasons to believe in this conjecture:

(1) Using Waldspurger formula and the subconvex bound (3.7) , one can show that the conjecture holds for pairs $(d, [\mathfrak{a}]_d)$ such that $\mathrm{Nr}([\mathfrak{a}]_d)$ is not too big compared to d: ie.

$$\mathrm{Nr}([\mathfrak{a}]_d) \leq |d|^\theta \tag{4.2}$$

for $\theta > 0$ an explicitable (small) constant; moreover under a suitable *Generalized Riemann Hypothesis* any fixed $\theta < 1/4$ should work.

(2) In [**EMV13**], using Linnik's ergodic method, we proved that the conjecture hold for pairs $(d, [\mathfrak{a}]_d)$ such that
   - $\mathbb{Q}(\sqrt{-d})$ is split at a fixed prime $p \neq 2$ (a Linnik type congruence condition),
   - the condition (4.2) holds for some fixed $\theta < 1/2$.

(3) In [**ST17**], Shende and Tsimerman formulated and studied a function field analog of the mixing conjecture and reduced the problem to a purely geometrical one: that of bounding (exponentially in the genus of the curve), the dimension of the cohomology of intersections of two translates of loci of theta divisors in the Jacobian of an hyperelliptic curve $\mathcal{C}$ (the function field of that curve is the analog of the quadratic extension K_d). Moreover they could prove that such bound hold in characteristic 0. Very recently, W. Sawin established these bounds (hence the mixing conjecture) in any sufficiently large positive characteristic [**Saw21**].

Perhaps, the most compelling evidence to date is the work of I. Khayutin [**Kha19**]:

THEOREM. *Let q_1, q_2 be two distinct primes and $([\mathfrak{a}])_d$ be a sequence of ideal classes indexed by the sequence of squarefree negative discriminants satisfying*

- *The field $K_d = \mathbb{Q}(\sqrt{d})$ splits at q_1, q_2,*
- *The Dedekind zeta function $\zeta_{K_d}(s)$ has no Siegel zero (say no zero in the interval $[1 - 10^{-1000}/\log|d|, 1]$).*

If $\mathrm{Nr}([\mathfrak{a}]_d) \to \infty$ as $d \to \infty$, the set $\mathcal{H}_{d,\mathfrak{a}}$ becomes equidistributed on $Y_0(1) \times Y_0(1)$ for the product of hyperbolic measures.

PROOF. (very rough sketch) This is an equidistribution problem for the product

$$\mathbf{G} = \mathrm{PGL}_2 \times \mathrm{PGL}_2 .$$

By Duke's theorem and the classification of joinings any (ergodic components of any) weak-$\star$ limit is algebraic and there are basically two possibilities for the group $\mathbf{L}$, either

$$\mathbf{L} = \mathrm{PGL}_2 \times \mathrm{PGL}_2$$

or $\mathbf{L}$ is PGL_2 embedded diagonally into $\mathrm{PGL}_2 \times \mathrm{PGL}_2$ (again to be more correct PGL_2 should be replaced by PSL_2).

The bulk of Khayutin's work consists in excluding the second part of this alternative (which as we have seen may occur if $\mathrm{Nr}([\mathfrak{a}]_d)$ remains constant). The idea is to show that the probability measures (or rather their adelic versions)

$$\mu_{d,[\mathfrak{a}]_d} = \frac{1}{|\mathcal{H}_d|} \sum_{[a,b,c]} \delta_{(z_{[a,b,c]}, z_{[\mathfrak{a}]\star[a,b,c]})}$$

do not accumulate too much mass in small (shrinking) neighborhoods Ω_d of any embedded modular curve

$$Y_0(N) \hookrightarrow Y_0(1) \times Y_0(1), N \geq 1.$$

Khayutin relates this question to evaluating the geometric side of the relative trace formula for the quotient

$$\mathrm{T}\backslash\mathrm{G} \times \mathrm{G}/\mathrm{G}$$

(where T and G are diagonally embedded) for highly concentrated test functions. In turn this can be related to another classical problem of analytic number theory: a specific instance of the *shifted convolution problem* (see [**Mic07**]). However due to the concentration of the test function, the shifted convolution problem is particularly difficult and cannot be analyzed via the "usual" methods (involving automorphic forms which would amount to switching to the spectral side of the above trace formula). Instead, Khayutin uses coarser put powerful methods from classical analytic number theory like *sieve methods* and the theory of *arithmetic multiplicative functions* in a way similar to the works of Holowinsky and Soundararajan on the Quantum Unique Ergodicity conjecture [**Hol09, Hol10, HS10**]; the requirement that the Siegel zero does not exists occurs for similar reasons as in [**Hol09, Hol10, HS10**] (which concern the values at 1 of symetric square L-functions) and is related to the fact that the savings produced by these methods are at best powers of $\log |d|$. □

Acknowledgments

This survey was written for the proceedings of the inaugural conference of the Linde hall housing the Mathematics department at Caltech. The conference was for me, the conclusion of three wonderful months spent in Caltech as an Olga Tausky-Todd distinguished visitor, during which I absolutely enjoyed the design of the new Linde Hall. I would like to thank Dinakar Ramakrishnan for his kind invitation, for our ongoing collaboration with many discussions related to Duke's equidistribution theorems and their existing and future applications as well as for his encouragements in writing this survey and his subsequent careful reading and helpful comments; many thanks are also due to my collaborators Manfred Einsiedler, Elon Lindenstrauss, Akshay Venkatesh as well as Menny Aka, Manuel Luethi and Andreas Wieser for our past and current collaborations around Duke's equidistribution theorems.

References

[AES16] Menny Aka, Manfred Einsiedler, and Uri Shapira, *Integer points on spheres and their orthogonal lattices*, Invent. Math. **206** (2016), no. 2, 379–396, DOI 10.1007/s00222-016-0655-7. MR3570295

[AEW19] M. Aka, M. Einsiedler, and A. Wieser, *Planes in four space and four associated CM points*, arXiv:1901.05833 (2019).

[ALMW20] M. Aka, E. Luethi, Ph. Michel, and A. Wieser, *Simultaneous supersingular reductions of cm elliptic curves*, arXiv:2005.01537 (2020).

[BB20] V. Blomer and F. Brumley, *Simultaneous equidistribution of toric periods and fractional moments of l-functions*, arXiv:2009.07093 (2020).

[CI00] J. B. Conrey and H. Iwaniec, *The cubic moment of central values of automorphic L-functions*, Ann. of Math. (2) **151** (2000), no. 3, 1175–1216, DOI 10.2307/121132. MR1779567

[Cog03] James W. Cogdell, *On sums of three squares* (English, with English and French summaries), J. Théor. Nombres Bordeaux **15** (2003), no. 1, 33–44. Les XXIIèmes Journées Arithmetiques (Lille, 2001). MR2018999

[Cor02] Christophe Cornut, *Mazur's conjecture on higher Heegner points*, Invent. Math. **148** (2002), no. 3, 495–523, DOI 10.1007/s002220100199. MR1908058

[CU05] Laurent Clozel and Emmanuel Ullmo, *Équidistribution de mesures algébriques* (French, with English summary), Compos. Math. **141** (2005), no. 5, 1255–1309, DOI 10.1112/S0010437X0500148X. MR2157138

[CV05] C. Cornut and V. Vatsal, *CM points and quaternion algebras*, Doc. Math. **10** (2005), 263–309. MR2148077

[DFI93] W. Duke, J. Friedlander, and H. Iwaniec, *Bounds for automorphic L-functions*, Invent. Math. **112** (1993), no. 1, 1–8, DOI 10.1007/BF01232422. MR1207474

[DSP90] William Duke and Rainer Schulze-Pillot, *Representation of integers by positive ternary quadratic forms and equidistribution of lattice points on ellipsoids*, Invent. Math. **99** (1990), no. 1, 49–57, DOI 10.1007/BF01234411. MR1029390

[Duk88] W. Duke, *Hyperbolic distribution problems and half-integral weight Maass forms*, Invent. Math. **92** (1988), no. 1, 73–90, DOI 10.1007/BF01393993. MR931205

[EL19] Manfred Einsiedler and Elon Lindenstrauss, *Joinings of higher rank torus actions on homogeneous spaces*, Publ. Math. Inst. Hautes Études Sci. **129** (2019), 83–127, DOI 10.1007/s10240-019-00103-y. MR3949028

[ELMV11] Manfred Einsiedler, Elon Lindenstrauss, Philippe Michel, and Akshay Venkatesh, *Distribution of periodic torus orbits and Duke's theorem for cubic fields*, Ann. of Math. (2) **173** (2011), no. 2, 815–885, DOI 10.4007/annals.2011.173.2.5. MR2776363

[ELMV12] Manfred Einsiedler, Elon Lindenstrauss, Philippe Michel, and Akshay Venkatesh, *The distribution of closed geodesics on the modular surface, and Duke's theorem*, Enseign. Math. (2) **58** (2012), no. 3-4, 249–313, DOI 10.4171/LEM/58-3-2. MR3058601

[EMV13] Jordan S. Ellenberg, Philippe Michel, and Akshay Venkatesh, *Linnik's ergodic method and the distribution of integer points on spheres*, Automorphic representations and L-functions, Tata Inst. Fundam. Res. Stud. Math., vol. 22, Tata Inst. Fund. Res., Mumbai, 2013, pp. 119–185. MR3156852

[EOY05] Noam Elkies, Ken Ono, and Tonghai Yang, *Reduction of CM elliptic curves and modular function congruences*, Int. Math. Res. Not. **44** (2005), 2695–2707, DOI 10.1155/IMRN.2005.2695. MR2181309

[GF87] E. P. Golubeva and O. M. Fomenko, *Asymptotic distribution of lattice points on the three-dimensional sphere* (Russian), Zap. Nauchn. Sem. Leningrad. Otdel. Mat. Inst. Steklov. (LOMI) **160** (1987), no. Anal. Teor. Chisel i Teor. Funktsiĭ. 8, 54–71, 297, DOI 10.1007/BF02342921; English transl., J. Soviet Math. **52** (1990), no. 3, 3036–3048. MR906844

[HM06] Gergely Harcos and Philippe Michel, *The subconvexity problem for Rankin-Selberg L-functions and equidistribution of Heegner points. II*, Invent. Math. **163** (2006), no. 3, 581–655, DOI 10.1007/s00222-005-0468-6. MR2207235

[HM19] Peter Humphries and Radziwill Maksym, *Optimal small scale equidistribution of lattice points on the sphere, heegner points, and closed geodesics*, arXiv:1910.01360 (2019).

[HMRL21] Sebastián Herrero, Ricardo Menares, and Juan Rivera-Letelier, *p-adic distribution of CM points and Hecke orbits I: Convergence towards the Gauss point*, Algebra Number Theory **14** (2020), no. 5, 1239–1290, DOI 10.2140/ant.2020.14.1239. MR4129386

[Hol09] Roman Holowinsky, *A sieve method for shifted convolution sums*, Duke Math. J. **146** (2009), no. 3, 401–448, DOI 10.1215/00127094-2009-002. MR2484279

[Hol10] Roman Holowinsky, *Sieving for mass equidistribution*, Ann. of Math. (2) **172** (2010), no. 2, 1499–1516, DOI 10.4007/annals.2010.172.1499. MR2680498

[HS10] Roman Holowinsky and Kannan Soundararajan, *Mass equidistribution for Hecke eigenforms*, Ann. of Math. (2) **172** (2010), no. 2, 1517–1528, DOI 10.4007/annals.2010.172.1517. MR2680499

[Iwa87] Henryk Iwaniec, *Fourier coefficients of modular forms of half-integral weight*, Invent. Math. **87** (1987), no. 2, 385–401, DOI 10.1007/BF01389423. MR870736

[Iwa97] Henryk Iwaniec, *Topics in classical automorphic forms*, Graduate Studies in Mathematics, vol. 17, American Mathematical Society, Providence, RI, 1997, DOI 10.1090/gsm/017. MR1474964

[Kha19] Ilya Khayutin, *Joint equidistribution of CM points*, Ann. of Math. (2) **189** (2019), no. 1, 145–276, DOI 10.4007/annals.2019.189.1.4. MR3898709

[KS93] Svetlana Katok and Peter Sarnak, *Heegner points, cycles and Maass forms*, Israel J. Math. **84** (1993), no. 1-2, 193–227, DOI 10.1007/BF02761700. MR1244668

[Lin68] Yu. V. Linnik, *Ergodic properties of algebraic fields*, Ergebnisse der Mathematik und ihrer Grenzgebiete, Band 45, Springer-Verlag New York Inc., New York, 1968. Translated from the Russian by M. S. Keane. MR0238801

[LMY13] Sheng-Chi Liu, Riad Masri, and Matthew P. Young, *Subconvexity and equidistribution of Heegner points in the level aspect*, Compos. Math. **149** (2013), no. 7, 1150–1174, DOI 10.1112/S0010437X13007033. MR3078642

[LR20] Stephen Lester and Maksym Radziwiłł, *Quantum unique ergodicity for half-integral weight automorphic forms*, Duke Math. J. **169** (2020), no. 2, 279–351, DOI 10.1215/00127094-2019-0040. MR4057145

[Mic04] P. Michel, *The subconvexity problem for Rankin-Selberg L-functions and equidistribution of Heegner points*, Ann. of Math. (2) **160** (2004), no. 1, 185–236, DOI 10.4007/annals.2004.160.185. MR2119720

[Mic07] Philippe Michel, *Analytic number theory and families of automorphic L-functions*, Automorphic forms and applications, IAS/Park City Math. Ser., vol. 12, Amer. Math. Soc., Providence, RI, 2007, pp. 181–295, DOI 10.1090/pcms/012/05. MR2331346

[MV10] Philippe Michel and Akshay Venkatesh, *The subconvexity problem for* GL_2, Publ. Math. Inst. Hautes Études Sci. **111** (2010), 171–271, DOI 10.1007/s10240-010-0025-8. MR2653249

[PY19] Ian Petrow and Matthew P. Young, *A generalized cubic moment and the Petersson formula for newforms*, Math. Ann. **373** (2019), no. 1-2, 287–353, DOI 10.1007/s00208-018-1745-1. MR3968874

[PY20] Ian Petrow and Matthew P. Young, *The Weyl bound for Dirichlet L-functions of cube-free conductor*, Ann. of Math. (2) **192** (2020), no. 2, 437–486, DOI 10.4007/annals.2020.192.2.3. MR4151081

[Rat83] Marina Ratner, *Horocycle flows, joinings and rigidity of products*, Ann. of Math. (2) **118** (1983), no. 2, 277–313, DOI 10.2307/2007030. MR717825

[Saw21] Will Sawin, *Bounds for the stalks of perverse sheaves in characteristic p and a conjecture of Shende and Tsimerman*, Invent. Math. **224** (2021), no. 1, 1–32, DOI 10.1007/s00222-020-01006-0. MR4228499

[Ser67] J.-P. Serre, *Complex multiplication*, Algebraic Number Theory (Proc. Instructional Conf., Brighton, 1965), Thompson, Washington, D.C., 1967, pp. 292–296. MR0244199

[Sil94] Joseph H. Silverman, *Advanced topics in the arithmetic of elliptic curves*, Graduate Texts in Mathematics, vol. 151, Springer-Verlag, New York, 1994, DOI 10.1007/978-1-4612-0851-8. MR1312368

[Sil09] Joseph H. Silverman, *The arithmetic of elliptic curves*, 2nd ed., Graduate Texts in Mathematics, vol. 106, Springer, Dordrecht, 2009, DOI 10.1007/978-0-387-09494-6. MR2514094

[SP04] Rainer Schulze-Pillot, *Representation by integral quadratic forms—a survey*, Algebraic and arithmetic theory of quadratic forms, Contemp. Math., vol. 344, Amer. Math. Soc., Providence, RI, 2004, pp. 303–321, DOI 10.1090/conm/344/06226. MR2060206

[ST17] Vivek Shende and Jacob Tsimerman, *Equidistribution in* $\mathrm{Bun}_2(\mathbb{P}^1)$, Duke Math. J. **166** (2017), no. 18, 3461–3504, DOI 10.1215/00127094-2017-0025. MR3732881

[Vat02] V. Vatsal, *Uniform distribution of Heegner points*, Invent. Math. **148** (2002), no. 1, 1–46, DOI 10.1007/s002220100183. MR1892842

[Ven10] Akshay Venkatesh, *Sparse equidistribution problems, period bounds and subconvexity*, Ann. of Math. (2) **172** (2010), no. 2, 989–1094, DOI 10.4007/annals.2010.172.989. MR2680486

[Wal85] J.-L. Waldspurger, *Sur les valeurs de certaines fonctions L automorphes en leur centre de symétrie* (French), Compositio Math. **54** (1985), no. 2, 173–242. MR783511

[Wal91] Jean-Loup Waldspurger, *Correspondances de Shimura et quaternions* (French), Forum Math. **3** (1991), no. 3, 219–307, DOI 10.1515/form.1991.3.219. MR1103429

[Wey16] Hermann Weyl, *Über die Gleichverteilung von Zahlen mod. Eins* (German), Math. Ann. **77** (1916), no. 3, 313–352, DOI 10.1007/BF01475864. MR1511862

[Wie85] Franck Wielonsky, *Séries d'Eisenstein, intégrales toroïdales et une formule de Hecke* (French), Enseign. Math. (2) **31** (1985), no. 1-2, 93–135. MR798908

[You17] Matthew P. Young, *Weyl-type hybrid subconvexity bounds for twisted L-functions and Heegner points on shrinking sets*, J. Eur. Math. Soc. (JEMS) **19** (2017), no. 5, 1545–1576, DOI 10.4171/JEMS/699. MR3635360

EPFL/MATH/TAN, Station 8, CH-1015 Lausanne, Switzerland
Email address: philippe.michel@epfl.ch

Proceedings of Symposia in Pure Mathematics
Volume **104**, 2021
https://doi.org/10.1090/pspum/104/01879

Gap and Type problems in Fourier analysis

A. Poltoratski

ABSTRACT. We survey history and recent progress in the area of Gap and Type problems of Fourier analysis. Among the applications we discuss problems on sampling sets for entire functions of zero exponential type and oscillations of Fourier integrals. We provide new simple examples illustrating gap ant type formulas.

1. Introduction

This note focuses on two old problems of Fourier analysis, the so-called Gap and Type problems. For a summable function f on the real line, $f \in L^1(\mathbb{R})$ we define its Fourier transform as

$$\hat{f}(s) = \frac{1}{\sqrt{2\pi}} \int_{\mathbb{R}} e^{-ist} f(t)dt.$$

From the classical Parseval-Plancherel identity we know that the Fourier transform can be extended as a unitary operator on $L^2(\mathbb{R})$. For a finite complex measure μ on $\mathbb{R}$ the Fourier transform can once again be defined directly as

$$\hat{\mu}(s) = \frac{1}{\sqrt{2\pi}} \int_{\mathbb{R}} e^{-ist} d\mu(t).$$

The function $\hat{f}(s)$ has various interpretations coming from physics and applied fields. One of the meanings of the Fourier transform is that it decomposes an arbitrary signal f into elementary waves $e^{ist} = \cos st + i \sin st$. Then $\hat{f}$ shows how much each elementary wave contributes to the original signal.

The following important principle was found by Norbert Wiener around 1925 and later became known as the uncertainty principle (UP) in harmonic analysis: A function (measure, distribution) and its Fourier transform cannot be simultaneously small.

As Wiener wrote in his memoir [**43**], problems of breaking a sound wave into elementary harmonics can be traced as far back as the time of Pythagoras who studied sounds produced by strings of a lute. In these settings Wiener's formulation of UP corresponds to the well-known property that a musical note cannot be both short in time and low in frequency.

According to [**43**], in 1925 Wiener delivered a lecture on his newfound principle at Göttingen, where it may have influenced the work of Heisenberg and Born on

2020 *Mathematics Subject Classification.* Primary 42A38.

The author was supported by NSF Grant DMS-1954085.

the foundations of quantum mechanics. The celebrated Heisenberg's inequality (see Section 2), which appeared several years later, became the original mathematical expression of UP.

Throughout the 20th century the mathematical study of UP involved a number of prominent mathematicians. Practically any definition of smallness (with few exceptions like smallness in the L^2-norm) inserted in Wiener's statement leads to a new deep problem of analysis. Many of such problems remain open to this day. Modern generalizations of UP spread from the original 1D Fourier transform into wide variety of mathematical areas including analysis on groups, frame expansions and abstract versions of the moment problem. The reader can find more information on the current state of UP, along with further references, in [**8**, **13**, **14**, **17**, **18**, **22**, **38**].

2. Forms of UP

In this section we recall several important results of UP relevant to our main subject.

As was mentioned before, Heisenberg's inequality, published by Weyl in 1928 and attributed to Pauli, can be viewed as the first mathematical form of UP:

$$\int_{\mathbb{R}} x^2|f(x)|^2 \int_{\mathbb{R}} x^2|\hat{f}(x)|^2 dx \geq \frac{||f||_2^4}{16\pi^2}.$$

Like in many results in that area the extreme case – the equation – occurs when f is a Gaussian. Viewing Heisenberg's inequality from the point of view of Wiener's UP statement, it says that the second moments of $|f|^2$ and $|\hat{f}|^2$ cannot both be small. In a sense of mass distribution, it says that both masses cannot be concentrated too close to zero.

A whole class of statements of UP concerns similar property but from the point of view of the support of the functions. The form of Wiener's statement which unites these problems is that the supports of f and $\hat{f}$ cannot both be small. Even within this class of problems the smallness of the support can be understood in a variety of ways.

It is an easy exercise at the level of a graduate analysis course to show that the supports of f and $\hat{f}$ cannot both be bounded. In fact, if $\operatorname{supp} f$, $f \in L^2$, is bounded then $\hat{f}$ is an entire function from the Paley-Wiener class and therefore $\operatorname{supp} \hat{f} = \mathbb{R}$. From a different part of the same course, if the support of f is semi-bounded, for instance lies in $\mathbb{R}_-$, then $\hat{f}$ belongs to the Hardy class of analytic functions in the upper half-plane $\mathbb{C}_+$ and its boundary values on $\mathbb{R}$ once again cannot vanish on a set of positive measure, i.e., $\operatorname{supp} \hat{f} = \mathbb{R}$.

From these elementary observations let us proceed to deep results of Amrein-Berthier (1977, [**1**]) and Benedicks (1985, [**3**]) which say that Lebesgue measures of the supports of f and $\hat{f}$ cannot both be finite. This statement holds not only on the line but also in $\mathbb{R}^n$.

The strongest result in this direction belongs to Nazarov (1993, [**36**]). It says that for any two sets $T, S \in \mathbb{R}^n$

$$||f||_{L^2(\mathbb{R}^n)} < C\exp\left(C|S||T|\right)(||f||_{L^2(S^c)} + ||\hat{f}||_{L^2(T^c)}),$$

where T^c, S^c denote the complements of T, S correspondingly. To relate this inequality with the last statement, notice that if the supports $S = \operatorname{supp} f$ and

$T = \operatorname{supp} \widehat{f}$ had finite measure then the right-hand side of the inequality would be equal to 0.

Nazarov's inequality gives a quantitative version to the previous results. Like many classical questions of UP, this line of problems is not completely closed. For instance, it is known that if one of the sets $S,\ T \in \mathbb{R}^n$ is convex, then the product $|S||T|$ in the right-hand side of the inequality may be raised into the power $1/n$, however it is unknown if the same is true in general.

To finish this brief sampling of results of UP, let us again return to Wiener's postulate and interpret smallness of f and $\widehat{f}$ in terms of fast decay near infinity. Here a classical result of Hardy (1933, [**16**]) says that both functions cannot decay faster than a Gaussian. More precisely, assume that

$$|f(x)| < C(1+|x|^n)e^{-a\pi x^2}, \text{ and } |\widehat{f}(x)| < C(1+|x|^n)e^{-b\pi x^2}$$

for some $a, b, C > 0$, $n \in \mathbb{N}$. If $ab > 1$ then $f \equiv 0$. If $ab = 1$ then $f = pe^{-a\pi x^2}$ for some polynomial p of degree at most n. Like many statements of UP this version was extensively studied – let us mention for instance further results by Morgan [**35**] and Dzhrbashyan [**11**].

Another classical result from the 'decay' part of UP is a theorem by Beurling (1959, [**4**]). It says that if $f \in L^1(\mathbb{R})$ and

$$\iint_{\mathbb{R}\times\mathbb{R}} |f(x)||\widehat{f}(y)|e^{|x||y|}dxdy < \infty$$

then $f \equiv 0$. Further extensions of this result were obtained by Hörmander (1991, [**20**]) and Hedenmalm (2012, [**19**]).)

3. The Gap problem

We now turn to one of the two problems mentioned in the title of this note. For a finite complex measure (function) μ on $\mathbb{R}$ we call the support of its Fourier transform, $\operatorname{supp} \widehat{\mu}$, *the Fourier spectrum of* μ. Any open interval in the complement of $\operatorname{supp} \widehat{\mu}$ is called a gap in the Fourier spectrum. Presence of large gaps in the spectrum imply various properties of the original measure. Such measures appear in a number of problems in analysis and applications.

The Gap problem asks to find the maximal possible size of the gap in the spectrum of a measure satisfying given conditions. The Gap problem clearly belongs to the area of UP. A typical statement in this area says that if μ is small in the sense of decay at infinity, porosity of the support, etc., then $\widehat{\mu}$ cannot be small in the sense that its support cannot have a large gap.

Measures with spectral gaps are studied in spectral theory, in various versions of the moment problem, prediction theory and completeness problems. They correspond to 'high-pass signals' in electrical engineering and signal processing. Spectral gaps of difference of two measures appear in a number of applications concerning uniqueness and determinacy. Let us mention, for instance, that two Schroedinger operators on an interval have a common piece of potential at the beginning of the interval if and only if the difference of their spectral measures has a spectral gap, see [**32**].

Various forms of the Gap problem were studied by Beurling, Kolmogorov, Krein, Levinson, Wiener and many other prominent mathematicians starting from 1930s. Here let us present a short selection of classical results.

The following theorem belongs to the 'decay/support' part of UP statements where the measure itself is small in the sense that it decays fast near infinity and its Fourier transform is small in the sense that it has a gap in its support.

THEOREM 3.1 (Norman Levinson's Gap Theorem, 1940, [**29**]). *Let μ be a finite measure on $\mathbb{R}$. Denote $M(x) = |\mu|((x,\infty))$. Suppose that*

$$\int_0^\infty \frac{|\log M(x)|}{1+x^2}dx = \infty.$$

If $\widehat{\mu}$ vanishes on an interval then $\mu \equiv 0$.

One of the classical statements in the 'support/support' class of UP is the following theorem by A. Beurling.

A sequence of disjoint intervals $\{I_n\}$ on the real line is called short if

$$\sum_n \frac{|I_n|^2}{1+\text{dist}^2(0,I_n)} < \infty. \tag{3.1}$$

If the last sum diverges we call such a sequence long. Here and throughout the paper we use the notation $|\cdot|$ for the length of an interval.

THEOREM 3.2 (Arne Beurling's Gap Theorem, 1961, [**4**]). *Let μ be a finite complex measure on $\mathbb{R}$. If the complement of the support of μ is long then the support of $\widehat{\mu}$ does not have any gaps, unless $\mu \equiv 0$.*

A natural question is whether the gap can be replaced with an arbitrary set of positive measure in these statements. More precisely, instead of existence of a gap in the closed support of $\widehat{\mu}$ one can ask about existence of a set of positive Lebesgue measure such that $\widehat{\mu}$ vanishes on that set. Such strengthening turned out to be possible in some of the early statements concerning the gap problem but impossible in others. For instance, Levinson's theorem above was later improved by Beurling from a gap to a set of Lebesgue measure zero, while Beurling's theorem cannot be similarly improved (Koosis, 1988 [**23**]). Moreover, Beurling's longness condition is also sharp (Benedicks, 1985 [**3**]). More gap theorems and further references can be found in [**38**].

All classical gap theorems give conditions on the smallness of a measure which imply absence of a spectral gap. Modern methods of complex and harmonic analysis give more precise quantitative results in that area. For a measurable non-empty set $X \subset \mathbb{R}$ let us define the *gap characteristic* $\mathbf{G}(X)$ of X as

$$\mathbf{G}(X) = \sup\{a \mid \exists\ \mu \in M(X),\ \mu \not\equiv 0, \ \text{ such that } \widehat{\mu} = 0 \ \text{ on } \ [0,a]\},$$

where $M(X)$ denotes the set of all finite complex measures on X (i.e., such that $|\mu|(\mathbb{R}\setminus X) = 0$). Observe that since $\widehat{e^{icx}\mu} = \widehat{\mu}(x-c)$ the exact placement of the spectral gap is unimportant.

Our goal for the rest of this section will be to find a formula for $\mathbf{G}(X)$.

It is convenient in many situation to extend the definition of the gap characteristic to wider classes of measures. For a measurable $X \subset \mathbb{R}$ we denote by $M_p(X)$ the set of polynomially finite measures supported on X, i.e., the set of all complex measures on X satisfying

$$\int \frac{d|\mu|(x)}{1+|x|^N} < \infty$$

for some $N > 0$. Although the Fourier transform for general polynomially finite measures can only be defined in the sense of distributions, we can introduce the following analogue of the spectral gap.

The Fourier transform $\mathcal{F} : L^2(\mathbb{R}) \to L^2(\mathbb{R})$, $\mathcal{F}f = \widehat{f}$, maps the subspace of functions supported on $[-a, a]$ to the space of entire functions

$$F(z) = \frac{1}{\sqrt{2\pi}} \int_{-a}^{a} e^{-ist} f(t)dt.$$

The image $\mathcal{F}(L^2([-a, a]))$ is the Paley-Wiener space PW_a. Recall that an entire function F has exponential type at most a, $a \geq 0$ if for all $z \in \mathbb{C}$

$$|F(z)| < Ce^{a|z|}$$

for some $C > 0$. According to the classical Paley-Wiener theorem, the space PW_a can be alternatively defined as the space of all entire functions of exponential type at most a which belong to $L^2(\mathbb{R})$. Every PW_a is a Hilbert space with the inner product inherited from $L^2(\mathbb{R})$. In the corresponding norm the Fourier transform becomes a unitary operator from $L^2([-a, a])$ to PW_a.

For a polynomially finite measure μ we say that μ annihilates PW_a, and write $\mu \perp PW_a$, if

$$\int f d\mu = 0 \text{ for every } f \in PW_a \cap L^1(|\mu|).$$

Since the Schwartz class of entire functions decreasing along the real line faster than any power of $|x|$ is dense in any PW_a, the set $PW_a \cap L^1(|\mu|)$ is always non-empty (and in fact is dense in PW_a). Note that for a finite measure the property $\mu \perp PW_a$ is equivalent to having a spectral gap of the size $2a$ at $(-a, a)$.

Now let us consider a modified definition of the gap characteristic of a non-empty set $X \subset \mathbb{R}$ with polynomially finite measures:

$$\mathbf{G}^*(X) = 2\sup\{a \mid \exists\, \mu \in M_p(X),\ \mu \not\equiv 0,\ \mu \perp PW_a\},$$

when the set in the right-hand side is non-empty and $\mathbf{G}(X) = 0$ otherwise.

LEMMA 3.3. *For any $X \subset \mathbb{R}$*

$$\mathbf{G}(X) = \mathbf{G}^*(X).$$

PROOF. The inequality $\mathbf{G} \leq \mathbf{G}^*$ follows from the definition because $M(X) \subset M_p(X)$. For the other inequality, let $\mu \in M_p(x)$ satisfy $\mu \perp PW_a$. Let $f \in PW_\varepsilon$ be a function from the Schwartz class, $f \neq 0$ on $\mathbb{R}$. Then the measure ν defined as $d\nu = f d\mu$ is a finite measure. Since $fg \in PW_a$ for any $g \in PW_{a-\varepsilon}$,

$$\int g f d\mu = 0$$

and therefore $\nu \perp PW_{a-\varepsilon}$. It follows that $\mathbf{G} \geq \mathbf{G}^*$. □

The following well-known statement will also be useful for us, see for instance [**33**], Lemma 2. It translates the presence of a spectral gap for a measure into exponential decay of its Cauchy integral along the imaginary axis.

LEMMA 3.4. *Let μ be a finite complex measure. Then the Fourier transform of μ vanishes on $[0, a]$ ($[-a, 0]$) if and only if*

$$e^{xy} \int \frac{d\mu(t)}{t - iy} \to 0,$$

as $y \to \infty$ ($-\infty$) for every $x \in [0, a]$ ($[-a, 0]$).

Now let us illustrate the notion of the gap characteristic $\mathbf{G}(X)$ with the following simple examples.

Examples of gap characteristics.

- $\mathbf{G}(\mathbb{R}) = \infty$. For any $a > 0$ consider the measure $d\mu = e^{iax} f dx$ where f is any function from the Hardy space H^1 in the upper half-plane. The Cauchy integral of μ in the upper half plane is $e^{iax} f$ which decays faster than e^{-ay} along the positive imaginary axis. By the last lemma, $\widehat{\mu}$ vanishes on $[0, a]$.

  A faster but somewhat less formal way to see the same is to notice that the Fourier transform of the Lebesgue measure, in the sense of distributions, is a point mass at 0, i.e., has infinite gaps in its support.
- $\mathbf{G}(X) = 0$ for any semibounded X. As was discussed before, the Fourier transform of any finite measure sitting on such a set will produce an analytic function in the upper or lower half-plane whose boundary values cannot vanish on a set of positive measure. This argument uses an advanced uniqueness theorem and thus is not completely elementary. Is there a simpler way to see this property?
- $\mathbf{G}(\mathbb{Z}) = 2\pi$. By the Parseval theorem, PW_π is equal to $L^2(\eta)$, where

$$\eta = \sum_{n \in \mathbb{Z}} \delta_n$$

  is the counting measure on $\mathbb{Z}$. For any $a < \pi$ there exists a non-zero function $f \in PW_\pi$ from the Schwartz class (or just summable with respect to η) which is orthogonal to PW_a. The function f cannot vanish identically on $\mathbb{Z}$ by Parseval. Hence $\bar{f}\eta$ annihilates PW_a and has a gap of the size at least $2a$.

  Alternatively, consider the measure

$$\nu = \sum_{n \in \mathbb{Z}} (-1)^n \delta_n.$$

  Even though this measure is not finite, the Cauchy integral from the last lemma converges due to the alternation of masses. One can show that this integral is equal to $1/\sin \pi z$ and therefore decays as $e^{-\pi|y|}$ along both imaginary half-axes.

  One can also calculate the Fourier transform of ν directly (in the sense of distributions) using the well-known Poisson identity, which in our normalization has the form

$$\widehat{\eta} = \sum_{n \in \mathbb{Z}} \delta_{2\pi n}.$$

  Using standard re-scaling and shift properties of the Fourier transform,

$$\widehat{\nu} = \mathcal{F}(\eta(2x)) - \mathcal{F}(\eta(2(x-1))) =$$

$$\frac{1}{\sqrt{2}} \left(\sum_{n \in \mathbb{Z}} \delta_{\pi n} - e^{i\pi x} \sum_{n \in \mathbb{Z}} \delta_{\pi n} \right) = \sqrt{2} \sum_{n \in \mathbb{Z}} \delta_{\pi(2n+1)}.$$

  The last measure has gaps of the size of 2π (formally speaking, $2\pi - \varepsilon$) in its support.

It is a good exercise for a graduate course to show directly that $\nu \perp PW_{2\pi-\varepsilon}$.

Via the rescaling property of the Fourier transform, we can now construct a set of any finite positive characteristic since $\mathbf{G}(d\mathbb{Z}) = \frac{2\pi}{d}$ for any positive d.

- By Beurling's theorem if X has long gaps then $\mathbf{G}(X) = 0$

These initial examples bring up many natural questions. Are there uncountable sets of finite positive characteristic? In particular, can a union of intervals of fixed length have characteristic strictly between zero and infinity? How much can one perturb the points of $\mathbb{Z}$, or any set with a known characteristic, so that the characteristic does not change? We will answer these questions and provide several more examples after giving a formula for $\mathbf{G}(X)$. First, we need the following definitions.

Let

$$\cdots < a_{-2} < a_{-1} < a_0 = 0 < a_1 < a_2 < \ldots$$

be a two-sided sequence of real points $a_n \to \pm\infty$ as $n \to \pm\infty$. We say that the intervals $I_n = (a_n, a_{n+1}]$ form a short partition of $\mathbb{R}$ if $|I_n| \to \infty$ as $|n| \to \infty$ and the sequence $\{I_n\}$ is short, i.e., satisfies (3.1).

For a finite set of distinct points $\Lambda \subset \mathbb{R}$ define

$$E(\Lambda) = \sum_{\lambda,\kappa\in\Lambda,\ \lambda\neq\kappa} \log|\lambda - \kappa|.$$

According to the 2D Coulomb-gas formalism, this quantity may be interpreted as potential energy of a system of 'flat' electrons placed at the points of Λ.

Example: To illustrate our next definition, let us calculate the energy of an arithmetic progression on an interval. Let $I \subset \mathbb{R}$ be an interval and let

$$\Lambda = I \cap C\mathbb{Z} = \{n + C, n + 2C, \ldots, n + kC\}$$

for some $C > 0$. Then

$$E(\Lambda) = \sum_{1\leq m\leq k} \log\left[C^{k-1}(m-1)!(k-m)!\right] = k^2 \log|I| + O(|I|^2) \tag{3.2}$$

by Stirling's formula. Here the notation $O(|I|^2)$ corresponds to the direction $|I| \to \infty$ (with C remaining fixed).

Note that the energy of k points on I will never exceed the main term $k^2 \log|I|$ in the last equation as the energy is the sum of less than k^2 terms each no greater than $\log|I|$. Thus, even though the uniform distribution of points on the interval does not maximize the energy $E(\Lambda)$, it comes within $O(|I|^2)$ from the maximum, which is negligible for our purposes, see the next definition. It is interesting to observe that the true maximum for the energy of k electrons on I is achieved when they are placed at the endpoints of I and the zeros of the Jacobi $(1,1)$-polynomial of degree $k-2$, see for example [**21**].

Next we define D-uniform sequences which will be used through the rest of the paper.

DEFINITION 3.5. Let $\Lambda = \{\lambda_n\}_{n\in\mathbb{Z}}$ be a sequence of distinct real points. We say that Λ is D-uniform if there exists a short partition I_n of $\mathbb{R}$ such that

$$\Delta_n = D|I_n| + o(|I_n|) \quad \text{as} \quad n \to \pm\infty \ \textit{(density condition)}$$

and

$$\sum_n \frac{\Delta_n^2 \log |I_n| - E_n}{1 + \text{dist}^2(0, I_n)} < \infty, \quad \textit{(energy condition)}$$

where

$$\Delta_n = \#(\Lambda \cap I_n) \quad \text{and} \quad E_n = E(\Lambda \cap I_n) = \sum_{\lambda_k, \lambda_l \in I_n,\ \lambda_k \neq \lambda_l} \log |\lambda_k - \lambda_l|.$$

Note that the density condition simply says that on each interval of the partition Λ must have about the same number of points as the arithmetic progression of density D, $\frac{1}{D}\mathbb{Z}$. The series in the energy condition is positive, since the energy E_n on each interval is always under $\Delta_n^2 \log |I_n|$. The physical meaning of the difference in the numerator is, up to lower order terms, the work needed to transform Λ into an arithmetic progression on the n-th interval. The lower order terms do not exceed $|I_n|^2$ and, due to the shortness of partition, are negligible in the energy condition.

Now we are ready to give a formula for the gap characteristic of a set $X \subset \mathbb{R}$.

THEOREM 3.6 ([**39**]).

$$\mathbf{G}(X) = 2\pi \sup \{D \mid X \textit{ contains a } D\textit{-uniform sequence}\}.$$

The main steps of the proof are as follows. First the problems is discretized using an idea of de Branges. We notice that all measures of norm ≤ 1, supported on X, with a spectral gap of fixed size form a $*$-weakly compact convex set. By the Krein-Milman theorem it is spanned by its extreme points. One can show that extreme points are discrete measures (we call a sequence discrete if it does not have finite accumulation points and a measure is discrete if it is supported on a discrete sequence). Thus the problem reduces to finding the sup of the length of the gap over discrete measures supported on X. It turns out that a discrete set can support a measure with a spectral gap only if the counting function of the set is close, in some sense, to a function from the real Dirichlet space. This property translates into the density and energy conditions in the definition of D-uniform sequences.

In the opposite direction, if the counting function is close to the real Dirichlet space then the measure with the desired spectral gap supported on the sequence is constructed by solving an extremal problem in the Dirichlet space, following an idea of Beurling and Malliavin. See [**38, 39**] for details of this proof.

To finish our discussion of the Gap problem let us consider the following examples.

For a discrete sequence of real points Λ its Beurling-Malliavin interior density is defined as

$$D_*(\Lambda) = \inf\{d \mid \exists \text{ long } \{I_n\} \text{ such that } \#(\Lambda \cap I_n) \leqslant d|I_n|,\ \forall n\}. \tag{3.3}$$

Examples:

- Recall that by Beurling's Gap Theorem, if X has long gaps then $\mathbf{G}(X) = 0$. Let us deduce this statement from the gap formula. If X has long gaps then for any short partition $\{I_n\}$ infinitely many intervals of the partition will fall inside the gaps. Any sequence of points from X will have no points on these intervals and therefore will not satisfy the density condition in the definition of a D-uniform sequence with any $D > 0$. Hence the supremum in the gap formula is 0 and $\mathbf{G}(X) = 0$.

- Let us consider the following problem. Let ε_n, $0 < \varepsilon_n < C < 1/2$ be a sequence of numbers tending to 0. Consider a real sequence $\Lambda = \{\lambda_n\}_{n\in\mathbb{Z}}$ satisfying $|\lambda_n - n| = \varepsilon_n$. What is the gap characteristic of $\mathbb{Z}\cup\Lambda$? Clearly, if the points of Λ are too close to $\mathbb{Z}$ then every pair of points n, λ_n counts as one and
$$\mathbf{G}(\mathbb{Z}\cup\Lambda) = \mathbf{G}(\mathbb{Z}) = 2\pi.$$
If however λ_n are far enough from n, $\mathbf{G}(\mathbb{Z}\cup\Lambda)$ should double to 4π. How large should ε_n be for that to happen? What if some of ε_n are small and some are large? The answer to this and similar questions now follow from the gap formula.

  For any Λ satisfying the above conditions
$$\mathbf{G}(\mathbb{Z}\cup\Lambda) = 2\pi(1+\Delta)$$
where
$$\Delta = \sup\{D_*(N) \mid N\subset\mathbb{Z}, \sum_{n\in N}\frac{\log\varepsilon_n}{1+n^2} > -\infty\}.$$
- We call a real sequence Λ separated if $|\lambda_n - \lambda_k| > c > 0$ for all $n\neq k$. Note that if a separated sequence satisfies the density condition in the definition of a D-uniform sequence, then the sequence is D-uniform because the energy condition will follow from the separation property. It follows that the gap formula for separated sequences is simplified to
$$\mathbf{G}(\Lambda) = 2\pi D_*(\Lambda),$$
see **[33]**.

  In particular, if $\Lambda = \{\lambda_n\}$ is a separated sequence such that $\lambda_n - n = O(n^{\frac{1}{2}-\varepsilon})$ then $\mathbf{G}(\Lambda) = 2\pi$. Note that ε in this condition cannot be removed because that would allow long gaps and by Beurling's theorem the gap characteristic would become 0.
- To answer questions from the beginning of this section, let us first construct an uncountable set with a finite positive characteristic. Consider
$$X = \cup_{n\in\mathbb{Z}}(dn-\varepsilon_n, dn+\varepsilon_n)$$
where $d>0$, $\varepsilon_n = e^{-|n|}$. It is not difficult to deduce from the gap formula that then $\mathbf{G}(X) = \frac{2\pi}{d}$.

  If X is a union of disjoint intervals whose length is bounded from below, then $\mathbf{G}(X)$ is either zero (for instance, when X has long gaps) or infinity. This follows from the observation that if one can put a D-uniform sequence on such an X for some $D>0$ then one can put on X such a sequence with any $D>0$.

4. The Type problem

The second problem in the title of this note is a problem of completeness of harmonics in a space of functions. Let μ be a finite positive measure on the real line. For $a>0$ denote by $\mathcal{E}_a$ the family of exponential functions $\mathcal{E}_a = \{e^{ist} | \ s\in[0,a]\}$. The exponential type of μ is defined as
$$T_\mu = \inf\{a>0 | \ \mathcal{E}_a \text{ spans } L^2(\mu)\}$$
if the set of such a is non-empty and infinity otherwise.

Similarly to the Gap problem, the definition can be extended to infinite measures:

$$T_\mu = \inf\{a > 0|\ PW_a \cap L^2(\mu) \text{ is dense in } L^2(\mu)\}.$$

The type problem asks to find T_μ in terms of μ.

The Type problem is connected to the Gap problem via duality. The family $\mathcal{E}_a$ does not span $L^2(\mu)$ if and only if there exists a function $f \in L^2(\mu)$ orthogonal to all the exponentials from $\mathcal{E}_a$. This is equivalent to the property that $\widehat{f\mu} = 0$ on $[0, a]$. Hence the Type problem asks what maximal size of the spectral gap can be obtained when multiplying μ by an $L^2(\mu)$-density.

This question first appears in the work of Wiener, Kolmogorov and Krein in 1930-40s. If μ is a spectral measure of a stationary Gaussian process, the property that $\mathcal{E}_a$ is complete in $L^2(\mu)$ is equivalent to the property that the process can be completely predicted after it is observed during a time period of length a. The type of a spectral measure is the infimum of the observation time needed for prediction. In particular, type zero means that the process can be predicted after observing for a period of length ε for any $\varepsilon > 0$ and type infinity means that the process cannot be predicted based on any finite period of observation.

The problem is stated by Krein in [**26**], where he mentions that a problem on prediction and filtration contained in Wiener's book [**44**], which circulated as a preprint since the early 40s, is equivalent to the Type problem. In 1944 Krein has solved the problem of predicting the future of the process from its past, or equivalently found a criterion of completeness of exponentials with frequencies from the interval $(-\infty, 0)$ in $L^2(\mu)$. In regard to our problem he wrote [**26**]: "Naturally, the problems present most interest when the interval is finite. However, their study in that case becomes especially difficult." In this section we will discuss the case of a finite interval.

The Type Problem can also be restated in terms of Bernstein's weighted approximation, see [**23**] or [**38**]. In addition to Krein's work [**26, 27**], connections with spectral theory of second order differential operators were studied by Gelfand and Levitan [**15**].

Before giving a formula for T_μ let us mention some of the historic results and examples

Examples and theorems:

- A classical theorem by Krein (1945, [**25**]) says that if $d\mu = w(x)dx$ and $\log w(x)/(1+x^2)$ is summable then $T_\mu = \infty$. A partial inverse, proved by Levinson and McKean (1964, [**10**]), holds for even monotone w.

  To prove Krein's theorem notice that for any such μ, $L^2(\mu)$ contains a non-zero function $g = f/w$ for some f from the Hardy space H^1 in the upper half-plane. Then the measure $e^{iax}f$ has the form $e^{iax}g\mu$ and the spectral gap of size at least a. Since a is an arbitrary positive constant, our discussion above implies that $T_\mu = \infty$.
- A theorem by Duffin and Schaeffer (1945, [**9**]) says that if μ is a measure such that for any $x \in \mathbb{R}$,

  $$\mu([x - L, x + L]) > d$$

  for some $L, d > 0$ then

  $$T_\mu \geqslant 2\pi/L.$$

- For discrete measures, in the case $\operatorname{supp}\mu = \mathbb{Z}$, a deep result by Koosis [**24**] shows an analogue of Krein's result: if $\mu = \sum w(n)\delta_n$ where
$$\sum \frac{\log w(n)}{1+n^2} > -\infty$$
then
$$T_\mu = 2\pi.$$
- A more recent result by Borichev and Sodin (2010, [**6**]) says that exponentially small perturbations of weight or support do not change the type of a measure.

For a discrete sequence $\Lambda = \{\lambda_n\}$ let $\Lambda^* = \{\lambda_n^*\}$ be the sequence of disjoint intervals centered at the points of Λ:
$$\lambda_n^* = (\lambda_n - \varepsilon_n, \lambda_n + \varepsilon_n), \quad \varepsilon_n = \frac{1}{3}\operatorname{dist}(\lambda_n,\ \Lambda \setminus \{\lambda_n\}).$$
The type formula uses D-uniform sequences defined in the previous section. Other versions of the formula can be found in [**38, 40**].

THEOREM 4.1 ([**37**]). *Suppose that $T_\mu < \infty$. Then*
$$T_\mu = 2\pi \max\left\{D \;\middle|\; \exists\ D\text{-uniform } \Lambda = \{\lambda_n\} \text{ such that } \sum \frac{\log \mu(\lambda_n^*)}{1+n^2} > -\infty\right\}$$
if the set of such d is non-empty and $T_\mu = 0$ otherwise.

The proof uses some of the ideas applied to the proof of the gap formula and utilizes the Toeplitz approach to completeness problems developed in our joint work with Nikolai Makarov [**30, 31**]. This approach allows one to connect various completeness and spectral problems with problems on injectivity of Toeplitz operators.

One may ask what happens if $L^2(\mu)$ is replaced with $L^p(\mu)$, $1 \le p \le \infty$ in the definition of the type. It turns out that the problem has only two distinct cases: $p = 1$, in which case it becomes equivalent to the Gap problem discussed in the last section, and $p > 1$ where it can be shown that $T_\mu^p = T_\mu^2 = T_\mu$, see [**38, 40**]. The fact that in the latter case the answer does not depend on p was somewhat unexpected to the experts.

To illustrate the type formula let us discuss the following examples.

Examples:

- A positive measure μ on $\mathbb{R}$ is called a Frostman measure if there exist positive constants α and C such that
$$\mu((x-\epsilon, x+\epsilon)) < C\epsilon^\alpha$$
for all $\epsilon > 0, x \in \mathbb{R}$.

  It follows from the type formula that if μ is a Frostman measure then T_μ is either 0 or ∞, see [**37**].

  In particular, all absolutely continuous measures $d\mu(x) = w(x)dx$ with $w \in L^p(\mathbb{R})$, $p > 1$ are Frostman and therefore have types 0 or ∞. As was discussed before, if a stationary Gaussian process has a Frostman spectral measure then it can either be predicted after an arbitrarily short observation, or is unpredictable (after a finite observation).

- Next, let us consider the following extension of the result by Koosis discussed earlier in this section. Let $A = \{a_n\} \subset \mathbb{R}$ be a separated sequence and let $\mu = \sum w(n)\delta_{a_n}$ be a positive finite measure on A. Consider the set S of all subsequences $\{a_{n_k}\}$ of A satisfying
$$\sum_k \frac{\log w(n_k)}{1+{n_k}^2} > -\infty.$$
Then
$$T_\mu = 2\pi \sup_{B\in S} D_*(B).$$

5. Pólya sequences and oscillations of Fourier Integrals

In this section we present further applications of the gap formula and discuss solutions to two more old problems of complex and harmonic analysis. These results were obtained in our joint work with Mishko Mitkovski [**33**, **34**].

The first problem we are about to discuss concerns so-called Pólya sequences and entire functions of type zero. Recall that an entire function F is said to have exponential type zero if
$$\limsup_{|z|\to\infty} \frac{\log|F(z)|}{|z|} = 0.$$
A sequence is separated if it satisfies $|\lambda_n - \lambda_m| \geq \delta > 0$ for all $n \neq m$. The problem we are about to consider starts its history from the result of Valiron [**42**, 1925], who proved that any entire function of exponential type zero bounded on $\mathbb{Z}$ is a constant. Valiron's result presents a beautiful extension of the classical Liouville theorem. Later this statement was popularized by Pólya, who mistakenly posted it as an open problem in [**41**, 1931]. Subsequently many different proofs and generalizations were given (see for example section 21.2 of [**28**] or chapter 10 of [**5**] for such results and further references).

The natural problem is to describe sequences which can replace $\mathbb{Z}$ in Valiron's statement, i.e., to describe separated real sequences Λ such that any entire function of exponential type zero that is bounded on Λ is constant. Such Λ's are called *Pólya sequences.* The problem of description of Pólya sequences was studied by Levinson [**29**, 1940], de Branges [**7**, 1963] and many others. A full elementary description was recently obtained in [**33**, 2010] using a version of the gap formula for separated sequences.

It turns out that a separated real sequence is a Pólya sequence if and only if its gap characteristic is positive. As was discussed in Section 3, in the case of a separated sequence Λ the gap formula becomes especially simple:
$$\mathbf{G}(\Lambda) = 2\pi D_*(\Lambda),$$
where D_* denotes the interior Beurling-Malliavin density, see (3.3). Putting these statements together one obtains the following description. The statement once again uses the notion of long intervals, see (3.1).

THEOREM 5.1. *Let $\Lambda = \{\lambda_n\}_{n=-\infty}^{\infty}$ be a separated sequence of real numbers. Then Λ is a Pólya sequence if and only if for every long sequence of intervals $\{I_n\}$*
$$\frac{\#(\Lambda \cap I_n)}{|I_n|} \not\to 0.$$

***

The second problem we wanted to discuss in this section is a problem on oscillation of Fourier integrals. Its history starts with the following classical result

THEOREM 5.2 (Sturm, 1836; Hurwitz). *Let*

$$f(x) = \sum_{n\geq m} (c_n e^{inx} + \bar{c}_n e^{-inx})$$

be a smooth function. Then f has at least $2m$ sign changes on $[-\pi, \pi]$.

This statement shows that a smooth function with a spectral gap has mandatory sign changes in its domain. To pass to a more advanced version of the same problem, let us consider a function with a spectral gap, i.e., let $f \in L^1(\mathbb{R})$ be a real function such that $\widehat{f} = 0$ on $[0, a]$ for some $a > 0$. Such functions are studied in electrical engineering and signal processing as high-pass signals. A problem by Grinevich from 1964, also included in V. Arnold's list of problems (2000 [**2**]), asks how often should f change signs near infinity. The problem was solved by A. Eremenko and D, Novikov in [**12**].

If $f \in L^1(\mathbb{R})$ is real, denote by $s(f, r)$ the number of sign changes of f on $[0, r]$ in any reasonable sense. For instance, we may say that f changes signs at least once on an interval I if the sets $\{f > 0\} \cap I$ and $\{f < 0\} \cap I$ both have positive measure. This lower estimate for the number of sign changes is good enough for our purposes.

THEOREM 5.3 (Eremenko-Novikov, 2003). *Let $f \in L^1(\mathbb{R})$ be a real-valued function. Suppose f has a spectral gap, that is $\widehat{f}$ vanishes on $[0, a]$ for some $a > 0$. Then*

$$\liminf_{r\to\infty} \frac{s(f,r)}{r} \geq \frac{a}{2\pi}.$$

As was mentioned before, the theorem proves a conjecture by Grinevich (1964) and extends results by Krein, Levin, Ostrovski and others.

Our goal for the rest of this section is to make the result of the last theorem more precise by involving the tools applied to the Gap and Type problems.

Let $X, Y \subset \mathbb{R}$ be closed sets. Let $M_r(X)$ denote the space of finite real-valued measures on a measurable set X. Consider the gap characteristic of the pair of sets X and Y defined as

$$\mathbf{G}(X, Y) = \sup\{\, a \mid \exists\, \mu \in M_r(X \cup Y),\ \mu > 0 \text{ on } X,\ \mu < 0 \text{ on } Y,\ \widehat{\mu} = 0 \text{ on } [0, a]\}.$$

A formula for $\mathbf{G}(X, Y)$ will give a more precise asymptotics for the sign changes of high-pass signals and additional information on their spacing or clustering. The formula uses the notion of D-uniform sequences defined in Section 3. By default, discrete sequences are assumed to be enumerated in increasing order.

THEOREM 5.4 (2015, [**34**]).

$$\mathbf{G}(X, Y) = 2\pi \sup\{\, D \mid \exists\ D\text{-uniform } \{\lambda_n\},\ \{\lambda_{2n}\} \subset X,\ \{\lambda_{2n+1}\} \subset Y\}.$$

Similarly to functions, we say that a finite real measure μ on $\mathbb{R}$ changes signs on (a, b) if there exist sets $A, B \subset (a, b)$ such that $\mu(A) > 0$ and $\mu(B) < 0$. In regard to Grinevich' problem our formula implies the following.

COROLLARY 1. *If μ has a spectral gap of the size $2\pi D$ then there exists a D-uniform sequence $\{\lambda_n\}$ such that μ changes signs on each $(\lambda_n, \lambda_{n+1})$.*

Note that the density condition from the definition of D-uniform sequences gives an improved estimate on the asymptotics of the sign changes.

Acknowledgments

This note is based on my lecture at the Linde Hall inaugural symposium at Caltech on February 23rd, 2019. As a former Caltech graduate student, I am deeply grateful to the organizers of the symposium, Dinakar Ramakrishnan, Alexander Kechris, Nikolai Makarov and Xinwen Zhu, for giving me an opportunity to participate in that momentous event in the life of Caltech and the whole mathematical community.

References

[1] W. O. Amrein and A. M. Berthier, *On support properties of L^p-functions and their Fourier transforms*, J. Functional Analysis **24** (1977), no. 3, 258–267, DOI 10.1016/0022-1236(77)90056-8. MR0461025

[2] V. I. Arnol′d, *Zadachi Arnol′da* (Russian, with Russian summary), Izdatel′stvo FAZIS, Moscow, 2000. With a preface by M. B. Sevryuk and V. B. Filippov. MR1832295

[3] M. Benedicks, *On Fourier transforms of functions supported on sets of finite Lebesgue measure*, J. Math. Anal. Appl. **106** (1985), no. 1, 180–183, DOI 10.1016/0022-247X(85)90140-4. MR780328

[4] A. Beurling, *On quasianalyticity and general distributions*, Mimeographed lecture notes, Summer institute, Stanford University (1961)

[5] R. P. Boas Jr., *Entire functions*, Academic Press Inc., New York, 1954. MR0068627

[6] A. Borichev and M. Sodin, *Weighted exponential approximation and non-classical orthogonal spectral measures*, Adv. Math. **226** (2011), no. 3, 2503–2545, DOI 10.1016/j.aim.2010.08.019. MR2739783

[7] L. de Branges, *Some applications of spaces of entire functions*, Canadian J. Math. **15** (1963), 563–583, DOI 10.4153/CJM-1963-058-1. MR153840

[8] I. Daubechies, *Ten lectures on wavelets*, CBMS-NSF Regional Conference Series in Applied Mathematics, vol. 61, Society for Industrial and Applied Mathematics (SIAM), Philadelphia, PA, 1992, DOI 10.1137/1.9781611970104. MR1162107

[9] R. J. Duffin and A. C. Schaeffer, *Power series with bounded coefficients*, Amer. J. Math. **67** (1945), 141–154, DOI 10.2307/2371922. MR11322

[10] H. Dym and H. P. McKean, *Gaussian processes, function theory, and the inverse spectral problem*, Academic Press [Harcourt Brace Jovanovich, Publishers], New York-London, 1976. Probability and Mathematical Statistics, Vol. 31. MR0448523

[11] M. M. Dzhrbashyan, *Uniquencss theorems for Fourier transforms and infinitely differentiable func- tions.* Izv. AN ArmSSR, ser. Hz-mat, 10, N6-7 (1957) 21-30 (Russian).

[12] A. Eremenko and D. Novikov, *Oscillation of functions with a spectral gap*, Proc. Natl. Acad. Sci. USA **101** (2004), no. 16, 5872–5873, DOI 10.1073/pnas.0302874101. MR2048457

[13] G. B. Folland, *A course in abstract harmonic analysis*, Studies in Advanced Mathematics, CRC Press, Boca Raton, FL, 1995. MR1397028

[14] G. B. Folland and A. Sitaram, *The uncertainty principle: a mathematical survey*, J. Fourier Anal. Appl. **3** (1997), no. 3, 207–238, DOI 10.1007/BF02649110. MR1448337

[15] I. M. Gel′fand and B. M. Levitan, *On the determination of a differential equation from its spectral function*, Amer. Math. Soc. Transl. (2) **1** (1955), 253–304. MR0073805

[16] G. H. Hardy, *A Theorem Concerning Fourier Transforms*, J. London Math. Soc. **8** (1933), no. 3, 227–231, DOI 10.1112/jlms/s1-8.3.227. MR1574130

[17] Havin, V.P.*On the Uncertainty Principle in Harmonic Analysis*, lecture notes.

[18] V. Havin and B. Jöricke, *The uncertainty principle in harmonic analysis*, Ergebnisse der Mathematik und ihrer Grenzgebiete (3) [Results in Mathematics and Related Areas (3)], vol. 28, Springer-Verlag, Berlin, 1994, DOI 10.1007/978-3-642-78377-7. MR1303780

[19] H. Hedenmalm, *Heisenberg's uncertainty principle in the sense of Beurling*, J. Anal. Math. **118** (2012), no. 2, 691–702, DOI 10.1007/s11854-012-0048-9. MR3000695

[20] L. Hörmander, *A uniqueness theorem of Beurling for Fourier transform pairs*, Ark. Mat. **29** (1991), no. 2, 237–240, DOI 10.1007/BF02384339. MR1150375

[21] S. V. Kerov, *Equilibrium and orthogonal polynomials* (Russian), Algebra i Analiz **12** (2000), no. 6, 224–237; English transl., St. Petersburg Math. J. **12** (2001), no. 6, 1049–1059. MR1816518

[22] S. V. Kislyakov, *Classical themes of Fourier analysis [see MR0915768 (91b:42021)]*, Commutative harmonic analysis, I, Encyclopaedia Math. Sci., vol. 15, Springer, Berlin, 1991, pp. 113–165, DOI 10.1007/978-3-662-02732-5_2. MR1134137

[23] P. Koosis, *The logarithmic integral. I*, Cambridge Studies in Advanced Mathematics, vol. 12, Cambridge University Press, Cambridge, 1988, DOI 10.1017/CBO9780511566196. MR961844

[24] P. Koosis, *A local estimate, involving the least superharmonic majorant, for entire functions of exponential type*, Algebra i Analiz **10** (1998), no. 3, 45–64; English transl., St. Petersburg Math. J. **10** (1999), no. 3, 441–455. MR1628022

[25] M. Krein, *On a problem of extrapolation of A. N. Kolmogoroff*, C. R. (Doklady) Acad. Sci. URSS (N. S.) **46** (1945), 306–309. MR0012700

[26] M. G. Kreĭn, *On a basic approximation problem of the theory of extrapolation and filtration of stationary random processes* (Russian), Doklady Akad. Nauk SSSR (N.S.) **94** (1954), 13–16. MR0062980

[27] M. G. Kreĭn, *On the transfer function of a one-dimensional boundary problem of the second order* (Russian), Doklady Akad. Nauk SSSR (N.S.) **88** (1953), 405–408. MR0058072

[28] B. Ya. Levin, *Lectures on entire functions*, Translations of Mathematical Monographs, vol. 150, American Mathematical Society, Providence, RI, 1996. In collaboration with and with a preface by Yu. Lyubarskii, M. Sodin and V. Tkachenko; Translated from the Russian manuscript by Tkachenko, DOI 10.1090/mmono/150. MR1400006

[29] N. Levinson, *Gap and Density Theorems*, American Mathematical Society Colloquium Publications, Vol. 26, American Mathematical Society, New York, 1940. MR0003208

[30] N. Makarov and A. Poltoratski, *Meromorphic inner functions, Toeplitz kernels and the uncertainty principle*, Perspectives in analysis, Math. Phys. Stud., vol. 27, Springer, Berlin, 2005, pp. 185–252, DOI 10.1007/3-540-30434-7_10. MR2215727

[31] N. Makarov and A. Poltoratski, *Beurling-Malliavin theory for Toeplitz kernels*, Invent. Math. **180** (2010), no. 3, 443–480, DOI 10.1007/s00222-010-0234-2. MR2609247

[32] N. Makarov and A. Poltoratski, *Two-spectra theorem with uncertainty*, J. Spectr. Theory **9** (2019), no. 4, 1249–1285, DOI 10.4171/jst/276. MR4033521

[33] M. Mitkovski and A. Poltoratski, *Pólya sequences, Toeplitz kernels and gap theorems*, Adv. Math. **224** (2010), no. 3, 1057–1070, DOI 10.1016/j.aim.2009.12.014. MR2628803

[34] M. Mitkovski and A. Poltoratski, *On the determinacy problem for measures*, Invent. Math. **202** (2015), no. 3, 1241–1267, DOI 10.1007/s00222-015-0588-6. MR3425390

[35] G. W. Morgan, *A Note on Fourier Transforms*, J. London Math. Soc. **9** (1934), no. 3, 187–192, DOI 10.1112/jlms/s1-9.3.187. MR1574180

[36] F. L. Nazarov, *Local estimates for exponential polynomials and their applications to inequalities of the uncertainty principle type* (Russian, with Russian summary), Algebra i Analiz **5** (1993), no. 4, 3–66; English transl., St. Petersburg Math. J. **5** (1994), no. 4, 663–717. MR1246419

[37] A. Poltoratski, *Type alternative for Frostman measures*, Adv. Math. **349** (2019), 348–366, DOI 10.1016/j.aim.2019.04.018. MR3940942

[38] A. Poltoratski, *Toeplitz approach to problems of the uncertainty principle*, CBMS Regional Conference Series in Mathematics, vol. 121, Published for the Conference Board of the Mathematical Sciences, Washington, DC; by the American Mathematical Society, Providence, RI, 2015, DOI 10.1090/cbms/121. MR3309830

[39] A. Poltoratski, *Spectral gaps for sets and measures*, Acta Math. **208** (2012), no. 1, 151–209, DOI 10.1007/s11511-012-0076-4. MR2910798

[40] A. Poltoratski, *A problem on completeness of exponentials*, Ann. of Math. (2) **178** (2013), no. 3, 983–1016, DOI 10.4007/annals.2013.178.3.4. MR3092474

[41] G. Pólya, *Jahresbericht der Deutchen Mathematiker-Vereinigung*, Vol. 40 (1931), Problem 105

[42] G. Valiron, *Sur la formule d'interpolation de Lagrange,* Bull. Sci. Math. 49 (1925), 181-192, 203-224

[43] N. Wiener, *I am a mathematician,* The M.I.T. Press, 1964

[44] N. Wiener, *Extrapolation, Interpolation, and Smoothing of Stationary Time Series,* M.I.T. press, 1949.

Department of Mathematics, University of Wisconsin, Van Vleck Hall, 480 Lincoln Drive, Madison, Wisconsin 53706

Email address: poltoratski@wisc.edu

Proceedings of Symposia in Pure Mathematics
Volume **104**, 2021
https://doi.org/10.1090/pspum/104/01874

Quantitative bounds for critically bounded solutions to the Navier-Stokes equations

Terence Tao

Abstract. We revisit the regularity theory of Escauriaza, Seregin, and Šverák for solutions to the three-dimensional Navier-Stokes equations which are uniformly bounded in the critical $L^3_x(\mathbb{R}^3)$ norm. By replacing all invocations of compactness methods in these arguments with quantitative substitutes, and similarly replacing unique continuation and backwards uniqueness estimates by their corresponding Carleman inequalities, we obtain quantitative bounds for higher regularity norms of these solutions in terms of the critical L^3_x bound (with a dependence that is triple exponential in nature). In particular, we show that as one approaches a finite blowup time T_*, the critical L^3_x norm must blow up at a rate $(\log\log\log \frac{1}{T_*-t})^c$ or faster for an infinite sequence of times approaching T_* and some absolute constant $c > 0$.

1. Introduction

This paper is concerned with quantitative bounds for solutions $u : [0,T]\times\mathbb{R}^3 \to \mathbb{R}^3$, $p : [0,T]\times\mathbb{R}^3 \to \mathbb{R}$ to the Navier-Stokes equations

$$\begin{aligned} \partial_t u + (u\cdot\nabla)u &= \Delta u - \nabla p \\ \nabla\cdot u &= 0. \end{aligned} \tag{1.1}$$

Here we have normalised the viscosity to equal one for simplicity. To avoid technicalities, we shall restrict attention to *classical* solutions, by which we mean solutions that are smooth and such that all derivatives of u, p lie in the space $L^\infty_t L^2_x([0,T]\times\mathbb{R}^3)$. As our bounds are quantitative and do not depend on any smooth norms of the solution, it is possible to extend the results here to weaker notions of solution, such as mild solutions of Kato [**K**], the weak Leray-Hopf solutions studied in [**ESS2**], or the suitable weak solutions from [**CKN**], by using the regularity theory of such solutions; we leave the details to the interested reader. As is well known, such solutions have a maximal Cauchy development $u : [0,T_*)\times\mathbb{R}^3 \to \mathbb{R}^3$, $p : [0,T_*)\times\mathbb{R}^3 \to \mathbb{R}$ for some $0 < T_* \le \infty$, with the restriction to $[0,T]\times\mathbb{R}^3$ a classical solution for all $T < T_*$, but for which no smooth extension to time T_* is possible if $T_* < \infty$. We refer to T_* as the *maximal time of existence* of such a classical solution.

2020 *Mathematics Subject Classification.* Primary 35Q35, 37N10, 76B99.
Key words and phrases. Navier-Stokes, blowup criterion.
The author is supported by NSF grant DMS-1266164 and by a Simons Investigator Award. We also thank Stan Palasek and Jiayan Wu for corrections.

The Navier-Stokes system enjoys the scaling symmetry $(u, p, T) \mapsto (u^\lambda, p^\lambda, \lambda^2 T)$ for any $\lambda > 0$, where

$$u^\lambda(t,x) := \lambda u(\lambda^2 t, \lambda x)$$

and

$$p^\lambda(t,x) := \lambda^2 p(\lambda^2 t, \lambda x),$$

Among other things, this means that the norm

$$\|u\|_{L^\infty_t L^3_x([0,T]\times\mathbb{R}^3)}$$

is scale-invariant (or *critical*) for this equation. In **[ESS2]** it was shown that as long as this norm stays bounded, solutions to Navier-Stokes remain regular. In particular, they showed an endpoint of the classical Prodi-Serrin-Ladyshenskaya blowup criterion [**Pr**], [**S2**], [**La**] or the Leray blowup criterion [**Le**]:

THEOREM 1.1 (Qualitative blowup criterion). [**ESS2**] *Suppose (u,p) is a classical solution to Navier-Stokes whose maximal time of existence T_* is finite. Then*

$$\limsup_{t\to T_*^+} \|u(t)\|_{L^3_x(\mathbb{R}^3)} = +\infty.$$

There are now many proofs, variants and generalisations [**ESS2**], [**KK**], [**GKP**], [**GKP2**], [**S1**], [**Ph**], [**GIP**] [**DD**], [**A**], [**BS**], [**SS**], [**WZ**] of this theorem, including extensions to higher dimensions or other domains than Euclidean spaces, replacing L^3 with another critical Besov or Lorenz space, or replacing the limit superior by a limit. However, in contrast to the more quantitative arguments of Leray, Prodi, Serrin and Ladyshenskaya, the proofs in the above references all rely at some point on a compactness argument to extract a limiting profile solution to which qualitative results such as unique continuation and backwards uniqueness for heat equations (as established in particular in [**ESS**]) can be applied. As such, the above proofs do not easily give any quantitative rate of blowup for the L^3 norm.

On the other hand, the proofs of unique continuation and backwards uniqueness rely on explicit Carleman inequalities which are fully quantitative in nature. Thus, one would expect it to be possible, at least in principle, to remove the reliance on compactness methods and obtain a quantitative version of Theorem 1.1. This is the purpose of the current paper. More precisely, in Section 6 we will establish the following two results:

THEOREM 1.2 (Quantitative regularity for critically bounded solutions). *Let $u : [0,T] \times \mathbb{R}^3 \to \mathbb{R}^3$, $p : [0,T] \times \mathbb{R}^3 \to \mathbb{R}$ be a classical solution to the Navier-Stokes equations with*

$$\|u\|_{L^\infty_t L^3_x([0,T]\times\mathbb{R}^3)} \leq A \tag{1.2}$$

for some $A \geq 2$. Then we have the derivative bounds

$$|\nabla^j_x u(t,x)| \leq \exp\exp\exp(A^{O(1)}) t^{-\frac{j+1}{2}}$$

and

$$|\nabla^j_x \omega(t,x)| \leq \exp\exp\exp(A^{O(1)}) t^{-\frac{j+2}{2}}$$

whenever $0 < t \leq T$, $x \in \mathbb{R}^3$, and $j = 0, 1$, where $\omega := \nabla \times u$ is the vorticity field. (See Section 2 for the asymptotic notation used in this paper.)

REMARK 1.3. It is not difficult to iterate using Schauder estimates in Hölder spaces and extend the above regularity bounds to higher values of j than $j = 0, 1$ (allowing the implied constants in the $O()$ notation to depend on j), and also control time derivatives (conceding a factor of t^{-1} for each time derivative); we leave this extension of Theorem 1.2 to the interested reader.

THEOREM 1.4 (Quantitative blowup criterion). *Let* $u : [0, T_*) \times \mathbb{R}^3 \to \mathbb{R}^3$, $p : [0, T_*) \times \mathbb{R}^3 \to \mathbb{R}$ *be a classical solution to the Navier-Stokes equations which blows up at a finite time* $0 < T_* < \infty$. *Then*

$$\limsup_{t \to T_*^-} \frac{\|u(t)\|_{L^3_x(\mathbb{R}^3)}}{(\log \log \log \frac{1}{T_* - t})^c} = +\infty$$

for an absolute constant $c > 0$.

We now discuss the method of proof of these theorems, which uses many of the same key inputs as in previous arguments (most notably the Carleman estimates used to prove backwards uniqueness and unique continuation), but also introduces some other ingredients in order to avoid having to make some rather delicate results from the qualitative theory (such as profile decompositions) quantitative, as doing so would almost certainly lead to much poorer bounds than the ones given here.

The main estimate focuses on bounding the scale-invariant quantity

$$N_0^{-1} |P_{N_0} u(t_0, x_0)| \tag{1.3}$$

for various points (t_0, x_0) in spacetime, and various frequencies N_0, where P_{N_0} is a Littlewood-Paley projection operator to frequencies $\sim N_0$ (see Section 2 for a precise definition). Using (1.2) and the Bernstein inequality, one can bound this quantity by $O(A)$. It is well known that if one could improve this bound somewhat for sufficiently large N_0, for instance to $O(A^{-C_0})$ for a large constant C_0, then (assuming A is large enough) the L^3 norm becomes sufficiently "dispersed" in space and frequency that one could adapt the local well-posedness theory for the Navier-Stokes equation (or the local regularity theory from [**CKN**]) to obtain good bounds. Hence we will focus on establishing such a bound for (1.3) for N_0 large[1] enough (see Theorem 5.1 for a precise statement).

The first step in doing so is to observe (basically from the Duhamel formula and some standard Littlewood-Paley theory) that if the quantity (1.3) is large for some N_0, t_0, x_0 with t_0 not too close to the initial time 0, then the quantity

$$N_1^{-1} |P_{N_1} u(t_1, x_1)| \tag{1.4}$$

is also large (with exactly the same lower bound) for some (t_1, x_1) a little bit to the past of (t_0, x_0) (but more or less within the "parabolic domain of dependence", in the sense that $x_1 = x_0 + O((t_0 - t_1)^{1/2})$) and with N_1 comparable to N_0; see Proposition 3.1(iv) for a precise statement. If one takes care to have exactly the same lower bounds for both (1.3) and (1.4), then this claim can be iterated, creating a chain of "bubbles of concentration" at various points (t_n, x_n) and frequencies N_n, propagating backwards in time, and for which

$$N_n^{-1} |P_{N_n} u(t_n, x_n)|$$

[1]Strictly speaking, it is the scale-invariant quantity $N_0^2 T$ that needs to be large, rather than N_0 itself, where T is the amount of time to the past of x_0 for which the solution exists and obeys the bounds (1.2).

is bounded from below uniformly in n. Furthermore, by using a "bounded total speed" property first observed in [**T**], one can ensure that (t_n, x_n) stays in the "parabolic domain of dependence" in the sense that $x_n = x_0 + O((t_0 - t_n)^{1/2})$. Due to the well known fact (dating back to the classical work of Leray [**Le**]) that solutions to Navier-Stokes enjoy large "epochs of regularity" in which one has control of high regularity norms of the solution in large time intervals outside of a small dimensional singular set of times (see Proposition 3.1(iii) for a precise quantification of this statement), one can show that there are a large number of points (t_n, x_n) for which the frequency N_n is basically as small as possible, in the sense that

$$N_n \sim |t_0 - t_n|^{-1/2}.$$

The (Littlewood-Paley component $P_{N_n} u$ of) the solution u is large near (t_n, x_n), and it is not difficult to then obtain analogous lower bounds on the vorticity

$$\omega := \nabla \times u$$

near (t_n, x_n). The importance of working with the vorticity comes from the fact that it obeys the *vorticity equation*

$$\partial_t \omega = \Delta\omega - (u \cdot \nabla)\omega + (\omega \cdot \nabla)u \tag{1.5}$$

which can be viewed as a variable coefficient heat equation (in which the lower order coefficients $u, \nabla u$ depend on the velocity field) for which the non-local effects of the pressure p do not explicitly appear. Using a quantitative version of unique continuation for backwards parabolic equations (see Proposition 4.3 for a precise statement) that can be established using Carleman inequalities, one can then obtain exponentially small, but still non-trivial, lower bounds[2] for enstrophy-type quantities such as

$$\int_{I_n} \int_{R_n \leq |x - x_n| \leq R'_n} |\omega(t,x)|^2 \, dxdt$$

for various cylindrical annuli $I_n \times \{x : R_n \leq |x - x_n| \leq R'_n\}$ surrounding (t_n, x_n), with R'_n a large multiple of R_n. Crucially, one can set R_n to be as large as one pleases (although the lower bound exhibits Gaussian decay in R_n). In order to apply the Carleman inequalities, it is important that the time interval I lies within one of the "epochs of regularity" in which one has good L^∞ estimates for $u, \nabla u, \omega, \nabla\omega$, but this can be accomplished without much difficulty (mainly thanks to the energy dissipation term in the energy inequality).

For many choices of scale R_n (a bit larger than $|t_0 - t_n|^{1/2}$), one can use an "energy pigeonholing argument" (as used for instance by Bourgain [**B**]) to make the energy (or more precisely, a certain component of the enstrophy) small in an annular region $\{x : R_n \leq |x - x_n| \leq R'_n\}$ at some time t'_n a little bit to the past of t_n; by modifying the somewhat delicate analysis of local enstrophies from [**T**] that again takes advantage of the "bounded total speed" property, one can then propagate this smallness forward in time (at the cost of shrinking the annular region $\{R_n \leq |x - x_n| \leq R'_n\}$ slightly), and in particular back up to time t_0, and parabolic regularity theory can then be used to obtain good L^∞ estimates for $u, \nabla u, \omega, \nabla\omega$ in these regions. This allows us to again use Carleman inequalities. Specifically,

---

[2]One can think of this as applying (a quantitative version) of unique continuation "in the contrapositive". Similarly for the invocation of backwards uniqueness below. Actually in practice the Carleman inequalities also require an additional term such as $|\nabla\omega(t,x)|^2$ in the integrand, but we ignore this term for sake of discussion.

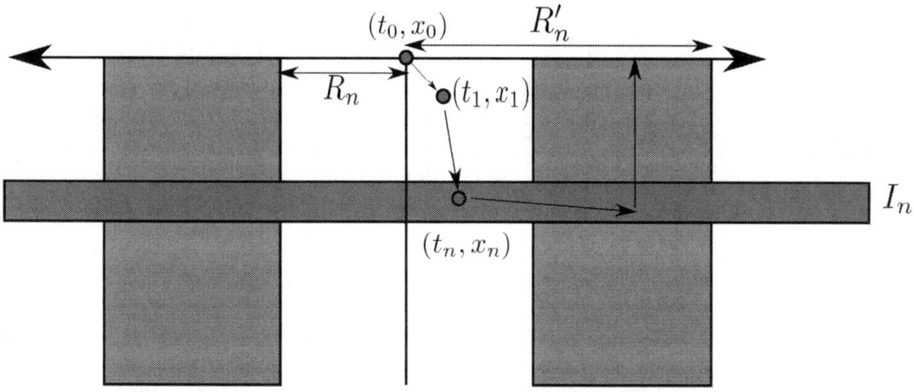

FIGURE 1. A schematic depiction of the main argument. Starting with a concentration of critical norm at a point (t_0, x_0) in spacetime, one propagates this concentration backwards in time to generate concentrations at further points (t_n, x_n) in spacetime. Restricting attention to an epoch of regularity $I_n \times \mathbb{R}^3$ (depicted here in purple), Carleman estimates are then used to establish lower bounds on the vorticity at other locations in space, and in particular where the epoch intersects an "annulus of regularity" (depicted in green) arising from an energy (or enstrophy) pigeonholing argument. A further application of Carleman estimates are then used to establish a lower bound on the vorticity (or velocity) in the annular region at time $t = t_0$, thus demonstrating a lack of compactness of the solution at this time which can be used to obtain a contradiction when N_0 (or more precisely the scale-invariant quantity $N_0^2 T$, where T is the lifespan of the solution) is large enough, by letting n vary.

by using the Carleman inequalities used to prove the backwards uniqueness result in [**ESS2**] (see Section 4 for precise statements), one can then propagate the lower bounds on $I_n \times \{x : R_n \leq |x - x_n| \leq R'_n\}$ forward in time until one returns to the original time t_0 of interest, eventually obtaining a small but nontrivial lower bound for quantities such as

$$\int_{R_n \leq |x - x_n| \leq R'_n} |\omega(t_0, x)|^2 \, dx$$

(ignoring for this discussion some slight adjustments to the scales R_n, R'_n that occur during this argument), which after some routine manipulations (and using the fact that (t_n, x_n) lies in the parabolic domain of dependence of (t_0, x_0)) also gives a lower bound on quantities like

$$\int_{R_n \leq |x - x_0| \leq R'_n} |u(t_0, x)|^3 \, dx.$$

Crucially, this lower bound is uniform in n. If one now lets n vary, the annuli $\{R_n \leq |x - x_0| \leq R'_n\}$ end up becoming disjoint for widely separated n, and one can eventually contradict (1.2) at time $t = t_0$ if N_0 is large enough.

REMARK 1.5. The triply exponential nature of the bounds in Theorem 1.2 (which is of course closely tied to the triply logarithmic improvement to Theorem 1.1 in Theorem 1.4) can be explained as follows. One exponential factor comes from the Bourgain energy pigeonholing argument to locate a good spatial scale R. A second exponential factor arises from the Carleman inequalities. The third exponential arises from locating enough disjoint spatial scales R_n to contradict (1.2). It seems that substantially new ideas would be needed in order to improve significantly upon this triple exponential bound.

REMARK 1.6. Of course, by Sobolev embedding, the $L^3_x(\mathbb{R}^3)$ norm in the above theorems can be replaced by the critical homogeneous Sobolev norm $\dot H^{1/2}_x(\mathbb{R}^3)$. It is likely that the arguments here can also be adapted to handle other critial Besov or Lorentz spaces (as long as the secondary exponent of such spaces is finite, so that the critical norm cannot simultaneously have a substantial presence at an unbounded number of scales), but we will not pursue this question here; based on Theorem 1.4, it is also reasonable to conjecture that the Orlicz norm $\|u(t)\|_{L^3(\log\log\log L)^{-c}(\mathbb{R}^3)}$ of u also must blow up as $t \to T_*^-$ for some absolute constant $c > 0$. On the other hand, our argument relies heavily in many places on the fact that we are working in three dimensions. It may be possible to obtain a higher-dimensional analogue of our results by finding quantitative versions of the argument in [**DD**], but we do not pursue this question here. Similarly, our arguments do not directly allow us to replace the limit superior in Theorem 1.1 with a limit, as is done in [**S1**] (see also [**A**]); again, it may be possible to also find quantitative analogues of these results, but we do not pursue this matter here.

2. Notation

We use the notation $X = O(Y)$, $X \lesssim Y$, or $Y \gtrsim X$ to denote the bound $|X| \leq CY$ for some absolute constant $C > 0$. If we need the implied constant C to depend on parameters we shall indicate this by subscripts, for instance $X \lesssim_j Y$ denotes the bound $|X| \leq C_j Y$ where C_j depends only on j.

Throughout this paper we will need a sufficiently large absolute constant C_0, which will remain fixed throughout the paper. For instance $C_0 = 10^5$ would suffice throughout our paper, if one worked out all the implied constants in the exponents carefully.

If $I \subset \mathbb{R}$ is a time interval, we use $|I|$ to denote its length. If $x_0 \in \mathbb{R}^3$ and $R > 0$, we use $B(x_0, R)$ to denote the ball $\{x \in \mathbb{R}^3 : |x - x_0| \leq R\}$, and if $B = B(x_0, R)$ is such a ball, we use $kB = B(x_0, kR)$ to denote its dilates for any $k > 0$.

We use the mixed Lebesgue norms

$$\|u\|_{L^q_t L^r_x(I \times \mathbb{R}^3)} := \left(\int_I \|u(t)\|^q_{L^r_x(\mathbb{R}^3)}\, dt \right)^{1/q}$$

where

$$\|u(t)\|_{L^r_x(\mathbb{R}^3)} := \left(\int_{\mathbb{R}^3} |u(t,x)|^r \, dx \right)^{1/r}$$

with the usual modifications when $q = \infty$ or $r = \infty$. For any measurable subset $\Omega \subset I \times \mathbb{R}^3$, we write $\|u\|_{L^q_t L^r_x(\Omega)}$ for $\|u 1_\Omega\|_{L^q_t L^r_x(I \times \mathbb{R}^3)}$, where 1_Ω is the indicator function of Ω.

Given a Schwartz function $f : \mathbb{R}^3 \to \mathbb{R}$, we define the Fourier transform

$$\hat{f}(\xi) := \int_{\mathbb{R}^3} f(x) e^{-2\pi i \xi \cdot x}\, dx$$

and then for any $N > 0$ we define the Littlewood-Paley projection $P_{\leq N}$ by the formula

$$\widehat{P_{\leq N} f}(\xi) := \varphi(\xi/N)\hat{f}(\xi)$$

where $\varphi : \mathbb{R}^3 \to \mathbb{R}$ is a fixed bump function supported on $B(0,1)$ that equals 1 on $B(0,1/2)$. We also define the companion Littlewood-Paley projections

$$\begin{aligned} P_N &:= P_N - P_{N/2} \\ P_{>N} &:= 1 - P_{\leq N} \\ \tilde{P}_N &:= P_{2N} - P_{N/4} \end{aligned}$$

where 1 denotes the identity operator; thus for instance $P_{\leq N} f = \sum_{k=0}^{\infty} P_{2^{-k}N} f$ and $P_{>N} f = \sum_{k=1}^{\infty} P_{2^k N} f$ for Schwartz f (with the convergence in a locally uniform sense). Also we have $P_N = P_N \tilde{P}_N$. These operators can also be applied to vector-valued Schwartz functions by working component by component. These operators commute with other Fourier multipliers such as the Laplacian Δ and its inverse Δ^{-1}, partial derivatives ∂_i, heat propagators $e^{t\Delta}$, and the Leray projection $\mathbb{P} := -\nabla \times \Delta^{-1} \nabla \times$ to divergence-free vector fields. To estimate such multipliers, we use the following general estimate:

LEMMA 2.1 (Multiplier theorem). *Let $N > 0$, and let $m : \mathbb{R}^3 \to \mathbb{C}$ be a smooth function supported on $B(0,N)$ that obeys the bounds*

$$|\nabla^j m(\xi)| \leq M N^{-j}$$

for all $0 \leq j \leq 100$ and some $M > 0$. Let T_m denote the associated Fourier multiplier, thus

$$\widehat{T_m f}(\xi) := m(\xi) f(\xi).$$

Then one has

$$\|T_m f\|_{L^q(\mathbb{R}^3)} \lesssim M N^{\frac{3}{p} - \frac{3}{q}} \|f\|_{L^p(\mathbb{R}^3)} \tag{2.1}$$

whenever $1 \leq p \leq q \leq \infty$ and $f : \mathbb{R}^3 \to \mathbb{R}$ is a Schwartz function. More generally, if $\Omega \subset \mathbb{R}^3$ is an open subset of $\mathbb{R}^3$, $A \geq 1$, and $\Omega_{A/N} := \{x \in \mathbb{R}^3 : \operatorname{dist}(x, \Omega) < A/N\}$ denotes the A/N-neighbourhood of Ω, then we have a local version

$$\|T_m f\|_{L^{q_1}(\Omega)} \lesssim M N^{\frac{3}{p_1} - \frac{3}{q_1}} \|f\|_{L^{p_1}(\Omega_{A/N})} + A^{-50} M |\Omega|^{\frac{1}{q_1} - \frac{1}{q_2}} N^{\frac{3}{p_2} - \frac{3}{q_2}} \|f\|_{L^{p_2}(\mathbb{R}^3)} \tag{2.2}$$

of the above estimate, whenever $1 \leq p_1 \leq q_1 \leq \infty$ and $1 \leq p_2 \leq q_2 \leq \infty$ are such that $q_2 \geq q_1$, and $|\Omega|$ denotes the volume of Ω.

By the usual limiting arguments, one can replace the hypothesis that f is Schwartz with the requirement that f lie in L^p. Also one can extend this theorem to vector-valued $f : \mathbb{R}^3 \to \mathbb{R}^3$ by working component by component. In practice, the A^{-50} factor will ensure that the second term on the right-hand side of (2.2) is negligible compared to the first, and can be ignored on a first reading.

PROOF. By homogeneity we can normalise $M = 1$; by scaling (or dimensional analysis) we may also normalise $N = 1$. We can write $T_m f$ as a convolution $T_m f = f * K$ of f with the kernel

$$K(x) := \int_{\mathbb{R}^3} m(\xi) e^{2\pi i \xi \cdot x}\, d\xi.$$

By repeated integration by parts we obtain the bounds $K(x) \lesssim (1 + |x|)^{-90}$ (say), so in particular $\|K\|_{L^r(\mathbb{R}^3)} \lesssim 1$ for all $1 \leq r \leq \infty$. From Young's convolution inequality we then conclude that

$$\|T_m f\|_{L^q(\mathbb{R}^3)} \lesssim \|f\|_{L^p(\mathbb{R}^3)}$$

giving (2.1). To prove (2.2), we see that the claim already follows from (2.1) when f is supported in $L^p(\Omega_A)$, so by the triangle inequality we may assume that f is supported on $\mathbb{R}^3 \backslash \Omega_A$. In this case we may replace the convolution kernel K by its restriction to the complement of $B(0, A)$, which allows us to improve the bound on the L^r norm of the kernel to (say) $O(A^{-50})$. The claim follows from Young's convolution inequality, after first using Hölder's inequality to bound $\|T_m f\|_{L^{q_1}(\Omega)} \leq |\Omega|^{\frac{1}{q_1} - \frac{1}{q_2}} \|T_m f\|_{L^{q_2}(\Omega)}$. □

Thus for instance, we have the Bernstein inequalities

$$\|\nabla^j f\|_{L^q(\mathbb{R}^3)} \lesssim_j N^{j + \frac{3}{p} - \frac{3}{q}} \|f\|_{L^p(\mathbb{R}^3)} \tag{2.3}$$

whenever $1 \leq p \leq q \leq \infty$, $j \geq 0$, and f is a Schwartz function whose Fourier transform is supported on $B(0, N)$, as can be seen by writing $f = P_{\leq 2N} f$ and applying Lemma 2.1. In a similar spirit, one has

$$\|P_N e^{t\Delta} \nabla^j f\|_{L^q(\mathbb{R}^3)} \lesssim_j \exp(-N^2 t/20) N^{j + \frac{3}{p} - \frac{3}{q}} \|f\|_{L^p(\mathbb{R}^3)} \tag{2.4}$$

for any $t > 0$ and any Schwartz f. Summing this, we obtain the standard heat kernel bounds

$$\|e^{t\Delta} \nabla^j f\|_{L^q(\mathbb{R}^3)} \lesssim_j t^{-\frac{j}{2} - \frac{3}{2p} + \frac{3}{2q}} \|f\|_{L^p(\mathbb{R}^3)}. \tag{2.5}$$

3. Basic estimates

The purpose of this section is to establish the following initial bounds for $L^\infty_t L^3_x$-bounded solutions to the Navier-Stokes equations.

PROPOSITION 3.1 (Initial estimates). *Let $u : [t_0 - T, t_0] \times \mathbb{R}^3 \to \mathbb{R}^3$, $p : [t_0 - T, t_0] \times \mathbb{R}^3 \to \mathbb{R}$ be a classical solution to Navier-Stokes that obeys the bound*

$$\|u\|_{L^\infty_t L^3_x([t_0 - T, t_0] \times \mathbb{R}^3)} \leq A. \tag{3.1}$$

for some $A \geq C_0$. We adopt the notation

$$A_j := A^{C_0^j}$$

for all j, thus $A_0 = A$ and $A_{j+1} = A_j^{C_0}$.

(i) *(Pointwise derivative estimates) For any $(t, x) \in [t_0 - T/2, t_0] \times \mathbb{R}^3$ and $N > 0$, we have*

$$P_N u(t,x) = O(AN); \quad \nabla P_N u(t,x) = O(AN^2); \quad \partial_t P_N u(t,x) = O(A^2 N^3); \tag{3.2}$$

similarly, the vorticity $\omega := \nabla \times u$ obeys the bounds

$$P_N \omega(t,x) = O(AN^2); \quad \nabla P_N \omega(t,x) = O(AN^3); \quad \partial_t P_N \omega(t,x) = O(A^2 N^4). \tag{3.3}$$

(ii) *(Bounded total speed) For any interval I in $[t_0 - T/2, t_0]$, one has*

$$\|u\|_{L^1_t L^\infty_x(I\times\mathbb{R}^3)} \lesssim A^4 |I|^{1/2}. \tag{3.4}$$

(iii) *(Epochs of regularity) For any interval I in $[t_0 - T/2, t_0]$, there is a subinterval $I' \subset I$ with $|I'| \gtrsim A^{-8}|I|$ such that*

$$\|\nabla^j u\|_{L^\infty_t L^\infty_x(I'\times\mathbb{R}^3)} \lesssim A^{O(1)}|I|^{-(j+1)/2}$$

and

$$\|\nabla^j \omega\|_{L^\infty_t L^\infty_x(I'\times\mathbb{R}^3)} \lesssim A^{O(1)}|I|^{-(j+2)/2}$$

for $j = 0, 1$.

(iv) *(Back propagation) Let $(t_1, x_1) \in [t_0 - T/2, t_0] \times \mathbb{R}^3$ and $N_1 \geq A_3 T^{-1/2}$ be such that*

$$|P_{N_1} u(t_1, x_1)| \geq A_1^{-1} N_1. \tag{3.5}$$

Then there exists $(t_2, x_2) \in [t_0 - T, t_1] \times \mathbb{R}^3$ and $N_2 \in [A_2^{-1} N_1, A_2 N_1]$ such that

$$A_3^{-1} N_1^{-2} \leq t_1 - t_2 \leq A_3 N_1^{-2}$$

and

$$|x_2 - x_1| \leq A_4 N_1^{-1}$$

and

$$|P_{N_2} u(t_2, x_2)| \geq A_1^{-1} N_2. \tag{3.6}$$

(v) *(Iterated back propagation) Let $x_0 \in \mathbb{R}^3$ and $N_0 > 0$ be such that*

$$|P_{N_0} u(t_0, x_0)| \geq A_1^{-1} N_0.$$

Then for every $A_4 N_0^{-2} \leq T_1 \leq A_4^{-1} T$, there exists

$$(t_1, x_1) \in [t_0 - T_1, t_0 - A_3^{-1} T_1] \times \mathbb{R}^3$$

and

$$N_1 = A_3^{O(1)} T_1^{-1/2}$$

such that

$$x_1 = x_0 + O(A_4^{O(1)} T_1^{1/2})$$

and

$$|P_{N_1} u(t_1, x_1)| \geq A_1^{-1} N_1.$$

(vi) *(Annuli of regularity) If $0 < T' < T/2$, $x_0 \in \mathbb{R}^3$, and $R_0 \geq (T')^{1/2}$, then there exists a scale*

$$R_0 \leq R \leq \exp(A_6^{O(1)}) R_0$$

such that on the region

$$\Omega := \{(t, x) \in [t_0 - T', t_0] \times \mathbb{R}^3 : R \leq |x - x_0| \leq A_6 R\}$$

we have

$$\|\nabla^j u\|_{L^\infty_t L^\infty_x(\Omega)} \lesssim A_6^{-2}(T')^{-(j+1)/2}$$

and

$$\|\nabla^j \omega\|_{L^\infty_t L^\infty_x(\Omega)} \lesssim A_6^{-2}(T')^{-(j+2)/2}$$

for $j = 0, 1$.

As C_0 is assumed large, any polynomial combination of $A = A_0, A_1, \dots, A_{j-1}$ will be dominated by A_j for any $j \geq 1$; we take advantage of this fact without comment in the sequel to simplify the estimates. The various numerical powers of A (or A_j) that appear in the above proposition are not of much significance, except that it is important for iterative purposes that the negative power A_1^{-1} appearing in (3.5) is exactly the same as the one appearing in (3.6).

In the remainder of this section t_0, T, A, u, p are as in Proposition 3.1. Our objective is now to establish the claims (i)-(vi).

We begin with the proof of (i). It suffices to establish (3.2), as (3.3) then follows from the Bernstein inequalities (2.3). The first two claims of (3.2) are immediate from (3.1) and (2.3). For the final claim, we first apply the Leray projection $\mathbb{P}$ to (1.1) to obtain the familiar equation

$$\partial_t u = \Delta u - \mathbb{P}\nabla \cdot (u \otimes u) \tag{3.7}$$

where the divergence $\nabla \cdot (u \otimes u)$ of the symmetric tensor $u \otimes u$ is expressed in coordinates as

$$(\nabla \cdot (u \otimes u))_i = \partial_j (u_i u_j)$$

with the usual summation conventions. We apply P_N to both sides of (3.7). From (3.1) and (2.3) we have

$$\|P_N \Delta u(t)\|_{L^\infty_x(\mathbb{R}^3)} \lesssim N^3 A.$$

From (3.1) and Hölder we have $\|u \otimes u(t)\|_{L^{3/2}_x(\mathbb{R}^3)} \lesssim A^2$, hence by Lemma 2.1 we have

$$\|P_N \mathbb{P}\nabla \cdot (u \otimes u)(t)\|_{L^\infty_x(\mathbb{R}^3)} \lesssim N^3 A^2,$$

and the final claim of (3.2) follows from the triangle inequality.

Now we prove (ii), (iii). It is not difficult to see that these estimates are invariant with respect to time translation (shifting I, t_0, u accordingly) and also rescaling (adjusting T, t_0, I, u accordingly). Hence we may assume without loss of generality that $I = [0,1] \subset [t_0 - T/2, t_0]$, which implies that $[-1,1] \subset [t_0 - T, t_0]$.

It will be convenient to remove[3] a linear component from u, as it is not well controlled in L^2_x type spaces. Namely, on $[-1,1] \times \mathbb{R}^3$ we split $u = u^{\mathrm{lin}} + u^{\mathrm{nlin}}$, where u^{lin} is the linear solution

$$u^{\mathrm{lin}}(t) := e^{(t+1)\Delta} u(-1) \tag{3.8}$$

and $u^{\mathrm{nlin}} := u - u^{\mathrm{lin}}$ is the nonlinear component. From (3.1) we have

$$\|u^{\mathrm{lin}}\|_{L^\infty_t L^3_x([-1,1]\times\mathbb{R}^3)}, \|u^{\mathrm{nlin}}\|_{L^\infty_t L^3_x([-1,1]\times\mathbb{R}^3)} \lesssim A. \tag{3.9}$$

From (3.7) and Duhamel's formula one has

$$u^{\mathrm{nlin}}(t) = -\int_{-1}^t e^{(t-t')\Delta} \mathbb{P}\nabla \cdot (u \otimes u)(t')\, dt'.$$

From (3.1), $u \otimes u$ has an $L^{3/2}_x(\mathbb{R}^3)$ norm of $O(A^2)$. From (2.5), the operator $e^{(t-t')\Delta}\mathbb{P}\nabla\cdot$ maps $L^{3/2}_x$ to L^2_x with an operator norm of $(t-t')^{-3/4}$. From Minkowski's inequality we conclude an energy bound for the nonlinear component:

$$\|u^{\mathrm{nlin}}\|_{L^\infty_t L^2_x([-1,1]\times\mathbb{R}^3)} \lesssim A^2. \tag{3.10}$$

[3]See also [C] for a similar technique to apply energy methods to Navier-Stokes solutions that lie in a function space other than L^2_x.

We now restrict attention to the slab $[-1/2,1]\times\mathbb{R}^3$. Here $t+1$ lies between $1/2$ and 2, and we can use (3.1), (3.8), and (2.5) to obtain very good bounds on u^{lin} (but only in spaces with an integrability exponent greater than or equal to 3). More precisely, we have

$$\|\nabla^j u^{\mathrm{lin}}\|_{L^\infty_t L^p_x([-1/2,1]\times\mathbb{R}^3)} \lesssim_j A \tag{3.11}$$

for any $3 \le p \le \infty$ and $j \ge 0$.

To exploit the bound (3.10), we use the energy method. Since u^{lin} solves the heat equation $\partial_t u^{\mathrm{lin}} = \Delta u^{\mathrm{lin}}$, we can subtract this from (1.1) to conclude that

$$\partial_t u^{\mathrm{nlin}} = \Delta u^{\mathrm{nlin}} - \nabla\cdot(u\otimes u) - \nabla p. \tag{3.12}$$

Taking inner products with u^{nlin}, which is divergence-free, and integrating by parts, we conclude that

$$\frac{1}{2}\partial_t \int_{\mathbb{R}^3} |u^{\mathrm{nlin}}|^2\, dx = -\int_{\mathbb{R}^3} |\nabla u^{\mathrm{nlin}}|^2\, dx + \int_{\mathbb{R}^3} (\nabla u^{\mathrm{nlin}})\cdot(u\otimes u)\, dx$$

where the quantity $(\nabla u^{\mathrm{nlin}})\cdot(u\otimes u)$ is defined in coordinates as

$$(\nabla u^{\mathrm{nlin}})\cdot(u\otimes u) := (\partial_i u^{\mathrm{nlin}}_j) u_i u_j.$$

From the divergence-free nature of u^{nlin} and integration by parts we have

$$\int_{\mathbb{R}^3} (\nabla u^{\mathrm{nlin}})\cdot(u^{\mathrm{nlin}}\otimes u^{\mathrm{nlin}})\, dx = 0$$

and hence

$$\frac{1}{2}\partial_t \int_{\mathbb{R}^3} |u^{\mathrm{nlin}}|^2\, dx = -\int_{\mathbb{R}^3} |\nabla u^{\mathrm{nlin}}|^2\, dx + \int_{\mathbb{R}^3} (\nabla u^{\mathrm{nlin}})\cdot(u\otimes u - u^{\mathrm{nlin}}\otimes u^{\mathrm{nlin}})\, dx.$$

Integrating this on $[-1/2,1]$ using (3.10) we conclude that

$$\int_{-1/2}^1\int_{\mathbb{R}^3} |\nabla u^{\mathrm{nlin}}|^2\, dxdt \lesssim A^2 + \int_{-1/2}^1\int_{\mathbb{R}^3} |\nabla u^{\mathrm{nlin}}||u\otimes u - u^{\mathrm{nlin}}\otimes u^{\mathrm{nlin}}|\, dxdt,$$

and hence by Young's inequality

$$\int_{-1/2}^1\int_{\mathbb{R}^3} |\nabla u^{\mathrm{nlin}}|^2\, dxdt \lesssim A^2 + \int_{-1/2}^1\int_{\mathbb{R}^3} |u\otimes u - u^{\mathrm{nlin}}\otimes u^{\mathrm{nlin}}|^2\, dxdt.$$

Splitting $u\otimes u - u^{\mathrm{nlin}}\otimes u^{\mathrm{nlin}} = u_l\otimes u + u^{\mathrm{nlin}}\otimes u_l$ and using (3.1), (3.9), (3.11) (with $p=6$, $j=0$) and Hölder's inequality, one has

$$\int_{-1/2}^1\int_{\mathbb{R}^3} |u\otimes u - u^{\mathrm{nlin}}\otimes u^{\mathrm{nlin}}|^2\, dxdt \lesssim A^4$$

and thus

$$\int_{-1/2}^1\int_{\mathbb{R}^3} |\nabla u^{\mathrm{nlin}}|^2\, dxdt \lesssim A^4. \tag{3.13}$$

By Plancherel's theorem this implies in particular that

$$\sum_N N^2 \|P_N u^{\mathrm{nlin}}\|^2_{L^2_t L^2_x([-1/2,1]\times\mathbb{R}^3)} \lesssim A^4 \tag{3.14}$$

where N ranges over powers of two. Also, from Sobolev embedding one has

$$\|u^{\mathrm{nlin}}\|_{L^2_t L^6_x([-1/2,1]\times\mathbb{R}^3)} \lesssim A^2. \tag{3.15}$$

We are now ready to establish the bounded total speed property (ii), which is a variant of [**T**, Proposition 9.1]. If $t \in [0,1]$ and $N \geq 1$ is a power of two, we see from (3.7) and Duhamel's formula that

$$P_N u^{\mathrm{nlin}}(t) = e^{(t+\frac{1}{2})\Delta} P_N u^{\mathrm{nlin}}\left(-\frac{1}{2}\right) - \int_{-1/2}^{t} P_N e^{(t-t')\Delta}\mathbb{P}\nabla \cdot \tilde{P}_N(u \otimes u)(t')\, dt'.$$

From (2.4) the operator $P_N e^{(t-t')\Delta}\mathbb{P}\nabla\cdot$ has an operator norm of $O(N\exp(-N^2(t-t')/20))$ on L^∞_x, while from (3.9), (2.4) we see that $e^{(t+\frac{1}{2})\Delta}P_N v(-\frac{1}{2})$ has an L^∞_x norm of $O(AN\exp(-N^2/20))$. Thus by Young's inequality

$$\|P_N u^{\mathrm{nlin}}\|_{L^1_t L^\infty_x([0,1]\times\mathbb{R}^3)} \lesssim AN\exp(-N^2/20) + N^{-1}\|\tilde{P}_N(u\otimes u)\|_{L^1_t L^\infty_x([-1/2,1]\times\mathbb{R}^3)}.$$

We split $u \otimes u = u^{\mathrm{lin}} \otimes u^{\mathrm{lin}} + u^{\mathrm{lin}} \otimes u^{\mathrm{nlin}} + u^{\mathrm{nlin}} \otimes u^{\mathrm{lin}} + u^{\mathrm{nlin}} \otimes u^{\mathrm{nlin}}$. From (3.11) one has

$$\|\tilde{P}_N(u^{\mathrm{lin}} \otimes u^{\mathrm{lin}})\|_{L^1_t L^\infty_x([-1/2,1]\times\mathbb{R}^3)} \lesssim A^2.$$

From (2.3), Hölder's inequality, and (3.11), (3.15) one has

$$\begin{aligned}\|\tilde{P}_N(u^{\mathrm{lin}} \otimes u^{\mathrm{nlin}})\|_{L^1_t L^\infty_x([-1/2,1]\times\mathbb{R}^3)} &\lesssim N^{1/2}\|u^{\mathrm{lin}} \otimes u^{\mathrm{nlin}}\|_{L^1_t L^6_x([-1/2,1]\times\mathbb{R}^3)} \\ &\lesssim A^3 N^{1/2}.\end{aligned}$$

Similarly with $u^{\mathrm{lin}} \otimes u^{\mathrm{nlin}}$ replaced by $u^{\mathrm{nlin}} \otimes u^{\mathrm{lin}}$. We then split $u^{\mathrm{nlin}} \otimes u^{\mathrm{nlin}} = P_{\leq N}u^{\mathrm{nlin}} \otimes P_{\leq N}u^{\mathrm{nlin}} + P_{\leq N}u^{\mathrm{nlin}} \otimes P_{>N}u^{\mathrm{nlin}} + P_{>N}u^{\mathrm{nlin}} \otimes P_{\leq N}u^{\mathrm{nlin}} + P_{>N}u^{\mathrm{nlin}} \otimes P_{>N}u^{\mathrm{nlin}}$. We have from Hölder that

$$\begin{aligned}\|P_{\leq N}u^{\mathrm{nlin}} \otimes P_{\leq N}u^{\mathrm{nlin}}\|_{L^1_t L^\infty_x([-1/2,1]\times\mathbb{R}^3)} &\lesssim \|P_{\leq N}u^{\mathrm{nlin}}\|^2_{L^2_t L^\infty_x([-1/2,1]\times\mathbb{R}^3)} \\ \|P_{\leq N}u^{\mathrm{nlin}} \otimes P_{>N}u^{\mathrm{nlin}}\|_{L^1_t L^2_x([-1/2,1]\times\mathbb{R}^3)}&, \\ \|P_{>N}u^{\mathrm{nlin}} \otimes P_{\leq N}u^{\mathrm{nlin}}\|_{L^1_t L^2_x([-1/2,1]\times\mathbb{R}^3)} &\lesssim \|P_{\leq N}u^{\mathrm{nlin}}\|_{L^2_t L^\infty_x([-1/2,1]\times\mathbb{R}^3)} \\ &\quad \times \|P_{>N}u^{\mathrm{nlin}}\|_{L^2_t L^2_x([-1/2,1]\times\mathbb{R}^3)} \\ \|P_{>N}u^{\mathrm{nlin}} \otimes P_{>N}u^{\mathrm{nlin}}\|_{L^1_t L^1_x([-1/2,1]\times\mathbb{R}^3)} &\lesssim \|P_{>N}u^{\mathrm{nlin}}\|^2_{L^2_t L^2_x([-1/2,1]\times\mathbb{R}^3)}\end{aligned}$$

and hence by (2.3), the triangle inequality, and Young's inequality

$$\begin{aligned}&\|\tilde{P}_N(u\otimes u)\|_{L^1_t L^\infty_x([-1/2,1]\times\mathbb{R}^3)} \\ &\quad \lesssim \|P_{\leq N}u^{\mathrm{nlin}}\|^2_{L^2_t L^\infty_x([-1/2,1]\times\mathbb{R}^3)} + N^3\|P_{>N}u^{\mathrm{nlin}}\|^2_{L^2_t L^2_x([-1/2,1]\times\mathbb{R}^3)}.\end{aligned}$$

Putting all this together, we conclude that

$$\begin{aligned}\|P_N u^{\mathrm{nlin}}\|_{L^1_t L^\infty_x([0,1]\times\mathbb{R}^3)} &\lesssim A^3 N^{-1/2} \\ &\quad + N^{-1}\|P_{\leq N}u^{\mathrm{nlin}}\|^2_{L^2_t L^\infty_x([-1/2,1]\times\mathbb{R}^3)} \\ &\quad + N^2\|P_{>N}u^{\mathrm{nlin}}\|^2_{L^2_t L^2_x([-1/2,1]\times\mathbb{R}^3)}.\end{aligned}$$

By (2.3) and Cauchy-Schwarz we have

$$\begin{aligned}\|P_{\leq N}u^{\mathrm{nlin}}\|^2_{L^2_t L^\infty_x([-1/2,1]\times\mathbb{R}^3)} &\lesssim \left(\sum_{N'\leq N}(N')^{3/2}\|P_{N'}u^{\mathrm{nlin}}\|_{L^2_t L^2_x([-1/2,1]\times\mathbb{R}^3)}\right)^2 \\ &\lesssim N^{1/2}\sum_{N'\leq N}(N')^2\|P_{N'}u^{\mathrm{nlin}}\|^2_{L^2_t L^2_x([-1/2,1]\times\mathbb{R}^3)}\end{aligned}$$

where N' ranges over powers of two, while from Plancherel's theorem one has

$$\|P_{>N}u^{\mathrm{nlin}}\|^2_{L^2_tL^2_x([-1/2,1]\times\mathbb{R}^3)} \lesssim \sum_{N'>N} \|P_{N'}u^{\mathrm{nlin}}\|^2_{L^2_tL^2_x([-1/2,1]\times\mathbb{R}^3)}.$$

Summing in N, and using the triangle inequality followed by (3.14), we conclude that

$$\|P_{\geq 1}u^{\mathrm{nlin}}\|_{L^1_tL^\infty_x([0,1]\times\mathbb{R}^3)} \lesssim A^3 + \sum_{N'}(N')^2\|P_{N'}u^{\mathrm{nlin}}\|^2_{L^2_tL^2_x([-1/2,1]\times\mathbb{R}^3)} \lesssim A^4.$$

From (3.9) and (2.3) we also have

$$\|u^{\mathrm{lin}}\|_{L^1_tL^\infty_x([0,1]\times\mathbb{R}^3)}, \|P_{<1}u^{\mathrm{nlin}}\|_{L^1_tL^\infty_x([0,1]\times\mathbb{R}^3)} \lesssim A$$

and we conclude

$$\|u\|_{L^1_tL^\infty_x([0,1]\times\mathbb{R}^3)} \lesssim A^4$$

which gives (ii).

Now we establish (iii). For $t \in [0,1]$ we define the enstrophy-type quantity

$$E(t) := \frac{1}{2}\int_{\mathbb{R}^3} |\nabla u^{\mathrm{nlin}}(t,x)|^2\, dx,$$

Taking the gradient of (3.12) and then taking the inner product with ∇u^{nlin}, we see upon integration by parts that

$$\partial_t E(t) = -\int_{\mathbb{R}^3} |\nabla^2 u^{\mathrm{nlin}}|^2\, dx + \int_{\mathbb{R}^3} \Delta u^{\mathrm{nlin}} \cdot (\nabla\cdot(u\otimes u))\, dx$$

and hence by Young's inequality

$$\partial_t E(t) \leq -\frac{1}{2}\|\nabla^2 u^{\mathrm{nlin}}\|^2_{L^2_x(\mathbb{R}^3)} + O\left(\|\nabla\cdot(u\otimes u)\|^2_{L^2_x(\mathbb{R}^3)}\right).$$

By the Leibniz rule and Hölder's inequality, one has

$$\|\nabla\cdot(u\otimes u)\|_{L^2_x(\mathbb{R}^3)} \lesssim \|u\|_{L^6_x(\mathbb{R}^3)}\|\nabla u\|_{L^3_x(\mathbb{R}^3)}.$$

From (3.11) and the triangle inequality one has

$$\|u\|_{L^6_x(\mathbb{R}^3)} \lesssim A + \|u^{\mathrm{nlin}}\|_{L^6_x(\mathbb{R}^3)}$$

and

$$\|\nabla u\|_{L^3_x(\mathbb{R}^3)} \lesssim A + \|\nabla u^{\mathrm{nlin}}\|_{L^3_x(\mathbb{R}^3)}$$

while from Sobolev embedding and Hölder one has

$$\|u^{\mathrm{nlin}}\|_{L^6_x(\mathbb{R}^3)} \lesssim \|\nabla u^{\mathrm{nlin}}\|_{L^2_x(\mathbb{R}^3)} \lesssim E(t)^{1/2}$$

and

$$\|\nabla u^{\mathrm{nlin}}\|_{L^3_x(\mathbb{R}^3)} \lesssim \|\nabla u^{\mathrm{nlin}}\|^{1/2}_{L^2_x(\mathbb{R}^3)}\|\nabla^2 u^{\mathrm{nlin}}\|^{1/2}_{L^2_x(\mathbb{R}^3)} \lesssim E(t)^{1/4}\|\nabla^2 u^{\mathrm{nlin}}\|^{1/2}_{L^2_x(\mathbb{R}^3)}.$$

We conclude that

$$\partial_t E(t) \leq -\frac{1}{2}\|\nabla^2 u^{\mathrm{nlin}}\|^2_{L^2_x(\mathbb{R}^3)} + O\left((A^2+E(t))(A^2+E(t)^{1/2}\|\nabla^2 u^{\mathrm{nlin}}\|_{L^2_x(\mathbb{R}^3)})\right)$$

and hence by Young's inequality

$$\partial_t E(t) \leq -\frac{1}{4}\|\nabla^2 u^{\mathrm{nlin}}\|^2_{L^2_x(\mathbb{R}^3)} + O((A^2+E(t))A^2 + (A^2+E(t))^2E(t)). \tag{3.16}$$

In particular we have

$$\partial_t E(t) \leq O(A^4 + A^4E(t) + E(t)^3). \tag{3.17}$$

From (3.13) we have

$$\int_0^1 E(t)\,dt \lesssim A^4,$$

and hence by the pigeonhole principle, we can find a time $t_1 \in [0,1/2]$ such that

$$E(t_1) \lesssim A^4.$$

A standard continuity argument using (3.17) then gives $E(t) \lesssim A^4$ for $t \in [t_1, t_1 + cA^{-8}] = [\tau(0), \tau(1)]$, where $\tau(s) := t_1 + scA^{-8}$ and $c > 0$ is a small absolute constant. Inserting this back into (3.16) one has

$$\partial_t E(t) \leq -\frac{1}{4}\|\nabla^2 u^{\mathrm{nlin}}\|_{L^2_x(\mathbb{R}^3)}^2 + O(A^{12})$$

and hence by the fundamental theorem of calculus

$$\int_{\tau(0)}^{\tau(1)}\int_{\mathbb{R}^3} |\nabla^2 u^{\mathrm{nlin}}|^2\,dxdt \lesssim A^4. \tag{3.18}$$

Thus we have

$$\|\nabla u^{\mathrm{nlin}}\|_{L^\infty_t L^2_x([\tau(0),\tau(1)]\times\mathbb{R}^3)} + \|\nabla^2 u^{\mathrm{nlin}}\|_{L^2_t L^2_x([\tau(0),\tau(1)]\times\mathbb{R}^3)} \lesssim A^2. \tag{3.19}$$

From the Gagliardo-Nirenberg inequality

$$\|u^{\mathrm{nlin}}\|_{L^\infty_x} \lesssim \|\nabla u^{\mathrm{nlin}}\|_{L^2_x}^{1/2}\|\nabla^2 u^{\mathrm{nlin}}\|_{L^2_x}^{1/2} \tag{3.20}$$

and Hölder's inequality, one concludes in particular that

$$\|u^{\mathrm{nlin}}\|_{L^4_t L^\infty_x([\tau(0),\tau(1)]\times\mathbb{R}^3)} \lesssim A^2$$

and hence by (3.11)

$$\|u\|_{L^4_t L^\infty_x([\tau(0),\tau(1)]\times\mathbb{R}^3)} \lesssim A^2; \tag{3.21}$$

also from Sobolev embedding and (3.19) one has

$$\|\nabla u^{\mathrm{nlin}}\|_{L^2_t L^6_x([\tau(0),\tau(1)]\times\mathbb{R}^3)} \lesssim A^2$$

and hence by (3.11)

$$\|\nabla u\|_{L^2_t L^6_x([\tau(0),\tau(1)]\times\mathbb{R}^3)} \lesssim A^2. \tag{3.22}$$

These are subcritical regularity estimates and can now be iterated to obtain even higher regularity. For $t \in [\tau(0.1), \tau(1)]$, we see from (3.7) that

$$u(t) = e^{(t-\tau(0))\Delta}u(\tau(0)) - \int_{\tau(0)}^t e^{(t-t')\Delta}\mathbb{P}\nabla\cdot(u\otimes u)(t')\,dt'. \tag{3.23}$$

From (2.5) the operator $e^{(t-t')\Delta}\mathbb{P}\nabla\cdot$ has norm $O((t-t')^{-1/2})$ on L^∞_x, while $e^{(t-\tau(0))\Delta}$ maps L^3_x to L^∞_x with norm $O((t-\tau(0))^{-1/2}) = O(A^{O(1)})$. We conclude from (3.1) that

$$\|u(t)\|_{L^\infty_x(\mathbb{R}^3)} \lesssim A^{O(1)} + \int_{\tau(0)}^t (t-t')^{-1/2}\|u(t)\|_{L^\infty_x(\mathbb{R}^3)}^2\,dt'.$$

From (3.21) and Young's convolution inequality, we conclude that

$$\|u\|_{L^8_t L^\infty_x([\tau(0.1),\tau(1)]\times\mathbb{R}^3)} \lesssim A^{O(1)}$$

Repeating the above argument, we now also see for $t \in [\tau(0.2), \tau(1)]$ that

$$\|u(t)\|_{L^\infty_x(\mathbb{R}^3)} \lesssim A^{O(1)} + \int_{\tau(0.1)}^t (t-t')^{-1/2}\|u(t)\|_{L^\infty_x(\mathbb{R}^3)}^2\,dt'$$

so from Hölder's inequality we conclude that

$$\|u\|_{L_t^\infty L_x^\infty([\tau(0.2),\tau(1)]\times\mathbb{R}^3)} \lesssim A^{O(1)}. \tag{3.24}$$

Now we differentiate (3.7) to conclude that

$$\nabla u(t) = \nabla e^{(t-\tau(0.2))\Delta} u(\tau(0.2)) - \int_{\tau(0.2)}^t \nabla e^{(t-t')\Delta}\mathbb{P}\nabla\cdot(u\otimes u)(t')\,dt'$$

for $t \in [\tau(0.3),\tau(1)]$. From (3.1), the first term $\nabla e^{(t-\tau(0.2))\Delta}u(\tau(0.2))$ has an L_x^∞ norm of $O(A^{O(1)})$. From (2.5), the operator $\nabla e^{(t-t')\Delta}\mathbb{P}$ maps L_x^6 to L_x^∞ with norm $O((t-t')^{-3/4})$, thus

$$\|\nabla u(t)\|_{L_x^\infty(\mathbb{R}^3)} \lesssim A^{O(1)} + \int_{\tau(0.2)}^t (t-t')^{-3/4}\|\nabla\cdot(u\otimes u)(t')\|_{L_x^6(\mathbb{R}^3)}\,dt'.$$

From (3.22), (3.24), Leibniz and Hölder one has

$$\|\nabla\cdot(u\otimes u)\|_{L_t^2 L_x^6([\tau(0.2),\tau(1)]\times\mathbb{R}^3)} \lesssim A^{O(1)}$$

and hence by fractional integration

$$\|\nabla u\|_{L_t^4 L_x^\infty([\tau(0.3),\tau(1)]\times\mathbb{R}^3)} \lesssim A^{O(1)}.$$

From this, (3.24), Leibniz, and Hölder one has

$$\|\nabla\cdot(u\otimes u)\|_{L_t^4 L_x^\infty([\tau(0.3),\tau(1)]\times\mathbb{R}^3)} \lesssim A^{O(1)}.$$

By (2.5), $\nabla e^{(t-t')\Delta}\mathbb{P}$ has an operator norm of $O((t-t')^{-1/2})$ on L_x^∞, thus

$$\|\nabla u(t)\|_{L_x^\infty(\mathbb{R}^3)} \lesssim A^{O(1)} + \int_{\tau(0.3)}^t (t-t')^{-1/2}\|\nabla\cdot(u\otimes u)(t')\|_{L_x^\infty(\mathbb{R}^3)}\,dt'$$

for $t \in [\tau(0.4),\tau(1)]$, and hence by Hölder's inequality

$$\|\nabla u\|_{L_t^\infty L_x^\infty([\tau(0.4),\tau(1)]\times\mathbb{R}^3)} \lesssim A^{O(1)}.$$

From the vorticity equation (1.5), we now have

$$\partial_t\omega = \Delta\omega + O(A^{O(1)}(|\omega| + |\nabla\omega|))$$

on $[\tau(0.4),\tau(1)]\times\mathbb{R}^3$, and also $\omega = O(A^{O(1)})$ on this slab. Standard parabolic regularity estimates (see e.g., [**LSU**]) then give

$$\|\nabla\omega\|_{L_t^\infty L_x^\infty([\tau(0.5),\tau(1)]\times\mathbb{R}^3)} \lesssim A^{O(1)}.$$

Setting $I' := [\tau(0.5),\tau(1)]$, we obtain the claim (iii). We remark that it is also possible to control higher derivatives $\nabla^j u, \nabla^j\omega$ with $j > 1$, for instance by using parabolic Schauder estimates in Hölder spaces, but we will not need to do so here.

Now we establish (iv). Let t_1, x_1, N_1 be as in that part of the proposition. By rescaling we may normalise $N_1 = 1$, and by translation invariance we may normalise $(t_1,x_1) = (0,0)$, so that $t_0 - T \le -\frac{T}{2} \le -\frac{A_3^2}{2}$, so in particular $[-2A_3, 0] \subset [t_0 - T, t_0]$. From (3.5) we have

$$|P_1 u(0,0)| \ge A_1^{-1}. \tag{3.25}$$

Assume for contradiction that the claim fails, then we have

$$\|P_N u\|_{L_t^\infty L_x^\infty([-A_3,-A_3^{-1}]\times B(0,A_4))} \le A_1^{-1}N$$

for all $A_2^{-1} \leq N \leq A_2$. From (3.2) and the fundamental theorem of calculus in time, we can enlarge the time interval to reach $t = 0$, so that

$$\|P_N u\|_{L_t^\infty L_x^\infty([-A_3,0]\times B(0,A_4))} \lesssim A_1^{-1} N. \tag{3.26}$$

Suppose now that $N \geq A_2^{-1}$. For $t \in [-A_3, 0]$, we can use Duhamel's formula, (3.7), and the triangle inequality to write

$$\begin{aligned}\|P_N u(t)\|_{L_x^{3/2}(B(0,A_4))} &\leq \|e^{(t+2A_3)\Delta} P_N u(-2A_3)\|_{L_x^{3/2}(B(0,A_4))} \\ &\quad + \int_{-2A_3}^t \|e^{(t-t')\Delta} P_N \nabla \cdot (u(t') \otimes u(t'))\|_{L_x^{3/2}(\mathbb{R}^3)}\, dt'.\end{aligned}$$

From (2.4), $e^{(t+2A_3)\Delta} P_N$ has an operator norm of $O(\exp(-N^2 A_3/20))$ on L_x^3, and $e^{(t-t')\Delta} P_N \nabla \cdot$ similarly has an operator norm of $O(N \exp(-N^2(t-t')/20))$ on $L_x^{3/2}$. Applying (3.1) and Hölder's inequality, we conclude that

$$\|P_N u(t)\|_{L_x^{3/2}(B(0,A_4))} \lesssim A A_4 \exp(-N^2 A_3/20) + A^2 N^{-1}$$

and hence in the range $N \geq A_2^{-1}$ we have

$$\|P_N u\|_{L_t^\infty L_x^{3/2}([-A_3,0]\times B(0,A_4))} \lesssim A^2 N^{-1}. \tag{3.27}$$

Now suppose that $N \geq A_2^{-1/2}$. For $t \in [-A_3/2, 0]$, we again use Duhamel's formula, (3.7) and the triangle inequality to write

$$\begin{aligned}\|P_N u(t)\|_{L_x^1(B(0,A_4/2))} &\leq \|e^{(t+A_3)\Delta} P_N u(-A_3)\|_{L_x^1(B(0,A_4/2))} \\ &\quad + \int_{-A_3}^t \|e^{(t-t')\Delta} P_N \nabla \cdot \tilde P_N (u(t') \otimes u(t'))\|_{L_x^1(B(0,A_4/2))}\, dt'.\end{aligned}$$

From (2.4), (3.1), and Hölder as before we have

$$\|e^{(t+A_3)\Delta} P_N u(-A_3)\|_{L_x^1(B(0,A_4/2))} \lesssim A A_4^2 \exp(-N^2 A_3/40).$$

From (2.2) one has

$$\begin{aligned}&\|e^{(t-t')\Delta} P_N \nabla \cdot \tilde P_N (u(t') \otimes u(t'))\|_{L_x^1(B(0,A_4/2))} \lesssim N \exp(-N^2(t-t')/20) \\ &\times \left(\|\tilde P_N(u(t') \otimes u(t'))\|_{L_x^1(B(0,3A_4/4)} + A_4^{-50} A_4^{1/2} \|\tilde P_N (u(t') \otimes u(t'))\|_{L_x^{3/2}(\mathbb{R}^3)}\right)\end{aligned}$$

and hence by (3.1)

$$\begin{aligned}&\|P_N u\|_{L_t^\infty L_x^1([-A_3/2,0]\times B(0,A_4/2))} \\ &\quad \lesssim A_4^{-40} + N^{-1} \|\tilde P_N (u(t') \otimes u(t'))\|_{L_t^\infty L_x^1([-A_3,0]\times B(0,3A_4/4))}.\end{aligned}$$

Since $\tilde P_N(P_{\leq N/100} u(t') \otimes P_{\leq N/100} u(t'))$ vanishes, we can write

$$\tilde P_N(u(t') \otimes u(t')) = \tilde P_N(P_{>N/100} u(t') \otimes u(t')) + \tilde P_N(P_{\leq N/100} u(t') \otimes P_{>N/100} u(t')). \tag{3.28}$$

From (2.2), (3.1) we have

$$\begin{aligned}&\|\tilde P_N(P_{>N/100} u(t') \otimes u(t'))\|_{L_t^\infty L_x^1([-A_3,0]\times B(0,3A_4/4))} \\ &\quad \lesssim \|P_{>N/100} u(t') \otimes u(t')\|_{L_t^\infty L_x^1([-A_3,0]\times B(0,A_4))} + A_4^{-40}.\end{aligned}$$

From (3.27) (and the triangle inequality) as well as (3.1) and Hölder's inequality, we thus have

$$\|\tilde P_N(P_{>N/100} u(t') \otimes u(t'))\|_{L_t^\infty L_x^1([-A_3,0]\times B(0,3A_4/4))} \lesssim A^3 N^{-1}.$$

Similarly for the other component of (3.28). We conclude that

$$\|P_N u\|_{L_t^\infty L_x^1([-A_3/2,0]\times B(0,A_4/2))} \lesssim A^3 N^{-2} \tag{3.29}$$

for all $N \geq A_2^{-1/2}$.

Now suppose that $A_2^{-1/3} \leq N \leq A_2^{1/3}$. For $t \in [-A_3/3, 0]$, we again use Duhamel's formula, (3.7), and the triangle inequality as before to write

$$\|P_N u(t)\|_{L_x^2(B(0,A_4/4))} \leq \|e^{(t+A_3/2)\Delta} P_N u(-A_3/2)\|_{L_x^2(B(0,A_4/4))} + \int_{-A_3/2}^{t} \|e^{(t-t')\Delta} P_N \nabla \cdot \tilde{P}_N(u(t') \otimes u(t'))\|_{L_x^2(B(0,A_4/4))}\, dt'.$$

Arguing as before we have

$$\|e^{(t+A_3/2)\Delta} P_N u(-A_3/2)\|_{L_x^2(B(0,A_4/4))} \lesssim A A_4^{1/2} \exp(-N^2 A_3/120)$$

and

$$\|e^{(t-t')\Delta} P_N \nabla \cdot \tilde{P}_N(u(t') \otimes u(t'))\|_{L_x^2(B(0,A_4/4))} \lesssim N^{5/2} \exp(-N^2(t-t')/20) \left(\|\tilde{P}_N(u(t') \otimes u(t'))\|_{L_x^1(B(0,A_4/3))} + A_4^{-50} N^{-1} \|\tilde{P}_N(u(t') \otimes u(t'))\|_{L_x^{3/2}(\mathbb{R}^3)}\right)$$

and thus

$$\begin{aligned} &\|P_N u\|_{L_t^\infty L_x^2([-A_3/4,0]\times B(0,A_4/4))} \\ &\quad\lesssim A_4^{-40} + N^{1/2} \|\tilde{P}_N(u(t') \otimes u(t'))\|_{L_t^\infty L_x^1([-A_3/2,0]\times B(0,A_4/3))}. \end{aligned} \tag{3.30}$$

We can split $\tilde{P}_N(u(t') \otimes u(t'))$ into $O(1)$ paraproduct terms of the form $\tilde{P}_N(P_{N'} u(t') \otimes P_{\leq N/100} u(t'))$ where $N' \sim N$, $O(1)$ terms of the form $\tilde{P}_N(P_{\leq N/100} u(t') \otimes P_{N'} u(t'))$, and a sum of the form $\sum_{N_1 \sim N_2 \gtrsim N} \tilde{P}_N(P_{N_1} u(t') \otimes P_{N_2} u(t'))$. For the "high-low" term $\tilde{P}_N(P_{N'} u(t') \otimes P_{\leq N/100} u(t'))$, we observe from (3.26), (3.2) and the triangle inequality that

$$\|P_{\leq N/100} u\|_{L_t^\infty L_x^\infty([-A_3,0]\times B(0,A_4))} \lesssim A_1^{-1} N.$$

Using this, (2.2), (3.29) (for the high frequency factor $P_{N'} u(t')$), and Hölder's inequality, we conclude that the contribution of this term to (3.30) is $O(A^3 A_1^{-1} N^{-1/2})$. Similarly for the "low-high" term $\tilde{P}_N(P_{\leq N/100} u(t') \otimes P_{N'} u(t'))$. Finally, to control the "high-high" term $\sum_{N_1 \sim N_2 \gtrsim N} \tilde{P}_N(P_{N_1} u(t') \otimes P_{N_2} u(t'))$, we use (2.2), the triangle inequality, Hölder, and (3.29) to control this contribution by

$$\lesssim A_4^{-40} + N^{1/2} \sum_{N_1 \sim N_2 \gtrsim N} A^3 N_1^{-2} \|P_{N_2} u\|_{L_t^\infty L_x^{3/2}([-A_3,0]\times B(0,A_4))}.$$

Using (3.27) when $N_2 \leq A_2$ and (3.2) otherwise, we see that this term also contributes $O(A^3 A_1^{-1} N^{-1/2})$. We have thus shown that

$$\|P_N u\|_{L_t^\infty L_x^2([-A_3/4,0]\times B(0,A_4/4))} \lesssim A^3 A_1^{-1} N^{-1/2} \tag{3.31}$$

for $A_2^{-1/3} \leq N \leq A_2^{1/3}$.

We now return once again to Duhamel's formula to estimate

$$|P_1 u(0,0)| \leq |e^{A_3 \Delta/4} P_1 u(-A_3/4)|(0) + \int_{-A_3/4}^{0} |e^{(t-t')\Delta} P_1 \nabla \cdot \tilde{P}_1(u(t') \otimes u(t'))|(0)\, dt'.$$

From (2.4), (3.1), the first term is $O(A_3 \exp(-A_3^2/320))$, thus from (3.25) we have

$$\int_{-A_3/4}^0 |e^{(t-t')\Delta} P_1 \nabla \cdot \tilde{P}_1(u(t') \otimes u(t'))|(0)\, dt' \gtrsim A_1^{-1}.$$

From (2.2), (3.1) one has

$$\begin{aligned} &|e^{(t-t')\Delta} P_1 \nabla \cdot \tilde{P}_1(u(t') \otimes u(t'))|(0) \\ &\lesssim \exp(-(t-t')/20)(\|\tilde{P}_1(u(t') \otimes u(t'))\|_{L^1_x(B(0,A_1))} + A_1^{-50}) \end{aligned}$$

and hence by the pigeonhole principle we have

$$\|\tilde{P}_1(u(t') \otimes u(t'))\|_{L^1_x(B(0,A_1))} \gtrsim A_1^{-1}.$$

for some $-A_3/4 \leq t' \leq 0$.

Fix this t'. As before, we can split $\tilde{P}_1(u(t') \otimes u(t'))$ into the sum of $O(1)$ "low-high" terms $\tilde{P}_1(P_{N'}u(t') \otimes P_{\leq 1/100}u(t'))$ and "high-low" terms $\tilde{P}_1(P_{\leq 1/100}u(t') \otimes P_{N'}u(t'))$ with $N' \sim 1$, plus a "high-high" term $\sum_{N_1 \sim N_2 \gtrsim 1} \tilde{P}_1(P_{N_1}u(t') \otimes P_{N_2}u(t'))$. For the first two types of terms, we use (2.2) (for frequencies larger than $A_2^{-1/3}$), (3.1), and Hölder to conclude that

$$\|P_{\leq 100}u(t')\|_{L^2_x(B(0,2A_1))} \lesssim A^3 A_1^{-1}$$

and then from (3.31), (2.2) (and (3.1) to control the global contribution of (2.2)) we see that the contribution of those two types of terms is $O(A^6 A_1^{-2})$. For the high-high terms with $N_1, N_2 \leq A_2^{1/3}$, we again use (3.31), (2.2), (3.1) to again obtain a bound of $O(A^6 A_1^{-2})$. For the cases when $N_1, N_2 \gtrsim A_2^{1/3}$, we use (3.27), (3.1) to obtain a much better bound $O(A^3 A_2^{-1/3})$. Putting all this together we obtain

$$A_1^{-1} \lesssim A^6 A_1^{-2}$$

giving the required contradiction. This establishes (iv).

Now we prove (v). We may assume that $A_4 N_0^{-2} \leq A_4^{-1} T$, since the claim is trivial otherwise. Thus we have $N_0 \geq A_4 T^{-1/2}$.

By iteratively applying (iv), we may find a sequence $(t_0, x_0), (t_1, x_1), \ldots, (t_n, x_n) \in [t_0 - T, t_0]$ and $N_0, N_1, \ldots, N_n > 0$ for some $n \geq 1$, with the properties

$$|P_{N_i} u(t_i, x_i)| \geq A_1^{-1} N_i \tag{3.32}$$

$$A_2^{-1} N_{i-1} \leq N_i \leq A_2 N_{i-1} \tag{3.33}$$

$$A_3^{-1} N_{i-1}^{-2} \leq t_{i-1} - t_i \leq A_3 N_{i-1}^{-2} \tag{3.34}$$

$$|x_i - x_{i-1}| \leq A_4 N_{i-1}^{-1} \tag{3.35}$$

for all $i = 1, \ldots, n$, with $t_i \in [t_0 - T/2, t_0]$ and $N_i \geq A_3 T^{-1/2}$ for $i = 0, \ldots, n-1$ and either $t_n \in [t_0 - T, t_0 - T/2]$ or $N_n < A_3 T^{-1/2}$. To see that this process terminates at a finite n, observe from the classical nature of u that the $P_{N_i} u(t_i, x_i)$ are uniformly bounded in i, which by (3.32) implies that the N_i are uniformly bounded above, and hence by (3.34) $t_{i-1} - t_i$ are uniformly bounded below; since t_i must stay above $t_0 - T$, we obtain the required finite time termination. By (3.34), the first time t_1 after t_0 lies in the interval

$$t_1 \in [t_0 - A_2 N_0^{-2}, t_0 - A_2^{-1} N_0^{-2}].$$

If $N_n < A_3 T^{-1/2}$, then by (3.34), (3.33)

$$t_{n-1} - t_n \geq A_3^{-1} N_{n-1}^{-2} \geq A_3^{-2} N_n^{-2} \geq A_3^{-4}(t_{n-1} - t_n) \leq A_3^{-4} T$$

so in particular $t_n \leq t_0 - A_3^{-4} T$. Of course this inequality also holds if $t_n \in [t_0 - T, t_0 - T/2]$. In either case, we see from the hypothesis $A_4 N_0^{-2} \leq T_1 \leq A_4^{-1} T$ that

$$t_n < t - T_1 \leq t_1.$$

Let m be the largest index for which $t_m \geq t - T_1$, thus $1 \leq m \leq n-1$ and $t_{m+1} > t - T_1$. By telescoping (3.34), we conclude that

$$\sum_{i=0}^{m} A_3 N_i^{-2} = \sum_{i=1}^{m+1} A_3 N_{i-1}^{-2} \geq t - t_{m+1} \geq T_1. \tag{3.36}$$

On the other hand, from (3.32) and (3.2) we have

$$|P_{N_i} u(t, x_i)| \gtrsim A_1^{-1} N_i$$

for $t \in [t_i - A_1^{-2} N_i^{-2}, t_i]$; as P_{N_i} is bounded on L^∞ by (2.3), this implies that

$$\|P_{N_i} u(t)\|_{L^\infty_x(\mathbb{R}^3)} \gtrsim A_1^{-1} N_i$$

for such t. From (3.34) we see that the time intervals $[t_i - A_1^{-2} N_i^{-2}, t_i]$ are disjoint and lie in $[t - T_1, t]$ for $i = 0, \ldots, m-1$. Applying (3.4), we conclude that

$$\sum_{i=0}^{m-1} A_1^{-1} N_i \times A_1^{-2} N_i^{-2} \lesssim A^4 T_1^{1/2}$$

and thus

$$\sum_{i=0}^{m-1} N_i^{-1} \lesssim A_1^4 T_1^{1/2}.$$

Using (3.33) to extend this sum to the final index m, we conclude that

$$\sum_{i=0}^{m} N_i^{-1} \lesssim A_2^2 T_1^{1/2}. \tag{3.37}$$

Comparing this with (3.36), we conclude that there exists $i = 0, \ldots, m$ such that

$$N_i^{-1} \gtrsim A_3^{-2} T_1^{1/2}.$$

Since $A_4 N_0^{-2} \leq A_4^{-1} T$, i cannot be zero, thus $1 \leq i \leq m$. From (3.34), (3.33) we have

$$\begin{aligned} t_0 - t_i &\geq t_{i-1} - t_i \\ &\geq A_3^{-1} N_{i-1}^{-2} \\ &\geq A_3^{-2} N_i^{-2} \\ &\gtrsim A_3^{-6} T_1. \end{aligned}$$

Since $t_0 - t_i$ is also bounded by T_1, we also have from (3.34) that $A_3^{-1} N_i^{-2} \leq T_1$, thus $N_i \geq A_3^{-1/2} T_1^{-1/2}$. Finally, from telescoping (3.35) and using (3.37), we conclude that

$$|x_i - x_0| \lesssim A_4^2 T_1^{1/2},$$

and the claim follows.

Finally, we prove (vi), which is the most difficult estimate. The claim is invariant with respect to time translation and rescaling, so we may assume that $[t_0 - T', t_0] = [0, 1]$. In particular $[-1, 1] \subset [t_0 - T, t_0]$, so we may decompose $u = u^{\rm lin} + u^{\rm nlin}$ as before with the estimates (3.10), (3.11), (3.13).

From (3.13) we can find a time $t_1 \in [-1/2, 0]$ such that

$$\int_{\mathbb{R}^3} |\nabla u^{\rm nlin}(t_1, x)|^2 \, dx \lesssim A^4.$$

Fix this time t_1. From (3.11) we thus have

$$\int_{\mathbb{R}^3} |\nabla u^{\rm nlin}(t_1, x)|^2 + \sum_{j=0}^{4} |\nabla^j u^{\rm lin}(t_1, x)|^3 \, dx \lesssim A^4.$$

By the pigeonhole principle, we can thus find a scale

$$A_6^{100} R_0 \leq R \leq \exp(A_6^{O(1)}) R_0 \tag{3.38}$$

such that

$$\int_{A_6^{-10} R \leq |x| \leq A_6^{10} R} |\nabla u^{\rm nlin}(t_1, x)|^2 + \sum_{j=0}^{4} |\nabla^j u^{\rm lin}(t_1, x)|^3 \, dx \lesssim A_6^{-10}. \tag{3.39}$$

Fix this R. We now propagate this estimate forward in time to $[t_1, 1]$. We first achieve this for the linear component $u^{\rm lin}$, which is straightforward. From Sobolev embedding we have

$$\sup_{A_6^{-9} R \leq |x| \leq A_6^9 R} |\nabla^j u^{\rm lin}(t_1, x)| \lesssim A_6^{-3}$$

for $j = 0, 1, 2$. Since $\nabla^j u^{\rm lin}$ solves the linear heat equation, we conclude from this, (2.2), and (3.11) that

$$\sup_{t_1 \leq t \leq 1} \sup_{A_6^{-8} R \leq |x| \leq A_6^8 R} |\nabla^j u^{\rm lin}(t, x)| \lesssim A_6^{-3} \tag{3.40}$$

for $j = 0, 1, 2$. This estimate (when combined with (3.11)) will suffice to control all the terms involving the linear component $u^{\rm lin}$ of the velocity (or the analogous component $\omega^{\rm lin} := \nabla \times u^{\rm lin}$ of the vorticity).

The vorticity $\omega := \nabla \times u$ obeys the vorticity equation (1.5). On $[t_1, 1] \times \mathbb{R}^3$, we decompose $\omega = \omega^{\rm lin} + \omega^{\rm nlin}$, where $\omega^{\rm lin} := \nabla \times u^{\rm lin}$ is the linear component of the vorticity and $\omega^{\rm nlin} := \nabla \times u^{\rm nlin}$ is the nonlinear component. As $\omega^{\rm lin}$ solves the heat equation, we have

$$\partial_t \omega^{\rm nlin} = \Delta \omega^{\rm nlin} - (u \cdot \nabla)\omega + (\omega \cdot \nabla) u. \tag{3.41}$$

As in [**T**, §10], we apply the energy method to this equation with a carefully chosen time-dependent cutoff function. Namely, let

$$R_- \in [A_6^{-8} R, 2A_6^{-8} R]; \quad R_+ \in [A_6^8 R/2, A_6^8 R] \tag{3.42}$$

be scales to be chosen later, and define the time-dependent radii

$$R_-(t) := R_- + C_0 \int_{t_1}^{t} (A_6 + \|u(t)\|_{L^\infty_x(\mathbb{R}^3)}) \, dt$$

$$R_+(t) := R_+ - C_0 \int_{t_1}^{t} (A_6 + \|u(t)\|_{L^\infty_x(\mathbb{R}^3)}) \, dt$$

that start at R_-, R_+ respectively, and contract inwards at a rate faster than the velocity field u. From the bounded total speed property (3.4), (3.38), and the hypothesis $R_0 \geq 1$, we conclude that

$$R_-(t) \in [A_6^{-8} R, 3A_6^{-8} R]; \quad R_+(t) \in [A_6^8 R/3, A_6^8 R]$$

for all $t \in [t_1, 1]$.

For $t \in [t_1, 1]$, we define the local enstrophy

$$E(t) := \frac{1}{2} \int_{\mathbb{R}^3} |\omega^{\text{nlin}}(t,x)|^2 \eta(t,x)\, dx$$

where η is the time-varying cutoff

$$\eta(t,x) := \max(\min(A_6, |x| - R_-(t), R_+(t) - |x|), 0),$$

thus η is supported in the annulus $\{R_-(t) \leq |x| \leq R_+(t)\}$, is Lipschitz with norm 1, and equals A_6 in the smaller annulus $\{R_-(t) + A_6 \leq |x| \leq R_+(t) - A_6\}$. From (3.39) we have the initial bound

$$E(t_1) \lesssim A_6^{-9}. \tag{3.43}$$

Now we control the time derivative $\partial_t E(t)$ for $t \in [t_1, 1]$. From (3.41) and integration by parts we have

$$\partial_t E(t) = -Y_1(t) - Y_2(t) + Y_3(t) + Y_4(t) + Y_5(t) + Y_6(t) + Y_7(t) + Y_8(t) + Y_9(t)$$

where Y_1 is the dissipation term

$$Y_1(t) := \int_{\mathbb{R}^3} |\nabla \omega^{\text{nlin}}(t,x)|^2\, dx,$$

$Y_2(t)$ is the recession term

$$Y_2(t) := -\frac{1}{2} \int_{\mathbb{R}^3} |\omega^{\text{nlin}}(t,x)|^2 \partial_t \eta(t,x)\, dx,$$

$Y_3(t)$ is the heat flux term

$$Y_3(t) := \frac{1}{2} \int_{\mathbb{R}^3} |\omega^{\text{nlin}}(t,x)|^2 \Delta \eta(t,x)\, dx,$$

$Y_4(t)$ is the transport term

$$Y_4(t) := \frac{1}{2} \int_{\mathbb{R}^3} |\omega^{\text{nlin}}(t,x)|^2 u(t,x) \cdot \nabla \eta(t,x)\, dx,$$

$Y_5(t)$ is a correction to the transport term arising from ω^{lin},

$$Y_5(t) := -\int_{\mathbb{R}^3} \omega^{\text{nlin}}(t,x) \cdot (u(t,x) \cdot \nabla) \omega^{\text{lin}}(t,x)\, \eta(t,x)\, dx,$$

$Y_6(t)$ is the main nonlinear term

$$Y_6(t) := \int_{\mathbb{R}^3} \omega^{\text{nlin}}(t,x) \cdot (\omega^{\text{nlin}}(t,x) \cdot \nabla) u^{\text{nlin}}(t,x)\, \eta(t,x)\, dx$$

and $Y_7(t), Y_8(t), Y_9(t)$ are corrections to the transport term arising from the u^{lin} and ω^{lin},

$$Y_7(t) := \int_{\mathbb{R}^3} \omega^{\mathrm{nlin}}(t,x) \cdot (\omega^{\mathrm{nlin}}(t,x) \cdot \nabla) u^{\mathrm{lin}}(t,x)\, \eta(t,x)\, dx$$

$$Y_8(t) := \int_{\mathbb{R}^3} \omega^{\mathrm{nlin}}(t,x) \cdot (\omega^{\mathrm{lin}}(t,x) \cdot \nabla) u^{\mathrm{nlin}}(t,x)\, \eta(t,x)\, dx$$

$$Y_9(t) := \int_{\mathbb{R}^3} \omega^{\mathrm{nlin}}(t,x) \cdot (\omega^{\mathrm{lin}}(t,x) \cdot \nabla) u^{\mathrm{lin}}(t,x)\, \eta(t,x)\, dx.$$

Here all derivatives of the Lipschitz function η are interpreted in a distributional sense. We now aim to control $Y_3(t), \ldots, Y_9(t)$ in terms of $Y_1(t), Y_2(t), E(t)$, and some other quantities that are well controlled. From definition of η we see that

$$-\partial_t \eta(t,x) = C_0(A_6 + \|u(t)\|_{L^\infty_x(\mathbb{R}^3)}) |\nabla \eta(t,x)|$$

so in particular we have that $Y_2(t)$ is non-negative and

$$Y_4(t) \leq C_0^{-1} Y_2(t).$$

A direct computation of $\Delta\eta$ in polar coordinates yields the bound

$$\begin{aligned} Y_3(t) \lesssim & \int_{|x| \in [R_-(t), R_-(t)+A_6] \cup [R_+(t)-A_6, R_+(t)]} \frac{|\omega^{\mathrm{nlin}}(t,x)|^2}{|x|}\, dx \\ & + \sum_{r = R_-(t), R_-(t)+A_6, R_+(t)-A_6, R_+(t)} r^2 \int_{S^2} |\omega^{\mathrm{nlin}}(t, r\theta)|\, d\theta \end{aligned}$$

where $d\theta$ is surface measure on the sphere (in fact the $r = R_-(t) + A_6, R_+(t) - A_6$ terms are non-positive and could be discarded if desired). This expression is difficult to estimate for fixed choices of R_-, R_+. However, if selects R_-, R_+ uniformly at random from the range (3.42), we see from Fubini's theorem that the expected value $\mathbb{E}|Y_3|$ of $|Y_3|$ can be estimated by

$$\mathbb{E}|Y_3(t)| \lesssim A_6 \int_{|x| \in [A_6^{-8} R, 3A_6^{-8} R] \cup [A_6^8 R/3, A_6^8 R]} \frac{|\omega^{\mathrm{nlin}}(t,x)|^2}{|x|^2}\, dx$$

and hence by (3.13), (3.38)

$$\mathbb{E} \int_{t_1}^1 |Y_3(t)|\, dt \lesssim A_6^{-10}$$

(say). Thus we can select R_-, R_+ so that

$$\int_{t_1}^1 |Y_3(t)|\, dt \lesssim A_6^{-10} \tag{3.44}$$

and we shall now do so.

To treat $Y_5(t)$, we use Young's inequality to bound

$$Y_5(t) \lesssim E(t) + \int_{\mathbb{R}^3} |(u \cdot \nabla) \omega^{\mathrm{lin}}|^2\, \eta\, dx.$$

Using (3.1), (3.11), (3.40), Hölder's inequality, we then have

$$Y_5(t) \lesssim E(t) + A_6^{-2}$$

(say).

In a similar vein, from (3.40) and Hölder's inequality one has

$$Y_7(t) \lesssim E(t)$$

(with plenty of room to spare) and from Young's inequality one has

$$Y_9(t) \lesssim E(t) + \int_{\mathbb{R}^3} |(\omega^{\mathrm{lin}} \cdot \nabla) u^{\mathrm{lin}}|^2 \, \eta(t,x) \, dx$$

and hence by (3.1), (3.11), (3.40), and Hölder

$$Y_9(t) \lesssim E(t) + A_6^{-2}.$$

For Y_8, we again use Young's inequality to bound

$$Y_8(t) \lesssim E(t) + \int_{\mathbb{R}^3} |(\omega^{\mathrm{lin}} \cdot \nabla) u^{\mathrm{nlin}}|^2 \, \eta(t,x) \, dx$$

and hence by (3.40)

$$Y_8(t) \lesssim E(t) + Y_{10}(t)$$

where

$$Y_{10}(t) := A_6^{-3} \int_{\mathbb{R}^3} |\nabla u^{\mathrm{nlin}}(t,x)|^2 \, dx.$$

Observe from (3.13) that

$$\int_{t_1}^{1} |Y_{10}(t)| \, dt \lesssim A_6^{-2}. \tag{3.45}$$

We are left with estimation of the most difficult term $Y_6(t)$. Following [**T**], we cover the annulus $\{R_-(t) \leq |x| \leq R_+(t)\}$ by a boundedly overlapping Whitney decomposition of balls $B = B(x_B, r_B)$, where the radius r_B of the ball is given as $r_B := \frac{1}{100}\eta(t, r_B)$. In particular, we have $\eta(t,x) \sim r_B$ on the dilate $10B = B(x_B, r_B)$ of the ball. We can then write

$$Y_6(t) \sim \sum_B r_B \int_B |\omega^{\mathrm{nlin}}|^2 |\nabla u^{\mathrm{nlin}}| \, dx$$

where we suppress the explicit dependence on t, x for brevity. Similarly one has

$$E(t) \sim \sum_B r_B \int_{10B} |\omega^{\mathrm{nlin}}|^2 \, dx \tag{3.46}$$

and

$$Y_1(t) \sim \sum_B r_B \int_{10B} |\nabla \omega^{\mathrm{nlin}}|^2 \, dx \tag{3.47}$$

To control $Y_6(t)$, we need to control ∇u^{nlin}. The Biot-Savart law suggests that this function has comparable size to ω^{nlin}, but we need to localise this intuition to the ball B and thus must address the slightly non-local nature of the Biot-Savart law. Fortunately this can be handled using standard cutoff functions. Namely, we have $\Delta u^{\mathrm{nlin}} = -\nabla \times \omega^{\mathrm{nlin}}$, hence if we let ψ_B be a smooth cutoff adapted to $3B$ that equals 1 on $2B$, then

$$u^{\mathrm{nlin}} = -\Delta^{-1}(\nabla \times (\omega^{\mathrm{nlin}} \psi_B)) + v$$

where v is harmonic on $2B$. From Sobolev embedding and Hölder one has

$$\|v\|_{L^2_x(2B)} \lesssim \|\omega^{\mathrm{nlin}} \psi_B\|_{L^{6/5}_x(\mathbb{R}^3)} + \|u^{\mathrm{nlin}}\|_{L^2_x(2B)} \lesssim r_B^{3/2} \|\omega^{\mathrm{nlin}}\|_{L^3_x(3B)} + \|u^{\mathrm{nlin}}\|_{L^2_x(2B)}$$

and hence by elliptic regularity for harmonic functions

$$\|\nabla v\|_{L^\infty_x(B)} \lesssim r_B^{-5/2}\|v\|_{L^2_x(2B)} \lesssim r_B^{-1}\|\omega^{\rm nlin}\|_{L^3_x(3B)} + r_B^{-5/2}\|u^{\rm nlin}\|_{L^2_x(2B)}.$$

We conclude the pointwise estimate

(3.48)

$$\nabla u^{\rm nlin} = -\nabla\Delta^{-1}(\nabla\times(\omega^{\rm nlin}\psi_B)) + O(r_B^{-1}\|\omega^{\rm nlin}\|_{L^3_x(3B)}) + O(r_B^{-5/2}\|u^{\rm nlin}\|_{L^2_x(2B)})$$

on B. By elliptic regularity, $\nabla\Delta^{-1}(\nabla\times(\omega^{\rm nlin}\psi_B))$ has an $L^3_x(B)$ norm of $O(\|\omega^{\rm nlin}\|_{L^3_x(3B)})$. From Hölder's inequality we thus have

$$\int_B |\omega^{\rm nlin}|^2|\nabla u^{\rm nlin}|\,dx \lesssim \|\omega^{\rm nlin}\|^3_{L^3_x(3B)} + r_B^{-5/2}\|\omega^{\rm nlin}\|^2_{L^2_x(3B)}\|u^{\rm nlin}\|_{L^2_x(3B)}$$

and hence $Y_6(t) \lesssim Y_{6,1}(t) + Y_{6,2}(t)$, where

$$Y_{6,1}(t) := \sum_B r_B\|\omega^{\rm nlin}\|^3_{L^3_x(3B)}$$

and

$$Y_{6,2}(t) := \sum_B r_B^{-3/2}\|\omega^{\rm nlin}\|^2_{L^2_x(3B)}\|u^{\rm nlin}\|_{L^2_x(3B)}.$$

For $Y_{6,2}(t)$, we first consider the contribution of the large balls in which $r_B \geq A^{10}$. Here we simply use (3.10) to bound $\|u^{\rm nlin}\|_{L^2_x(3B)} \lesssim A^2$. Since $r_B^{-3/2}A^2 \lesssim r_B$ for large balls B, the contribution of this case is $O(E(t))$ thanks to (3.46). Now we look at the small balls in which $r_B < A^{10}$. Here we use Hölder to bound

$$\|u^{\rm nlin}\|_{L^2_x(3B)} \lesssim r_B^{3/2}\|u^{\rm nlin}\|_{L^\infty_x(\mathbb{R}^3)} \lesssim r_B^{3/2}(A^2 + \|u\|_{L^\infty_x(\mathbb{R}^3)})$$

so the contribution of this case is bounded by

$$\sum_{B:r_B<A^{10}} \|\omega^{\rm nlin}\|^2_{L^2_x(3B)}(A^2 + \|u\|_{L^\infty_x(\mathbb{R}^3)}).$$

For small balls B, $3B$ is completely contained inside the region in which $\partial_t\eta \gtrsim C_0(A_6+\|u\|_{L^\infty_x(\mathbb{R}^3)})$, so the contribution of this case can be bounded by $O(C_0^{-1}Y_2(t))$. Thus

$$Y_{6,2}(t) \lesssim E(t) + C_0^{-1}Y_2(t).$$

Now we control $Y_{6,1}(t)$. For each ball B, define the mean vorticity ω_B by

$$\omega_B := \frac{\int_{\mathbb{R}^3}\omega^{\rm nlin}\psi_B\,dx}{\int_{\mathbb{R}^3}\psi_B\,dx}.$$

From the Poincaré inequality, Sobolev embedding, and the triangle inequality we have

$$\|\omega^{\rm nlin} - \omega_B\|_{L^6_x(3B)} \lesssim \|\nabla\omega^{\rm nlin}\|_{L^2_x(10B)}$$

and similarly

$$|\omega_B - \omega_{B'}| \lesssim r_B^{-1/2}\|\nabla\omega^{\rm nlin}\|_{L^2_x(10B)} \tag{3.49}$$

when B, B' are overlapping Whitney balls. We can now use Hölder's inequality to bound

$$\begin{aligned} Y_{6,1}(t) &\lesssim \sum_B r_B^4\omega_B^3 + \sum_B r_B\|\omega^{\rm nlin} - \omega_B\|^3_{L^3_x(3B)} \\ &\lesssim \sum_B r_B^4\omega_B^3 + \sum_B r_B\|\omega^{\rm nlin} - \omega_B\|^{3/2}_{L^2_x(3B)}\|\nabla\omega^{\rm nlin}\|^{3/2}_{L^2_x(10B)}. \end{aligned}$$

By Young's inequality and (3.47), we then have

$$Y_{6,1}(t) \leq \frac{1}{2} Y_1(t) + O(\sum_B r_B^4 \omega_B^3 + \sum_B r_B \|\omega^{\mathrm{nlin}} - \omega_B\|_{L^2_x(3B)}^6).$$

From the triangle inequality and Cauchy-Schwarz, and (3.46), one has

$$\|\omega^{\mathrm{nlin}} - \omega_B\|_{L^2_x(3B)} \lesssim \|\omega^{\mathrm{nlin}}\|_{L^2_x(3B)} \lesssim r_B^{-1/2} E(t)^{1/2}$$

and hence from Hölder, (3.47)

$$\sum_B r_B \|\omega^{\mathrm{nlin}} - \omega_B\|_{L^2_x(3B)}^6 \lesssim \sum_B r_B \|\omega^{\mathrm{nlin}} - \omega_B\|_{L^6_x(3B)}^2 E(t)^2 \lesssim E(t)^2 Y_1(t).$$

Now we estimate $\sum_B r_B^4 \omega_B^3$. We can arrange the Whitney decomposition so that all the radii r_B are powers of 1.001, and that every ball B of radius less than (say) $A_6/100$ has a "parent" ball $p(B)$ that overlaps B and has radius $1.001 r_B$. From the triangle inequality we have

$$|\omega_B| \leq |\omega_{p^k(B)}| + \sum_{i=0}^{k-1} |\omega_{p^i(B)} - \omega_{p^{i+1}(B)}|$$

for any Whitney ball B, where $k = k_B$ is the first natural number for which the iterated parent $p^k(B)$ has radius larger than $A_6^{1/2}$. By Hölder we then have

$$|\omega_B|^3 \lesssim |\omega_{p^k(B)}|^3 + \sum_{i=0}^{k-1} (1+i)^{10} |\omega_{p^i(B)} - \omega_{p^{i+1}(B)}|^3.$$

From a volume packing argment we see that for a given i, a Whitney ball B' is of the form $p^i(B)$ for at most $O((1.001)^{2i})$ choices of B. One can then sum the geometric series (exactly as in [**T**, §10]) and conclude that

$$\sum_B r_B^4 \omega_B^3 \lesssim \sum_{B: r_B \geq A_6/100} r_B^4 \omega_B^3 + \sum_{B: r_B < A_6^{1/2}} r_B^4 |\omega_B - \omega_{p(B)}|^3.$$

For the small balls in which $r_B < A_6/100$, we observe from (3.47) and Cauchy-Schwarz that

$$\omega_B, \omega_{p(B)} \lesssim r_B^{-2} E(t)^{1/2}$$

and thus from (3.49), (3.47)

$$\sum_{B: r_B < A_6/100} r_B^4 |\omega_B - \omega_{p(B)}|^3 \lesssim E(t)^{1/2} Y_1(t).$$

For the large balls in which $r_B \geq A_6/100$, we write $\omega^{\mathrm{nlin}} = \nabla \times u^{\mathrm{nlin}}$ and integrate by parts using Cauchy-Schwarz to find that

$$\omega_B \lesssim r_B^{-5/2} \|u^{\mathrm{nlin}}\|_{L^2_x(3B)}$$

and hence using (3.10) and the bounded overlap of the Whitney balls

$$\sum_{B: r_B \geq A_6/100} r_B^4 \omega_B^3 \lesssim \sum_{B: r_B \geq A_6/100} A^2 r_B^{-7/2} \|u^{\mathrm{nlin}}\|_{L^2_x(3B)}^2 \lesssim A^4 A_6^{-7/2} \lesssim A_6^{-2}.$$

Thus we have

$$Y_{6,1} \leq \frac{1}{2} Y_1(t) + O(E(t)^{1/2} Y_1(t) + A_6^{-2} + E(t)^2 Y_1(t)).$$

Putting all this together, we see that

$$\partial_t E(t) \leq -\frac{1}{2} Y_1(t) + O\left(E(t) + |Y_3(t)| + |Y_{10}(t)| + A_6^{-2} + E(t)^{1/2} Y_1(t) + E(t)^2 Y_1(t)\right).$$

A standard continuity argument using (3.43), (3.44), (3.45) then gives

$$E(t) \lesssim A_6^{-2} \tag{3.50}$$

for all $t_1 \leq t \leq 1$, and also

$$\int_{t_1}^1 Y_1(t)\, dt \lesssim A_6^{-2}. \tag{3.51}$$

These are subcritical regularity estimates and can now be iterated[4] as in the proof of (iii) to obtain higher regularity. First we move from control of the vorticity back to control of the velocity. From (3.48) and elliptic regularity one has

$$\|\nabla u^{\mathrm{nlin}}\|_{L^2_x(B)}^2 \lesssim \|\omega^{\mathrm{nlin}}\|_{L^2_x(3B)}^2 + r_B^{-2} \|u^{\mathrm{nlin}}\|_{L^2_x(2B)}^2$$

for any ball B; summing this on balls of radius A_6^{10} (say) using (3.50), (3.10), we conclude that

$$\int_{A_6^{-7} R \leq |x| \leq A_6^7 R} |\nabla u^{\mathrm{nlin}}(t,x)|^2\, dx \lesssim A_6^{-2} \tag{3.52}$$

for all $t_1 \leq t \leq 1$. Similarly we have

$$\|\nabla^2 u^{\mathrm{nlin}}\|_{L^2_x(B)}^2 \lesssim \|\nabla \omega^{\mathrm{nlin}}\|_{L^2_x(3B)}^2 + r_B^{-2} \|\nabla u^{\mathrm{nlin}}\|_{L^2_x(2B)}^2$$

and using (3.52), (3.51) in place of (3.10), (3.50) we conclude that

$$\int_{t_1}^1 \int_{A_6^{-6} R \leq |x| \leq A_6^6 R} |\nabla^2 u^{\mathrm{nlin}}(t,x)|^2\, dx dt \lesssim A_6^{-2}.$$

Using the Gagliardo-Nirenberg inequality (3.20) as before we see that

$$\|u^{\mathrm{nlin}}\|_{L^4_t L^\infty_x([t_1,1] \times \{A_6^{-5} R \leq |x| \leq A_6^5 R\})} \lesssim A_6^{-2}$$

which when combined with (3.40) gives

$$\|u\|_{L^4_t L^\infty_x([t_1,1] \times \{A_6^{-5} R \leq |x| \leq A_6^5 R\})} \lesssim A_6^{-2}.$$

By repeating the arguments in (iii) (using (2.2) in place of (2.1) to handle the long-range components of the heat kernel, which can be controlled with extremely good bounds using (3.1)), one can then show iteratively that

$$\|u\|_{L^8_t L^\infty_x([t_1,1] \times \{A_6^{-4} R \leq |x| \leq A_6^4 R\})} \lesssim A_6^{-2},$$

then

$$\|u\|_{L^\infty_t L^\infty_x([t_1,1] \times \{A_6^{-3} R \leq |x| \leq A_6^3 R\})} \lesssim A_6^{-2},$$

then

$$\|\nabla u\|_{L^4_t L^\infty_x([t_1,1] \times \{A_6^{-2} R \leq |x| \leq A_6^2 R\})} \lesssim A_6^{-2},$$

then finally

$$\|\nabla u\|_{L^\infty_t L^\infty_x([t_1,1] \times \{A_6^{-1} R \leq |x| \leq 2 A_6 R\})} \lesssim A_6^{-2}$$

and

$$\|\nabla \omega\|_{L^\infty_t L^\infty_x([t_1,1] \times \{R \leq |x| \leq A_6 R\})} \lesssim A_6^{-2}$$

giving (vi).

---

[4] It is likely that one can also proceed at this point using the local regularity theory from [**CKN**].

4. Carleman inequalities for backwards heat equations

We will need some Carleman inequalities for backwards heat equations which are essentially contained in previous literature (most notably [**ESS2**], [**ESS**]), but made slightly more quantitative for our application (also it will be convenient to not demand that the functions involved vanish at the starting and final time). Following [**ESS2**], we shall reverse the direction of time and work here with backwards heat equations rather than forward ones.

Our main tool is the following general inequality (cf. [**ESS**, Lemma 2]):

LEMMA 4.1 (General Carleman inequality). *Let $[t_1, t_2]$ be a time interval, and let $u \in C_c^\infty([t_1,t_2] \times \mathbb{R}^d \to \mathbb{R}^m)$ be a (vector-valued) test function solving the backwards heat equation*

$$Lu = f$$

with L the backwards heat operator

$$L := \partial_t + \Delta, \tag{4.1}$$

and let $g : [t_1,t_2] \times \mathbb{R}^d \to \mathbb{R}$ be smooth. Let $F : [t_1,t_2] \times \mathbb{R}^d \to \mathbb{R}$ denote the function

$$F := \partial_t g - \Delta g - |\nabla g|^2.$$

Then we have the inequality

$$\partial_t \int_{\mathbb{R}^d} \left(|\nabla u|^2 + \frac{1}{2} F|u|^2 \right) e^g dx \geq \int_{\mathbb{R}^d} \left(\frac{1}{2}(LF)|u|^2 + 2D^2 g(\nabla u, \nabla u) - \frac{1}{2}|Lu|^2 \right) e^g dx$$

for all $t \in I$, where $D^2 g$ is the bilinear form expressed in coordinates as

$$D^2 g(v,w) := (\partial_i \partial_j g) v_i \cdot w_j$$

with the usual summation conventions. In particular, from the fundamental theorem of calculus one has

$$\begin{aligned} &\int_{t_1}^{t_2} \int_{\mathbb{R}^d} \left(\frac{1}{2}(LF)|u|^2 + 2D^2 g(\nabla u, \nabla u) \right) e^g dx dt \\ &\quad \leq \frac{1}{2} \int_{t_1}^{t_2} \int_{\mathbb{R}^d} |Lu|^2 \; e^g dx dt + \int_{\mathbb{R}^d} \left(|\nabla u|^2 + \frac{1}{2} F|u|^2 \right) e^g dx|_{t=t_1}^{t=t_2}. \end{aligned}$$

The above inequality is valid in all dimensions, but in this paper we will only need this lemma in the case $d = m = 3$.

PROOF. By breaking u into components, we may assume without loss of generality that we are in the scalar case $m = 1$.

We use the usual commutator method. Introducing the weighted (and time-dependent) inner product

$$\langle u, v \rangle := \int_{\mathbb{R}^n} uv \; e^g dx$$

for test functions $u, v : I \times \mathbb{R}^n \to \mathbb{R}$, we compute after differentiating under the integral sign and integrating by parts

$$\begin{aligned} \langle Lu, v \rangle + \langle u, Lv \rangle &= \int_{\mathbb{R}^n} (\partial_t(uv) + \Delta(uv) - 2\nabla u \cdot \nabla v) \; e^g dx \\ &= \partial_t \langle u, v \rangle + \int_{\mathbb{R}^n} \left(-(\partial_t g)uv + (\Delta g + |\nabla g|^2)uv - 2\nabla u \cdot \nabla v \right) \; e^g dx \\ &= \partial_t \langle u, v \rangle - \langle Fu, v \rangle - 2\langle \partial_i u, \partial_i v \rangle \end{aligned}$$

with the usual summation conventions. We can write

$$-\langle \partial_i u, \partial_i v\rangle - \frac{1}{2}\langle Fu, v\rangle = \langle Su, v\rangle \tag{4.2}$$

where S is the differential operator

$$Su := \Delta u + \nabla g \cdot \nabla u - \frac{1}{2}Fu$$

which is then formally self-adjoint with respect to the inner product $\langle , \rangle$; one can view S as the self-adjoint component of L. We can then rewrite the above identity as

$$\partial_t \langle u, v\rangle = \langle Lu, v\rangle + \langle u, Lv\rangle - 2\langle Su, v\rangle.$$

In particular, by the self-adjointness of S we have for any test functions u, v that

$$\begin{aligned}
\partial_t \langle Su, v\rangle &= \langle LSu, v\rangle + \langle Su, Lv\rangle - 2\langle Su, Sv\rangle \\
&= \langle [L, S]u, v\rangle + \langle SLu, v\rangle + \langle Su, Lv\rangle - 2\langle Su, Sv\rangle \\
&= \langle [L, S]u, v\rangle + \langle Lu, Sv\rangle + \langle Su, Lv\rangle - 2\langle Su, Sv\rangle \\
&= \langle [L, S]u, v\rangle + \frac{1}{2}\langle Lu, Lv\rangle - \frac{1}{2}\langle (L - 2S)u, (L - 2S)v\rangle.
\end{aligned}$$

Among other things, this shows that the differential operator $[L, S]$ (which does not involve any time derivatives) is formally self-adjoint with respect to the inner product $\langle , \rangle$. Specialising to the case $u = v$, we conclude in particular the inequality

$$\partial_t \langle Su, u\rangle \leq \langle [L, S]u, u\rangle + \frac{1}{2}\langle Lu, Lu\rangle. \tag{4.3}$$

Now we compute $[L, S]$. As previously noted, $[L, S]$ is a formally self-adjoint differential operator that does not involve any time derivatives. Since the second order operator L commutes with the second order component Δ of S, we see that $[L, S]$ is a second-order operator. The highest order terms can be easily computed in coordinates as

$$[L, S]u = 2(\partial_i \partial_j g)\partial_i \partial_j u + \text{l.o.t.}$$

and hence after integrating by parts the symmetric quadratic form $\langle [L, S]u, v\rangle$ must take the form

$$\langle [L, S]u, v\rangle = \int_{\mathbb{R}^d} (-2D^2 g(\nabla u, \nabla v) + Huv)\, e^g \, dx$$

for some function H; setting $u = 1$, we see that H must equal

$$H = [L, S]1 = LS1 = -\frac{1}{2}LF.$$

We conclude that

$$\langle [L, S]u, u\rangle = \int_{\mathbb{R}^d} (-2D^2 g(\nabla u, \nabla u) - \frac{1}{2}(LF)|u|^2)\, e^g \, dx.$$

Inserting this identity back into (4.3) and using (4.2), we obtain the claim. □

The inequality below is a quantitative variant of [**ESS**, Lemma 4].

PROPOSITION 4.2 (First Carleman inequality). *Let $T > 0$, $0 < r_- < r_+$, and let $\mathcal{A}$ denote the cylindrical annulus*

$$\mathcal{A} := \{(t, x) \in \mathbb{R} \times \mathbb{R}^3 : t \in [0, T]; r_- \leq |x| \leq r_+\}.$$

Let $u : \mathcal{A} \to \mathbb{R}^3$ *be a smooth function obeying the differential inequality*

$$|Lu| \leq C_0^{-1}T^{-1}|u| + C_0^{-1/2}T^{-1/2}|\nabla u| \tag{4.4}$$

on $\mathcal{A}$. *Assume the inequality*

$$r_-^2 \geq 4C_0T. \tag{4.5}$$

Then one has

$$\int_0^{T/4}\int_{10r_- \leq |x| \leq r_+/2} (T^{-1}|u|^2 + |\nabla u|^2)\; dxdt \lesssim C_0^2 e^{-\frac{r_-r_+}{4C_0T}}(X + e^{2r_+^2/C_0T}Y)$$

where

$$X := \int\int_{\mathcal{A}} e^{2|x|^2/C_0T}(T^{-1}|u(t,x)|^2 + |\nabla u(t,x)|^2)\; dxdt$$

and

$$Y := \int_{r_- \leq |x| \leq r_+} |u(0,x)|^2\; dx.$$

The key feature here is the gain of $e^{-\frac{r_-r_+}{4C_0T}}$, which can be compared against the trivial bound of $e^{-2r_-^2/C_0T}X$ that follows by lower bounding the factor $e^{2|x|^2/C_0T}$ appearing in X by $e^{2r_-^2/C_0T}$. Thus, this lemma becomes powerful when the ratio r_+/r_- is large. Informally, Proposition 4.2 asserts that if u solves (4.4) on $\mathcal{A}$, has some mild Gaussian decay as $|x| \to \infty$, and is extremely small at $t = 0$, then it is also very small in the interior of $\mathcal{A}$ near $t = 0$. The various numerical constants such as $1/4$ or 10 appearing in the above proposition can be modified (and optimised) if desired, but we fix a specific choice of constants for sake of concreteness. The weight $e^{2|x|^2/C_0T}$ in X is inconvenient, but it is negligible when compared against the "natural" decay rate of $e^{-|x|^2/4t}$ arising from the fundamental solution of the heat equation, and it can be managed in our application by using the second Carleman inequality given below. Specialising Proposition 4.2 the case $u(0,x) = 0$ (so that $Y = 0$) and sending r_+ to infinity, one recovers a variant of the backwards uniqueness result in [**ESS**, Lemma 4].

PROOF. We may assume that

$$r_+ \geq 20r_- \tag{4.6}$$

since the claim is vacuous otherwise. By the pigeonhole principle, one can find a time $T_0 \in [T/2, T]$ such that

$$\int_{r_- \leq |x| \leq r_+} e^{2|x|^2/C_0T}(T^{-1}|u(T_0,x)|^2 + |\nabla u(T_0,x)|^2)\; dx \lesssim T^{-1}X. \tag{4.7}$$

Fix this time T_0. In the discussion below we implicitly restrict (t,x) to the region $\mathcal{A} \cap ([0,T_0] \times \mathbb{R}^3)$. We set

$$\alpha := \frac{r_+}{2C_0T^2} \tag{4.8}$$

and observe from (4.5), (4.6) that

$$\alpha \geq \frac{40}{r_-T}. \tag{4.9}$$

Following [**ESS2**], we apply Lemma 4.1 on the interval $[0, T_0]$ with the weight

$$g := \alpha(T_0 - t)|x| + \frac{1}{C_0 T}|x|^2$$

and u replaced by ψu, where $\psi(x)$ is a smooth cutoff supported on the region $r_1 \leq |x| \leq r_2$ that equals 1 on $2r_1 \leq |x| \leq r_2/2$ and obeys the estimates $|\nabla^j \psi(x)| = O(1/|x|^j)$ for $j = 0, 1, 2$. Since $\alpha(T_0 - t)|x|$ is convex in x, we have

$$D^2 g(\nabla(\psi u), \nabla(\psi u)) \geq 2\frac{1}{C_0 T}|\nabla(\psi u)|^2.$$

The function F defined in Lemma 4.1 can be computed on $\mathcal{A}$ as

$$\begin{aligned} F &= -\alpha|x| - \frac{2\alpha(T_0 - t)}{|x|} - \frac{6}{C_0 T} - \left(\alpha(T_0 - t) + \frac{2|x|}{C_0 T}\right)^2 \\ &= -\alpha|x| - \frac{2\alpha(T_0 - t)}{|x|} - \frac{6}{C_0 T} - \alpha^2(T_0 - t)^2 - 4\alpha\frac{T_0 - t}{C_0 T}|x| - \frac{4|x|^2}{C_0^2 T^2}. \end{aligned}$$

In particular F is negative. We also calculate

$$LF = 2\alpha^2(T_0 - t) + \frac{4\alpha|x|}{C_0 T} - \frac{8\alpha(T_0 - t)}{C_0 T|x|} - \frac{24}{C_0^2 T^2}.$$

We see from (4.5) that $\frac{T_0 - t}{|x|} \leq \frac{1}{4}|x|$, so that

$$LF \geq \frac{2\alpha|x|}{C_0 T} - \frac{24}{C_0^2 T^2}.$$

By (4.9) we thus have $LF \geq \frac{56}{C_0^2 T^2}$. Applying Lemma 4.1 and discarding some terms, we conclude that

$$\begin{aligned} &\int_0^{T_0} \int_{2r_- \leq |x| \leq r_+/2} \left(28 C_0^{-2} T^{-2}|u|^2 + 4 C_0^{-1} T^{-1}|\nabla u|^2\right) e^g \, dx dt \\ &\leq \frac{1}{2} \int_0^{T_0} \int_{\mathbb{R}^3} |L(\psi u)|^2 \, e^g \, dx dt \\ &\quad + \int_{\mathbb{R}^3} |\nabla(\psi u)(T_0, x)|^2 \, e^{g(T_0, x)} dx + \int_{\mathbb{R}^3} |(\psi u)(0, x)|^2 |F(0, x)| \, e^{g(0, x)} \, dx. \end{aligned}$$

In the region $2r_- \leq |x| \leq r_+/2$ we have from (4.4) that

$$|L(\psi u)|^2 = |Lu|^2 \leq 2 C_0^{-2} T^{-2}|u|^2 + 2 C_0^{-1} T^{-1}|\nabla u|^2.$$

In the regions $r_- \leq |x| \leq 2r_-$ or $r_+/2 \leq |x| \leq r_+$, we have

$$|L(\psi u)|^2 \lesssim |Lu|^2 + |x|^{-2}|\nabla u|^2 + |x|^{-4}|u|^2 \lesssim C_0^{-2} T^{-2}|u|^2 + C_0^{-1} T^{-1}|\nabla u|^2$$

thanks to (4.4), (4.5). For all other x, $L(\psi u)$ vanishes. A similar calculation gives

$$|\nabla(\psi u)|^2 \lesssim |\nabla u|^2 + C_0^{-1} T^{-1}|u|^2.$$

We therefore have

$$\begin{aligned}\int_0^{T_0}\int_{2r_-\le|x|\le r_+/2}&\left(C_0^{-2}T^{-2}|u|^2+C_0^{-1}T^{-1}|\nabla u|^2\right)\ e^g\,dxdt\\ &\lesssim \int_0^{T_0}\int_{|x|\in[r_-,2r_-]\cup[r_+/2,r_+]}(C_0^{-2}T^{-2}|u|^2+C_0^{-1}T^{-1}|\nabla u|^2)\ e^g\,dxdt\\ &\quad+\int_{r_-\le|x|\le r_+}(C_0^{-1}T^{-1}|u|^2+|\nabla u|^2)(T_0,x)\ e^{g(T_0,x)}dx\\ &\quad+\int_{r_-\le|x|\le r_+}|u(0,x)|^2|F(0,x)|\ e^{g(0,x)}\ dx.\end{aligned}$$

From (4.7) one has

$$\int_{r_-\le|x|\le r_+}(C_0^{-1}T^{-1}|u|^2+|\nabla u|^2)(T_0,x)\ e^{g(T_0,x)}dx\lesssim T^{-1}X.$$

When $t\in[0,T_0]$ and $|x|\in[r_+/2,r_+]$, one has

$$e^g\le e^{\alpha T|x|-\frac{|x|^2}{C_0T}}e^{2|x|^2/C_0T}\le e^{2|x|^2/C_0T}$$

by (4.8). When instead $t\in[0,T_0]$ and $|x|\in[r_-,2r_-]$, one has

$$e^g\le e^{\alpha T|x|-\frac{|x|^2}{C_0T}}e^{2|x|^2/C_0T}\le e^{2\alpha Tr_-}e^{2|x|^2/C_0T}.$$

We conclude that

$$\begin{aligned}\int_0^{T_0}\int_{2r_-\le|x|\le r_+/2}&\left(C_0^{-2}T^{-2}|u|^2+C_0^{-1}T^{-1}|\nabla u|^2\right)\ e^g\,dxdt\\ &\lesssim e^{2\alpha Tr_-}T^{-1}X+\int_{r_-\le|x|\le r_+}|u(0,x)|^2|F(0,x)|\ e^{g(0,x)}\ dx.\end{aligned}$$

In the region $t\in[0,T/4]$, $10r_-\le|x|\le r_+/2$, one has

$$e^g\ge e^{\frac{\alpha T|x|}{4}+\frac{|x|^2}{C_0T}}\ge e^{\frac{5}{2}\alpha Tr_-}$$

and hence

$$\begin{aligned}\int_0^{T/4}\int_{10r_-\le|x|\le r_+/2}&\left(C_0^{-2}T^{-2}|u|^2+C_0^{-1}T^{-1}|\nabla u|^2\right)\ dxdt\\ &\lesssim e^{-\alpha Tr_-/2}\left(T^{-1}X+\int_{r_-\le|x|\le r_+}|u(0,x)|^2|F(0,x)|\ e^{g(0,x)}\ dx\right).\end{aligned}$$

From (4.8) we have

$$e^{-\alpha Tr_-/2}=e^{-\frac{r_-r_+}{4C_0T}}.$$

Finally, for $t=0$ and $r_-\le|x|\le r_+$ one has

$$e^g\le e^{\alpha T|x|+\frac{|x|^2}{C_0T}}\le e^{\frac{3r_+^2}{2C_0T}}$$

and

$$\begin{aligned}|F| &\lesssim \alpha r_+ + \alpha T/r_- + C_0^{-1}T^{-1} + \alpha^2 T^2 + C_0^{-1}\alpha r_+ + C_0^{-2}T^{-2}r_+^2\\ &\lesssim \frac{r_+^2}{T^2} + \frac{r_+}{r_- T} + \frac{1}{T} + \frac{r_+^2}{T^2} + \frac{r_+^2}{T^2} + \frac{r_+^2}{T^2}\\ &\lesssim \frac{r_+^2}{T}(T^{-1} + \frac{1}{r_+ r_-})\\ &\lesssim \frac{r_+^2}{T}T^{-1}\end{aligned}$$

since $\frac{1}{r_+r_-} \leq \frac{1}{r_-^2} \lesssim \frac{1}{T}$ by (4.6), (4.5). Bounding $\frac{r_+^2}{T}e^{\frac{3r_+^2}{2T}} \lesssim e^{\frac{2r_+^2}{T}}$ and multiplying by T, we conclude that

$$\begin{aligned}&\int_0^{T/4}\int_{10r_- \leq |x| \leq r_+/2} (C_0^{-2}T^{-1}|u|^2 + C_0^{-1}|\nabla u|^2)\ dxdt\\ &\quad \lesssim e^{-\frac{r_- r_+}{4C_0 T}}\left(X + e^{2r_+^2/T}\int_{r_- \leq |x| \leq r_+} |u(0,x)|^2\ dx\right)\end{aligned}$$

giving the claim. □

Our second application of Lemma 4.1 is the following quantitative version of standard parabolic unique continuation results.

PROPOSITION 4.3 (Second Carleman inequality). *Let $T, r > 0$, and let $\mathcal{C}$ denote the cylindrical region*

$$\mathcal{C} := \{(t,x) \in \mathbb{R} \times \mathbb{R}^3 : t \in [0,T]; |x| \leq r\}.$$

Let $u : \mathcal{C} \to \mathbb{R}^3$ be a smooth function obeying the differential inequality (4.4) on $\mathcal{C}$. Assume the inequality

$$r^2 \geq 4000T. \tag{4.10}$$

Then for any

$$0 < t_1 \leq t_0 < \frac{T}{1000} \tag{4.11}$$

one has

$$\int_{t_0}^{2t_0}\int_{|x| \leq r/2} (T^{-1}|u|^2 + |\nabla u|^2)\, e^{-|x|^2/4t}\ dxdt \lesssim e^{-\frac{r^2}{500t_0}} X + t_0^{3/2}(et_0/t_1)^{O(r^2/t_0)} Y$$

where

$$X := \int_0^T \int_{|x| \leq r} (T^{-1}|u|^2 + |\nabla u|^2)\ dxdt$$

and

$$Y := \int_{|x| \leq r} |u(0,x)|^2 t_1^{-3/2} e^{-|x|^2/4t_1}\ dx.$$

As with the previous inequality, the numerical constants here such as 1000, 500 can be optimised if desired, but this explicit choice of constants suffices for our application. The key feature here is the gain of $e^{-\frac{r^2}{500t_0}}$. Specialising to the case where u vanishes to infinite order at $(0,0)$, sending $t_1 \to 0$ (which sends $(et_0/t_1)^{O(\frac{r^2}{t_0})} Y$

to zero thanks to the infinite order vanishing), and then sending $t_0 \to 0$, we obtain a variant of a standard unique continuation theorem for backwards parabolic equations (see e.g., [**ESS2**, Theorem 4.1]).

PROOF. By the pigeonhole principle, we can select a time

$$\frac{T}{200} \leq T_0 \leq \frac{T}{100} \tag{4.12}$$

such that

$$\int_{|x|\leq r} (T^{-1}|u|^2 + |\nabla u|^2)\, dx \lesssim T^{-1}X. \tag{4.13}$$

We define

$$\alpha = \frac{r^2}{400 t_0} \tag{4.14}$$

so from (4.10), (4.11) we have

$$\alpha \geq 10. \tag{4.15}$$

We apply Lemma 4.1 on $[0, T_0] \times \mathbb{R}^3$ with the weight

$$g := -\frac{|x|^2}{4(t+t_1)} - \frac{3}{2}\log(t+t_1) - \alpha \log \frac{t+t_1}{T_0+t_1} + \alpha \frac{t+t_1}{T_0+t_1}$$

(which is a modification of the logarithm of the fundamental solution $\frac{1}{t^{3/2}} e^{-|x|^2/4t}$ of the heat equation) and u replaced by ψu, where $\psi(x)$ is a smooth cutoff supported on the region $|x| \leq r$ that equals 1 on $|x| \leq r/2$ and obeys the estimates

$$|\nabla^j \psi(x)| = O(r^{-j}) \tag{4.16}$$

for $r/2 \leq |x| \leq r$ and $j = 0, 1, 2$. Clearly

$$D^2 g(\nabla(\psi u), \nabla(\psi u)) = -\frac{1}{2(t+t_1)} |\nabla(\psi u)|^2.$$

We can calculate

$$\begin{aligned} F &= \frac{|x|^2}{4(t+t_1)^2} - \frac{3}{2(t+t_1)} - \frac{\alpha}{t+t_1} + \frac{\alpha}{T_0+t_1} + \frac{3}{2(t+t_1)} + \left|\frac{x}{2(t+t_1)}\right|^2 \\ &= \frac{\alpha}{T_0+t_1} - \frac{\alpha}{t+t_1} \end{aligned}$$

and hence

$$LF = \frac{\alpha}{(t+t_1)^2}.$$

From Lemma 4.1 we thus have

$$\begin{aligned} \partial_t \int_{\mathbb{R}^3} &\left(|\nabla(\psi u)|^2 - \frac{\alpha}{2(t+t_1)}|\psi u|^2 + \frac{\alpha}{2(T_0+t_1)}|\psi u|^2 \right) e^g\, dx \\ &\geq \int_{\mathbb{R}^3} \left(\frac{\alpha}{2(t+t_1)^2}|\psi u|^2 - \frac{1}{t+t_1}|\nabla(\psi u)|^2 - \frac{1}{2}|Lu|^2 \right) e^g\, dx. \end{aligned}$$

To exploit this differential inequality we use the method of integrating factors. If we introduce the energy

$$E(t) := \int_{\mathbb{R}^3} \left(|\nabla(\psi u)|^2 - \frac{\alpha}{2(t+t_1)}|\psi u|^2 + \frac{\alpha}{2(T_0+t_1)}|\psi u|^2 \right) e^g\, dx$$

then we conclude from the product rule that

$$\begin{aligned}
&\partial_t \left(\left((t+t_1) + \frac{(t+t_1)^2}{10T_0} \right) E(t) \right) \\
&\quad \geq \left(1 + \frac{t+t_1}{5(T_0+t_1)} \right) E(t) \\
&\qquad + (t+t_1+\frac{(t+t_1)^2}{10T_0}) \int_{\mathbb{R}^3} \left(\frac{\alpha}{2(t+t_1)^2} |\psi u|^2 + \frac{1}{t+t_1} |\nabla(\psi u)|^2 - \frac{1}{2} |L(\psi u)|^2 \right) e^g \, dx \\
&\quad = \int_{\mathbb{R}^3} \left(\frac{t+t_1}{10(T_0+t_1)} |\nabla(\psi u)|^2 + \alpha \frac{5(T_0+t_1)-(t+t_1)}{10(T_0+t_1)^2} |\psi u|^2 \right. \\
&\qquad\qquad \left. - \frac{1}{2}(t+t_1+\frac{(t+t_1)^2}{10(T_0+t_1)}) |L(\psi u)|^2 \right) e^g \, dx \\
&\quad \geq \int_{\mathbb{R}^3} \left(\frac{t+t_1}{10(T_0+t_1)} |\nabla(\psi u)|^2 + \frac{\alpha}{10(T_0+t_1)} |\psi u|^2 - (t+t_1) |L(\psi u)|^2 \right) e^g \, dx
\end{aligned}$$

and hence by the fundamental theorem of calculus

$$\begin{aligned}
&\int_0^{T_0} \int_{\mathbb{R}^3} \left(\frac{t+t_1}{10(T_0+t_1)} |\nabla(\psi u)|^2 + \frac{\alpha}{10(T_0+t_1)} |\psi u|^2 \right) e^g \, dx dt \\
&\quad \leq \int_0^{T_0} \int_{\mathbb{R}^3} (t+t_1) |L(\psi u)|^2 \, e^g \, dx dt + \left((t+t_1) + \frac{(t+t_1)^2}{10(T_0+t_1)} \right) E(t) |_{t=0}^{t=T_0}.
\end{aligned}$$

Discarding some terms, we conclude that

$$\begin{aligned}
&\int_0^{T_0} \int_{|x| \leq r/2} \left(\frac{t+t_1}{10(T_0+t_1)} |\nabla u|^2 + \frac{\alpha}{10(T_0+t_1)} |u|^2 \right) e^g \, dx dt \\
&\quad \leq \int_0^{T_0} \int_{\mathbb{R}^3} (t+t_1) |L(\psi u)|^2 \, e^g \, dx dt \\
&\qquad + O\left(T_0 \int_{\mathbb{R}^3} |\nabla(\psi u)(T_0,x)|^2 \, e^g \, dx + \alpha \int_{|x| \leq r} |u(0,x)|^2 \, e^g \, dx \right).
\end{aligned} \tag{4.17}$$

When $|x| \leq r/2$, one has

$$|L(\psi u)|^2 = |Lu|^2 \leq 2T^{-2}|u|^2 + 2T^{-1}|\nabla u|^2$$

thanks to (4.4). By (4.12), (4.15) the contribution of this case is less than half of the left-hand side of (4.17). When $r/2 \leq |x| \leq r$, we have from (4.16), (4.10) that

$$\begin{aligned}
|L(\psi u)|^2 &\lesssim T^{-2}|u|^2 + T^{-1}|\nabla u|^2 + r^{-4}|u|^2 + r^{-2}|\nabla|^2 \\
&\lesssim T^{-2}|u|^2 + T^{-1}|\nabla u|^2.
\end{aligned}$$

Finally, $L(\psi u)$ vanishes for $|x| > r$. Putting all this together, we conclude that

$$\begin{aligned}
&\int_0^{T_0} \int_{|x| \leq r/2} \left(\frac{t+t_1}{T_0+t_1} |\nabla u|^2 + \frac{\alpha}{T_0+t_1} |u|^2 \right) e^g \, dx dt \\
&\quad \lesssim \int_0^{T_0} \int_{r/2 \leq |x| \leq r} (t+t_1)(T^{-2}|u|^2 + T^{-1}|\nabla u|^2) \, e^g \, dx dt \\
&\qquad + T_0 \int_{\mathbb{R}^3} |\nabla(\psi u)(T_0,x)|^2 \, e^g \, dx + \alpha \int_{|x| \leq r} |u(0,x)|^2 \, e^g \, dx.
\end{aligned}$$

Restricting the left-hand integral to the region $t_0 \leq t \leq 2t_0$ and also bounding $t_1 \leq t_0 \leq T_0 \leq T$ in several places, we conclude that

$$\begin{aligned}
&\int_{t_0}^{2t_0} \int_{|x| \leq r/2} \left(\frac{t_0}{T_0} |\nabla u|^2 + \frac{\alpha}{T_0} |u|^2 \right) e^g \, dx dt \\
&\lesssim \int_0^{T_0} \int_{r/2 \leq |x| \leq r} (T^{-1}|u|^2 + |\nabla u|^2) \, e^g \, dx dt \\
&\quad + T \int_{\mathbb{R}^3} |\nabla(\psi u)(T_0, x)|^2 \, e^g \, dx + \alpha \int_{|x| \leq r} |u(0,x)|^2 \, e^g \, dx.
\end{aligned}$$

From elementary calculus we have the inequality

$$-\frac{a}{t} - b \log t \leq b \log \frac{b}{ae}$$

for any $a, b, t > 0$ (the left-hand side attains its maximum when $t = a/b$). When $r/2 \leq |x| \leq r$ and $0 \leq t \leq T_0$, we then have

$$\begin{aligned}
g &\leq -\frac{|x|^2}{4(t+t_1)} - \left(\alpha + \frac{3}{2}\right) \log(t + t_1) + \alpha \log(T_0 + t_1) + \alpha \\
&\leq \left(\alpha + \frac{3}{2}\right) \log \frac{4\left(\alpha + \frac{3}{2}\right)}{e|x|^2} + \alpha \log(T_0 + t_1) + \alpha \\
&\leq \left(\alpha + \frac{3}{2}\right) \log \frac{32\alpha}{er^2} + \alpha \log(T_0 + t_1) + \alpha \\
&\leq \alpha \log \frac{32\alpha(T_0 + t_1)}{r^2} + \frac{3}{2} \log \frac{32\alpha}{er^2}
\end{aligned}$$

and thus by (4.14)

$$\int_0^{T_0} \int_{r/2 \leq |x| \leq r} (T^{-1}|u|^2 + |\nabla u|^2) \, e^g \, dx dt \lesssim t_0^{-3/2} \exp(\alpha \log \frac{32\alpha(T_0 + t_1)}{r^2}) X.$$

When $|x| \leq r$ and $t = T_0$, then

$$g \leq -\frac{3}{2} \log t_0 + \alpha$$

and $\nabla(\psi u)$ is supported on the ball $\{|x| \leq r\}$ and obeys the estimate

$$|\nabla(\psi u)| \lesssim |\nabla u| + r^{-1}|u| \lesssim T^{-1}|u| + |\nabla u|$$

thanks to (4.10), and hence by (4.13)

$$T \int_{\mathbb{R}^3} |\nabla(\psi u)(T_0, x)|^2 \, e^g \, dx \lesssim t_0^{-3/2} \exp(\alpha) X.$$

From (4.10), (4.15) we have $\log \frac{32\alpha(T_0 + t_0)}{r^2} \geq 1$. Thus

$$\begin{aligned}
&\int_{t_0}^{2t_0} \int_{|x| \leq r/2} \left(\frac{t_0}{T_0} |\nabla u|^2 + \frac{\alpha}{T_0} |u|^2 \right) e^g \, dx dt \\
&\lesssim t_0^{-3/2} \exp(\alpha \log \frac{32\alpha(T_0 + t_1)}{r^2}) X \qquad + \alpha \int_{|x| \leq r} |u(0,x)|^2 \, e^g \, dx.
\end{aligned}$$

In the region $t_0 \leq t \leq 2t_0$, $|x| \leq r/2$, we have

$$g \geq -\frac{|x|^2}{4t} - \frac{3}{2} \log(3t_0) - \alpha \log \frac{3t_0}{T_0 + t_1}$$

so that

$$e^g \gtrsim t_0^{-3/2} e^{-|x|^2/4t} \exp\left(-\alpha \log \frac{3t_0}{T_0 + t_1}\right).$$

Finally, when $t = 0$ and $|x| \leq r$, we have

$$g \leq -\frac{|x|^2}{4t_1} - \frac{3}{2}\log t_1 - \alpha \log \frac{t_1}{T_0 + t_1} + \alpha$$

so that

$$e^g \leq t_1^{-3/2} e^{-|x|^2/4t_1} \exp\left(\alpha \log \frac{e(T_0 + t_1)}{t_1}\right).$$

We conclude that

$$\begin{aligned}
&\int_{t_0}^{2t_0} \int_{|x|\leq r/2} \left(\frac{t_0}{T_0}|\nabla u|^2 + \frac{\alpha}{T_0}|u|^2\right)\, dxdt \\
&\quad \lesssim \exp\left(\alpha \log \frac{96\alpha t_0}{r^2}\right) X \\
&\qquad + \alpha \exp\left(\alpha \log(\frac{3et_0}{t_1})\right) t_0^{3/2} \int_{|x|\leq r} |u(0,x)|^2 t_1^{-3/2} e^{-|x|^2/4t_1}\, dx.
\end{aligned}$$

From (4.14) we have $\log \frac{96\alpha t_0}{r^2} \leq -1$, while from (4.15), (4.12), (4.10), (4.14) we have $\frac{\alpha}{T_0} \gtrsim T^{-1}$ and $\frac{t_0}{T_0} \gtrsim \frac{t_0}{r^2} \gtrsim \alpha^{-1}$. We conclude that

$$\begin{aligned}
&\int_{t_0}^{2t_0} \int_{|x|\leq r/2} (|\nabla u|^2 + T^{-1}|u|^2)\, dxdt \\
&\quad \lesssim \alpha^2 e^{-\alpha} X + \exp\left(\alpha \log(\frac{3et_0}{t_1})\right) t_0^{3/2} \int_{|x|\leq r} |u(0,x)|^2 t_1^{-3/2} e^{-|x|^2/4t_1}\, dx.
\end{aligned}$$

From (4.14) we have $\alpha = O(r^2/t_0)$ and $\alpha^2 e^{-\alpha} \lesssim e^{-\frac{r^2}{500t_0}}$, and the claim follows. □

5. Main estimate

In this section we combine the estimates in Proposition 3.1 with the Carleman inequalities from the previous section to obtain

THEOREM 5.1 (Main estimate). *Let t_0, T, u, p, A obey the hypotheses of Proposition 3.1, and suppose that there exists $x_0 \in \mathbb{R}^3$ and $N_0 > 0$ such that*

$$|P_{N_0} u(t_0, x_0)| \geq A_1^{-1} N_0$$

where as before we set $A_j := A^{C_0^j}$. Then

$$TN_0^2 \leq \exp(\exp(\exp(A_6^{O(1)}))).$$

PROOF. After translating in time and space we may normalise $(t_0, x_0) = (0, 0)$. Let T_1 be an arbitrary time scale in the interval

$$A_4 N_0^{-2} \leq T_1 \leq A_4^{-1} T.$$

By Proposition 3.1(v), there exists

$$(t_1, x_1) \in [-T_1, -A_3^{-O(1)} T_1] \times B(0, A_4^{O(1)} T_1^{1/2})] \tag{5.1}$$

and

$$N_1 = A_3^{O(1)} T_1^{-1/2} \tag{5.2}$$

such that

$$|P_{N_1} u(t_1, x_1)| \geq A_1^{-1} N_1.$$

From the Biot-Savart law we have

$$P_{N_1} u(t_1, x_1) = -\Delta^{-1} P_{N_1} \nabla \times \tilde{P}_{N_1} \omega(t_1, x_1),$$

and hence by (2.2)

$$P_{N_1} u(t_1, x_1) \lesssim N_1^{-1} \|\tilde{P}_{N_1} \omega(t_1)\|_{L^\infty(B(x_1, A_1/N_1))} + A_1^{-50} N_1^{-1} \|\tilde{P}_{N_1} \omega(t_1)\|_{L^\infty(\mathbb{R}^3)}.$$

From (3.1), (2.3) one has

$$\|\tilde{P}_{N_1} \omega(t_1)\|_{L^\infty(\mathbb{R}^3)} \lesssim A N_1^2$$

and thus we have

$$|\tilde{P}_{N_1} \omega(t_1, x_1')| \gtrsim A_1^{-1} N_1^2$$

for some $x_1' = x_1 + O(A_1/N_1) = O(A_4^{O(1)} T_1^{1/2})$. By Proposition 3.1(i), one has

$$\nabla \tilde{P}_{N_1} \omega = O(A N_1^3); \quad \partial_t \tilde{P}_{N_1} \omega = O(A N_1^4)$$

and thus

$$|\tilde{P}_{N_1} \omega(t, x)| \gtrsim A_1^{-1} N_1^2 \tag{5.3}$$

for all $(t, x) \in [t_1, t_1 + A_1^{-2} N_1^{-2}] \times B(x_1', A_1^{-2} N_1^{-1})$. By Proposition 3.1(iii), there is an interval

$$I' \subset [t_1, t_1 + A_1^{-2} N_1^{-2}] \cap [-T_1, -A_3^{-O(1)} T_1]$$

with $|I'| = A_3^{-O(1)} T_1$ such that

$$u(t, x) = O(A_3^{O(1)} T_1^{-1/2}), \nabla u(t, x) = O(A_3^{O(1)} T_1^{-1})$$

and

$$\omega(t, x) = O(A_3^{O(1)} T_1^{-1}), \nabla \omega(t, x) = O(A_3^{O(1)} T_1^{-3/2}) \tag{5.4}$$

on $I' \times \mathbb{R}^3$. By shrinking I' as necessary, we may thus assume that

$$|u(t, x)| \leq C_0^{-1/2} |I'|^{-1/2}; \quad |\nabla u(t, x)| \leq C_0^{-1} |I'|^{-1}. \tag{5.5}$$

On the other hand, from (5.3), (5.1), (5.2) one has

$$\int_{B(0, A_4^{O(1)} T_1^{1/2})} |\tilde{P}_{N_1} \omega(t, x)|^2 \, dx \gtrsim A_3^{-O(1)} T_1^{-1/2}$$

for all $t \in I'$. From (2.2) and (5.5) this implies that

$$\int_{B(0, A_4^{O(1)} T_1^{1/2})} |\omega(t, x)|^2 \, dx \gtrsim A_3^{-O(1)} T_1^{-1/2} \tag{5.6}$$

for all $t \in I'$.

Write $I' := [t' - T', t']$, and let $x_* \in \mathbb{R}^3$ be any point with $|x_*| \geq A_5 T_1^{1/2}$. We apply Proposition 4.3 on the slab $[0, T'] \times \mathbb{R}^3$ with $r := A_5 |x_*|$, $t_0 := T'/2$, and $t_1 := A_5^{-4} T'$, and u replaced by the function

$$(t, x) \mapsto \omega(t' - t, x_* + x)$$

(so that the hypothesis (4.4) follows from the vorticity equation and (5.5)) to conclude that

$$Z \lesssim \exp(-A_5 |x_*|^2 / T') X + (T')^{3/2} \exp(O(A_5^3 |x_*|^2 / T')) Y$$

where

$$X := \int_{I'} \int_{B(x_*, A_5|x_*|)} ((T')^{-1}|\omega|^2 + |\nabla\omega|^2)\, dxdt$$

and

$$Y := (T')^{-3/2} \int_{B(x_*, A_5|x_*|)} |\omega(t', x)|^2 e^{-A_5^4|x-x_*|^2/4T'}\, dx$$

and

$$Z := \int_{t'-T'}^{t'-T'/2} \int_{B(x_*, A_5|x_*|/2)} (T')^{-1}|\omega|^2 e^{-|x-x_*|^2/4(t'-t)}\, dxdt.$$

From (5.6) we have

$$Z \gtrsim A_3^{-O(1)} \exp(-|x_*|^2/100T')(T')^{-1/2}.$$

From (5.4) we have

$$X \lesssim (T')^{-2} A_5^3 |x_*|^3 \lesssim A_5^3 \exp(|x_*|^2/T')(T')^{-1/2}$$

and hence the expression $\exp(-A_5|x_*|^2/T')X$ is negligible compared to Z. We conclude that

$$Y \gtrsim \exp(-O(A_5^3|x_*|^2/T'))(T')^{-2}.$$

Using (5.4), the contribution to Y outside of the ball $B(x_*, |x_*|/2)$ is negligible, thus

$$\int_{B(x_*, |x_*|/2)} |\omega(t', x)|^2\, dx \gtrsim \exp(-O(A_5^3|x_*|^2/T'))(T')^{-1/2}$$

and therefore

$$\int_{B(0,2R)\backslash B(0,R/2)} |\omega(t', x)|^2\, dx \gtrsim \exp(-O(A_5^3 R^2/T'))(T')^{-1/2}$$

whenever $R \geq A_5 T_1^{1/2}$. A similar argument holds with t' replaced by any time in $[t' - T'/4, t']$. We conclude in particular that we have the Gaussian lower bound

$$\int_{-T_1}^{-A_4^{-1}T_1} \int_{B(0,2R)\backslash B(0,R/2)} |\omega(t,x)|^2\, dxdt \gtrsim \exp(-A_5^4 R^2/T_1) T_1^{1/2} \tag{5.7}$$

for any time scale T_1 and spatial scale R with $A_4 N_0^{-2} \leq T_1 \leq A_4^{-1}T$ and $R \geq A_5 T_1^{1/2}$.

Now let T_2 be a scale for which

$$A_4^2 N_0^{-2} \leq T_2 \leq A_4^{-1}T. \tag{5.8}$$

By Proposition 3.1(vi), there exists a scale

$$A_6 T_2^{1/2} \leq R \leq \exp(A_6^{O(1)}) T_2^{1/2} \tag{5.9}$$

such that on the cylindrical annulus

$$\Omega := \{(t,x) \in [-T_2, 0] \times \mathbb{R}^3 : R \leq |x| \leq A_6 R\}$$

one has the estimates

$$\nabla^j u(t,x) = O(A_6^{-2} T_2^{-\frac{j+1}{2}}); \quad \nabla^j \omega(t,x) = O(A_6^{-2} T_2^{-\frac{j+2}{2}}) \tag{5.10}$$

for $j = 0, 1$. We apply Proposition 4.2 on the slab $[0, T_2/C_0] \times \mathbb{R}^3$ with $r_- := 10R$, $r_+ := A_6 R/10$, and u replaced by the function

$$(t, x) \mapsto \omega(-t, x)$$

(so that the hypothesis (4.4) follows from the vorticity equation and (5.10)) to conclude that

$$Z' \lesssim \exp(-A_6^{1/2} R^2/T_2) X' + \exp(\exp(A_6^{O(1)})) Y'$$

where

$$X' := \int_{-T_2/C_0}^{0} \int_{10R \leq |x| \leq A_6 R/10} e^{\frac{2|x|^2}{T_2}} (T_2^{-1} |\omega|^2 + |\nabla \omega|^2) \, dx dt$$

and

$$Y' := \int_{10R \leq |x| \leq A_6 R/10} |\omega(0,x)|^2 \, dx$$

and

$$Z' := \int_{-T_2/4C_0}^{0} \int_{100R \leq |x| \leq A_1 R/20} T_2^{-1} |\omega|^2 \, dx dt.$$

From (5.7) (with R replaced by $200R$) we have

$$Z' \gtrsim \exp(-A_5^5 R^2/T_2) T_2^{-1/2}.$$

Thus we either have

$$X' \gtrsim \exp(A_6^{1/3} R^2/T_2) T_2^{-1/2} \tag{5.11}$$

or

$$Y' \gtrsim \exp(-\exp(A_6^{O(1)})) T_2^{-1/2}. \tag{5.12}$$

Suppose for the moment that (5.11) holds. From the pigeonhole principle, we can then find a scale

$$10R \leq R' \leq A_6 R/10 \tag{5.13}$$

such that

$$\int_{-T_2/C_0}^{0} \int_{R' \leq |x| \leq 2R'} e^{\frac{2|x|^2}{T_2}} (T_2^{-1} |\omega|^2 + |\nabla \omega|^2) \, dx dt \gtrsim \exp(A_6^{1/4} R^2/T_2) T_2^{-1/2}$$

and thus

$$\int_{-T_2/C_0}^{0} \int_{R' \leq |x| \leq 2R'} (T_2^{-1} |\omega|^2 + |\nabla \omega|^2) \, dx dt \gtrsim \exp(-10(R')^2/T_2) T_2^{-1/2}$$

From (5.10) we see that the contribution to the left-hand side arising from those times t in the interval $[-\exp(-20(R')^2/T_2)T_2, 0]$ is negligible, thus

$$\int_{-T_2/C_0}^{-\exp(-20(R')^2/T_2)T_2} \int_{R' \leq |x| \leq 2R'} (T_2^{-1} |\omega|^2 + |\nabla \omega|^2) \, dx dt \gtrsim \exp(-10(R')^2/T_2) T_2^{-1/2}.$$

Thus by a further application of the pigeonhole principle, one can locate a time scale

$$\exp(-20(R')^2/T_2) T_2 \leq t_0 \leq T_2/C_0 \tag{5.14}$$

such that

$$\int_{-2t_0}^{-t_0} \int_{R' \leq |x| \leq 2R'} (T_2^{-1} |\omega|^2 + |\nabla \omega|^2) \, dx dt \gtrsim \exp(-10(R')^2/T_2) T_2^{-1/2}.$$

Covering the annulus $R' \leq |x| \leq 2R'$ by $O(\exp(O((R')^2/T_2))$ balls of radius $t_0^{1/2}$, one can then find x_* with $R' \leq |x_*| \leq 2R'$ such that

$$\int_{-2t_0}^{-t_0} \int_{B(x_*, t_0^{1/2})} (T_2^{-1}|\omega|^2 + |\nabla \omega|^2)\, dx dt \gtrsim \exp(-O((R')^2/T_2))T_2^{-1/2}. \tag{5.15}$$

Now we apply Proposition 4.3 on the slab $[0, 1000t_0] \times \mathbb{R}^3$ with

$$r := C_0^{1/4}(t_0/T_2)^{1/2}R' \leq |x_*|/10, t_1 := t_0,$$

and u replaced by the function

$$(t, x) \mapsto \omega(-t, x_* + x)$$

(so that the hypothesis (4.4) follows from the vorticity equation and (5.5)) to conclude that

$$Z'' \lesssim \exp(-C_0^{1/2}\frac{(R')^2}{500T_2})X'' + t_0^{3/2}\exp(O(C_0^{1/2}(R')^2/T_2))Y'' \tag{5.16}$$

where

$$X'' := \int_{-T_2}^{0} \int_{B(x_*, |x_*|/2)} (t_0^{-1}|\omega|^2 + |\nabla\omega|^2)\, dx dt$$

and

$$Y'' := t_0^{-3/2} \int_{B(x_*, |x_*|/2)} |\omega(0, x)|^2 e^{-|x - x'|^2/4t_0}\, dx$$

and

$$Z'' := \int_{-2t_0}^{[-t_0} \int_{B(x_*, t_0^{1/2})} (t_0^{-1}|\omega|^2 + |\nabla\omega|^2) e^{-|x-x_*|^2/4|t|}\, dx dt.$$

From (5.15), (5.14) one has

$$Z'' \gtrsim \exp(-O((R')^2/T_2))T_2^{-1/2}.$$

From (5.10), (5.14) one has

$$X'' \lesssim T_2^{-1}t_0^{-1}(R')^3 \lesssim \exp(O((R')^2/T_2))T_2^{-1/2}.$$

As C_0 is large, the first term on the right-hand side of (5.16) can thus be absorbed by the left-hand side, so we conclude that

$$Y'' \gtrsim \exp(-O(C_0^{1/2}(R')^2/T_2))T_2^{-2}$$

and hence

$$\int_{R'/2 \leq |x| \leq 2R'} |\omega(0, x)|^2\, dx \gtrsim \exp(-O(C_0^{1/2}(R')^2/T_2))T_2^{-2}t_0^{3/2}.$$

Using the bounds (5.13), (5.9), (5.14), we conclude in particular that

$$\int_{2R \leq |x| \leq A_6 R/2} |\omega(0, x)|^2\, dx \gtrsim \exp(-\exp(A_6^{O(1)}))T_2^{-1/2}. \tag{5.17}$$

Note that this bound is also implied by (5.12). Thus we have unconditionally established (5.17) for any scale T_2 obeying (5.8), and for a suitable scale R obeying (5.9) and the bounds (5.10).

We now convert this vorticity lower bound (5.17) to a lower bound on the velocity. The annulus $\{2R \leq |x| \leq A_6 R/2\}$ has volume $O(\exp(\exp(A_6^{O(1)}))T_2^{3/2})$

by (5.9), hence by the pigeonhole principle there exists a point x_* in this annulus for which

$$|\omega(0, x_*)| \gtrsim \exp(-\exp(A_6^{O(1)}))T_2^{-1}.$$

Comparing this with (5.10), we see that

$$|\int_{\mathbb{R}^3} \omega(0, x_* - ry)\varphi(y)\, dy| \gtrsim \exp(-\exp(A_6^{O(1)}))T_2^{-1}$$

for some bump function φ supported on $B(0,1)$, where r is a radius of the form $r = \exp(-\exp(A_6^{O(1)}))T_2^{1/2}$. Writing $\omega = \nabla \times u$ and integrating by parts, we conclude that

$$|\int_{\mathbb{R}^3} u(0, x_* - ry)\nabla \times \varphi(y)\, dy| \gtrsim \exp(-\exp(A_6^{O(1)}))T_2^{-1/2}$$

and hence by Hölder's inequality

$$\int_{B(0,1)} |u(0, x_* - ry)|^3\, dy \gtrsim \exp(-\exp(A_6^{O(1)}))T_2^{-3/2}$$

or equivalently

$$\int_{B(x_*,r)} |u(0, x)|^3\, dx \gtrsim \exp(-\exp(A_6^{O(1)})).$$

We conclude that for any scale T_2 obeying (5.8), we have

$$\int_{T_2^{1/2} \leq |x| \leq \exp(A_7)T_2^{1/2}} |u(0, x)|^3\, dx \gtrsim \exp(-\exp(A_6^{O(1)})).$$

Summing over a set of such scales T_2 increasing geometrically at ratio $\exp(A_7)$, we conclude that if $T \geq A_4^2 N_0^{-2}$, then

$$\int_{\mathbb{R}^3} |u(0, x)|^3\, dx \gtrsim \exp(-\exp(A_6^{O(1)})) \log(TN_0^2).$$

Comparing this with (3.1), one obtains the claim. □

6. Applications

Using the main estimate, we now prove the theorems claimed in the introduction.

We begin with Theorem 1.2. By increasing A as necessary we may assume that $A \geq C_0$, so that Theorem 5.1 applies. By rescaling it suffices to establish the claim when $t = 1$, so that $T \geq 1$. Applying Theorem 5.1 in the contrapositive, we see that

$$\|P_N u\|_{L_t^\infty L_x^\infty([1/2,1]\times\mathbb{R}^3)} \leq A_1^{-1} N \tag{6.1}$$

whenever $N \geq N_*$, where

$$N_* := \exp(\exp(\exp(A^{C_0^7}))).$$

We now insert this bound into the energy method. As before, we split $u = u^{\text{lin}} + u^{\text{nlin}}$ on $[1/2, 1] \times \mathbb{R}^3$, where

$$u^{\text{lin}}(t) := e^{t\Delta}u(0)$$

and $u^{\text{nlin}} := u - u^{\text{lin}}$, and similarly split $\omega = \omega^{\text{lin}} + \omega^{\text{nlin}}$. From (2.5), (3.1) we have

$$\|\nabla^j u^{\text{lin}}\|_{L_t^\infty L_x^p([1/2,1]\times\mathbb{R}^3)} \lesssim_j A \tag{6.2}$$

for all $j \geq 0$ and $3 \leq p \leq \infty$. We introduce the nonlinear enstrophy

$$E(t) := \frac{1}{2} \int_{\mathbb{R}^3} |\omega^{\mathrm{nlin}}(t,x)|^2 \, dx$$

for $t \in [1/2, 1]$, and compute the time derivative $\partial_t E(t)$. From the vorticity equation (3.41) and integration by parts we have

$$\partial_t E(t) = -Y_1(t) + Y_2(t) + Y_3(t) + Y_4(t) + Y_5(t) \tag{6.3}$$

where

$$\begin{aligned} Y_1(t) &= \int_{\mathbb{R}^3} |\nabla \omega^{\mathrm{nlin}}(t,x)|^2 \, dx \\ Y_2(t) &= -\int_{\mathbb{R}^3} \omega^{\mathrm{nlin}} \cdot (u \cdot \nabla) \omega^{\mathrm{lin}} \, dx \\ Y_3(t) &= \int_{\mathbb{R}^3} \omega^{\mathrm{nlin}} \cdot (\omega^{\mathrm{nlin}} \cdot \nabla) u^{\mathrm{nlin}} \, dx \\ Y_4(t) &= \int_{\mathbb{R}^3} \omega^{\mathrm{nlin}} \cdot (\omega^{\mathrm{nlin}} \cdot \nabla) u^{\mathrm{lin}} \, dx \\ Y_5(t) &= \int_{\mathbb{R}^3} \omega^{\mathrm{nlin}} \cdot (\omega^{\mathrm{lin}} \cdot \nabla) u^{\mathrm{nlin}} \, dx \\ Y_6(t) &= \int_{\mathbb{R}^3} \omega^{\mathrm{nlin}} \cdot (\omega^{\mathrm{lin}} \cdot \nabla) u^{\mathrm{lin}} \, dx. \end{aligned}$$

From Hölder, (6.2), (3.1) we have

$$Y_2(t), Y_6(t) \lesssim A^2 E(t)^{1/2} \lesssim A^4 + E(t)$$

and similarly

$$Y_4(t), Y_5(t) \lesssim A E(t),$$

using Plancherel's theorem to control

$$\|\nabla u^{\mathrm{nlin}}\|_{L^2_x(\mathbb{R}^3)} \lesssim \|\omega^{\mathrm{nlin}}\|_{L^2_x(\mathbb{R}^3)}. \tag{6.4}$$

For $Y_3(t)$ we apply a Littlewood-Paley decomposition to all three factors to bound

$$Y_3(t) \lesssim \sum_{N_1, N_2, N_3} \int_{\mathbb{R}^3} P_{N_1} \omega^{\mathrm{nlin}} \cdot (P_{N_2} \omega^{\mathrm{nlin}} \cdot \nabla) P_{N_3} u^{\mathrm{nlin}} \, dx$$

where N_1, N_2, N_3 range over powers of two. The integral vanishes unless two of the N_1, N_2, N_3 are comparable to each other, and the third is less than or comparable to the other two. Controlling the two highest frequency terms in L^2_x and the lower one in L^∞_x, and using the Littlewood-Paley localised version of (6.4), we conclude that

$$Y_3(t) \lesssim \sum_{N_1, N_2, N_3 : N_1 \sim N_2 \gtrsim N_3} \|P_{N_1} \omega^{\mathrm{nlin}}\|_{L^2_x(\mathbb{R}^3)} \|P_{N_2} \omega^{\mathrm{nlin}}\|_{L^2_x(\mathbb{R}^3)} \|P_{N_3} \omega^{\mathrm{nlin}}\|_{L^\infty_x(\mathbb{R}^3)}.$$

From (3.1), (2.3), the quantity $\|P_{N_3} \omega^{\mathrm{nlin}}\|_{L^\infty_x(\mathbb{R}^3)}$ is bounded by $O(AN_3^2)$; for $N_3 \geq N_*$, (2.3) we have the superior bound $O(A_1^{-1} N_3^2)$. We thus see that

$$\sum_{N_3 \lesssim N_2} \|P_{N_3} \omega^{\mathrm{nlin}}\|_{L^\infty_x(\mathbb{R}^3)} \lesssim A_1^{-1} N_2^2 + A N_*^2$$

and thus by Cauchy-Schwarz

$$Y_3(t) \lesssim \sum_{N_1} \|P_{N_1}\omega^{\mathrm{nlin}}\|^2_{L^2_x(\mathbb{R}^3)}(A_1^{-1}N_1^2 + AN_*^2).$$

On the other hand, from Plancherel's theorem we have

$$Y_1(t) \sim \sum_{N_1} \|P_{N_1}\omega^{\mathrm{nlin}}\|^2_{L^2_x(\mathbb{R}^3)} N_1^2$$

and

$$E(t) \sim \sum_{N_1} \|P_{N_1}\omega^{\mathrm{nlin}}\|^2_{L^2_x(\mathbb{R}^3)}$$

and hence

$$Y_3(t) \lesssim A_1^{-1}Y_1(t) + AN_*^2 E(t).$$

Putting all this together, we conclude that

$$\partial_t E(t) + Y_1(t) \lesssim AN_*^2 E(t) + A^4.$$

In particular, from Gronwall's inequality we have

$$E(t_2) \lesssim E(t_1) + A^4$$

whenever $1/2 \leq t_1 \leq t_2 \leq 1$ is such that $|t_2 - t_1| \leq A^{-1}N_*^{-2}$. On the other hand, from a (slightly rescaled) version of (3.13) we have

$$\int_{1/2}^1 E(t)\, dt \lesssim A^4$$

and hence on any time interval in $[1/2, 1]$ of length $A^{-1}N_*^{-2}$ there is at least one time t with $E(t) \lesssim A^5 N_*^2$. We conclude that

$$E(t) \lesssim A^5 N_*^2 \lesssim N_*^{O(1)},$$

for all $t \in [3/4, 1]$, which then also implies

$$\int_{3/4}^1 Y_1(t) \lesssim N_*^{O(1)}.$$

Iterating this as in the proof of Proposition 3.1(iii) (or Proposition 3.1(vi)), we now have the estimtes

$$|u(t,x)|, |\nabla u(t,x)|, |\omega(t,x)|, |\nabla\omega(t,x)| \lesssim N_*^{O(1)}$$

on $[7/8, 1] \times \mathbb{R}^3$. This gives Theorem 1.2.

REMARK 6.1. More generally, one would expect in view of Theorem 5.1 that any reasonable function space estimate obeyed by the linear heat equation with L^3_x initial data will now also hold for classical solutions to Navier-Stokes obeying (1.2), but with an additional loss of $\exp\exp\exp(A^{O(1)})$ in the estimates. It seems likely that a modification of the arguments above would be able to obtain such estimates, particularly if one replaces the linear estimates (3.11) (or (6.2)) by more refined estimates that involve the profile of the initial data $u(0)$, and in particular on how the Littlewood-Paley components $\|P_N u(0)\|_{L^3_x(\mathbb{R}^3)}$ of the L^3_x norm of that data vary with the frequency N. We will not pursue this question further here.

Now we prove Theorem 1.4. We may rescale $T_* = 1$. Let $c > 0$ be a sufficiently small constant, and suppose for contradiction that

$$\limsup_{t \to 1^+} \frac{\|u(t)\|_{L^3_x(\mathbb{R}^3)}}{(\log\log\log \frac{1}{1-t})^c} < +\infty,$$

thus we have

$$\|u(t)\|_{L^3_x(\mathbb{R}^3)} \leq M(\log\log\log(1000 + \frac{1}{1-t}))^c \tag{6.5}$$

for all $0 \leq t < 1$ and some constant M. Applying Theorem 1.2, we obtain (for c small enough) the bounds

$$\|u(t)\|_{L^\infty_x(\mathbb{R}^3)}, \|\nabla u(t)\|_{L^\infty_x(\mathbb{R}^3)}, \|\omega(t)\|_{L^\infty_x(\mathbb{R}^3)}, \|\nabla \omega(t)\|_{L^\infty_x(\mathbb{R}^3)} \lesssim_M (1-t)^{-1/10} \tag{6.6}$$

(say) for all $1/2 \leq t < 1$. In particular, u is bounded in $L^2_t L^\infty_x$, contradicting the classical Prodi-Serrin-Ladyshenskaya blowup criterion [**Pr**], [**S2**], [**La**]; one could also use the Beale-Kato-Majda criterion [**BKM**] and (6.6) to obtain the required contradiction. The claim follows.

References

[A] D. Albritton, *Blow-up criteria for the Navier-Stokes equations in non-endpoint critical Besov spaces*, Anal. PDE **11** (2018), no. 6, 1415–1456, DOI 10.2140/apde.2018.11.1415. MR3803715

[BS] T. Barker and G. Seregin, *A necessary condition of potential blowup for the Navier-Stokes system in half-space*, Math. Ann. **369** (2017), no. 3-4, 1327–1352, DOI 10.1007/s00208-016-1488-9. MR3713543

[BKM] J. T. Beale, T. Kato, and A. Majda, *Remarks on the breakdown of smooth solutions for the 3-D Euler equations*, Comm. Math. Phys. **94** (1984), no. 1, 61–66. MR763762

[B] J. Bourgain, *Global wellposedness of defocusing critical nonlinear Schrödinger equation in the radial case*, J. Amer. Math. Soc. **12** (1999), no. 1, 145–171, DOI 10.1090/S0894-0347-99-00283-0. MR1626257

[CKN] L. Caffarelli, R. Kohn, and L. Nirenberg, *Partial regularity of suitable weak solutions of the Navier-Stokes equations*, Comm. Pure Appl. Math. **35** (1982), no. 6, 771–831, DOI 10.1002/cpa.3160350604. MR673830

[C] C. P. Calderón, *Existence of weak solutions for the Navier-Stokes equations with initial data in L^p*, Trans. Amer. Math. Soc. **318** (1990), no. 1, 179–200, DOI 10.2307/2001234. MR968416

[DD] H. Dong and D. Du, *The Navier-Stokes equations in the critical Lebesgue space*, Comm. Math. Phys. **292** (2009), no. 3, 811–827, DOI 10.1007/s00220-009-0852-y. MR2551795

[ESS] L. Escauriaza, G. Seregin, and V. Šverák, *Backward uniqueness for parabolic equations*, Arch. Ration. Mech. Anal. **169** (2003), no. 2, 147–157, DOI 10.1007/s00205-003-0263-8. MR2005639

[ESS2] L. Iskauriaza, G. A. Serëgin, and V. Shverak, *$L_{3,\infty}$-solutions of Navier-Stokes equations and backward uniqueness* (Russian, with Russian summary), Uspekhi Mat. Nauk **58** (2003), no. 2(350), 3–44, DOI 10.1070/RM2003v058n02ABEH000609; English transl., Russian Math. Surveys **58** (2003), no. 2, 211–250. MR1992563

[GKP] I. Gallagher, G. S. Koch, and F. Planchon, *A profile decomposition approach to the $L^\infty_t(L^3_x)$ Navier-Stokes regularity criterion*, Math. Ann. **355** (2013), no. 4, 1527–1559, DOI 10.1007/s00208-012-0830-0. MR3037023

[GKP2] I. Gallagher, G. S. Koch, and F. Planchon, *Blow-up of critical Besov norms at a potential Navier-Stokes singularity*, Comm. Math. Phys. **343** (2016), no. 1, 39–82, DOI 10.1007/s00220-016-2593-z. MR3475661

[GIP] I. Gallagher, D. Iftimie, and F. Planchon, *Asymptotics and stability for global solutions to the Navier-Stokes equations* (English, with English and French summaries), Ann. Inst. Fourier (Grenoble) **53** (2003), no. 5, 1387–1424. MR2032938

[K] T. Kato, *Strong L^p-solutions of the Navier-Stokes equation in $\mathbf{R}^m$, with applications to weak solutions*, Math. Z. **187** (1984), no. 4, 471–480, DOI 10.1007/BF01174182. MR760047

[KK] C. E. Kenig and G. S. Koch, *An alternative approach to regularity for the Navier-Stokes equations in critical spaces* (English, with English and French summaries), Ann. Inst. H. Poincaré Anal. Non Linéaire **28** (2011), no. 2, 159–187, DOI 10.1016/j.anihpc.2010.10.004. MR2784068

[La] O. A. Ladyženskaja, *Uniqueness and smoothness of generalized solutions of Navier-Stokes equations* (Russian), Zap. Naučn. Sem. Leningrad. Otdel. Mat. Inst. Steklov. (LOMI) **5** (1967), 169–185. MR0236541

[LSU] O. A. Ladyženskaja, V. A. Solonnikov, and N. N. Ural′ceva, *Linear and quasilinear equations of parabolic type* (Russian), Translations of Mathematical Monographs, Vol. 23, American Mathematical Society, Providence, R.I., 1968. Translated from the Russian by S. Smith. MR0241822

[Le] J. Leray, *Sur le mouvement d'un liquide visqueux emplissant l'espace* (French), Acta Math. **63** (1934), no. 1, 193–248, DOI 10.1007/BF02547354. MR1555394

[Ph] N. C. Phuc, *The Navier-Stokes equations in nonendpoint borderline Lorentz spaces*, J. Math. Fluid Mech. **17** (2015), no. 4, 741–760, DOI 10.1007/s00021-015-0229-2. MR3412277

[Pr] G. Prodi, *Un teorema di unicità per le equazioni di Navier-Stokes* (Italian), Ann. Mat. Pura Appl. (4) **48** (1959), 173–182, DOI 10.1007/BF02410664. MR126088

[S1] G. Seregin, *A certain necessary condition of potential blow up for Navier-Stokes equations*, Comm. Math. Phys. **312** (2012), no. 3, 833–845, DOI 10.1007/s00220-011-1391-x. MR2925135

[SS] G. Seregin and V. Šverák, *On global weak solutions to the Cauchy problem for the Navier-Stokes equations with large L_3-initial data*, Nonlinear Anal. **154** (2017), 269–296, DOI 10.1016/j.na.2016.01.018. MR3614655

[S2] J. Serrin, *On the interior regularity of weak solutions of the Navier-Stokes equations*, Arch. Rational Mech. Anal. **9** (1962), 187–195, DOI 10.1007/BF00253344. MR136885

[T] T. Tao, *Localisation and compactness properties of the Navier-Stokes global regularity problem*, Anal. PDE **6** (2013), no. 1, 25–107, DOI 10.2140/apde.2013.6.25. MR3068540

[WZ] W. Wang and Z. Zhang, *Blow-up of critical norms for the 3-D Navier-Stokes equations*, Sci. China Math. **60** (2017), no. 4, 637–650, DOI 10.1007/s11425-016-0344-5. MR3629487

UCLA Department of Mathematics, Los Angeles, CA 90095-1555
Email address: tao@math.ucla.edu

Proceedings of Symposia in Pure Mathematics
Volume **104**, 2021
https://doi.org/10.1090/pspum/104/01875

The Continuum Hypothesis

W. Hugh Woodin

Abstract. This is essentially the annotated slides of the lecture given at the Linde Hall Inaugural Symposium in February of 2019. The intent of this lecture was to present one view on the current situation regarding Cantor's Continuum Hypothesis.

1. Introduction

This account, starting with the next section, is essentially the annotated slides of the lecture given at the Linde Hall Inaugural Symposium in February of 2019. The intent of this lecture was to present one view on the current situation regarding Cantor's Continuum Hypothesis.

Recall that the Continuum Hypothesis (CH) is the assertion that if $X \subseteq \mathbb{R}$ is infinite then either X is countable or there is a bijection of X with $\mathbb{R}$. By the results of Gödel from 1938 and the results 25 years later of Cohen, CH can neither be proved or refuted by the basic (ZFC) axioms of Set theory.

But what exactly did Gödel and Cohen prove? We give here in the introduction, an informal but precise summary of the results of Cohen and Gödel. This will be repeated in very slightly different terms, but at the same time perhaps obscured by the introduction of a number of other notions from Set Theory.

A model of set theory is a pair (M, E) such that

$$E \subset M \times M$$

is a binary relation on the nonempty set M. The points of M are the "sets" of the model and the binary relation E specifies when one set is an element of another. If

$$(M, E) \models \text{ZFC}$$

which is the condition that each formal axiom of ZFC holds when interpreted in (M, E), then (M, E) is a model of ZFC. The axioms ZF are the axioms ZFC without the Axiom of Choice, and so one also has the notion that (M, E) is a model of ZF.

The results of Gödel and Cohen on CH primarily concern both models of ZFC but also pairs of such models. Suppose (M_0, E_0) and (M_1, E_1) are models of ZFC. Then (M_0, E_0) is an *inner model* of (M_1, E_1), and (M_1, E_1) is an *outer model* of (M_0, E_0) if the following hold.

2020 *Mathematics Subject Classification.* Primary 03E65.
Key words and phrases. Set Theory, determinacy, large cardinals.
The author was supported in part by NSF Grant #1664764.

(1) $M_0 \subseteq M_1$ and $E_0 = E_1 \restriction M_0$.
(2) For all $p \in M_0$ and for all $q \in M_1$, if $(q, p) \in E_1$ then $q \in M_0$.
(3) $\text{Ord}^{(M_0,E_0)} = \text{Ord}^{(M_1,E_1)}$.

Therefore an inner model is a substructure (and an outer model is an extension) but which satisfies the key additional requirements (2) and (3). Thus in passing from (M_1, E_1) to (M_0, E_0), if a set of M_1 survives so must all of its elements. The third requirement is the requirement that the two models have the same *ordinals*, which is the requirement that the two models have the same *height*.

The definition of inner and outer models of ZF is exactly the same. We can now state Gödel's theorem.

THEOREM 1.1 (Gödel). *Suppose that $(M, E) \models$ ZF. Then there exists an inner model (M_0, E_0) of (M, E) such that*

$$(M_0, E_0) \models \text{ZFC} + \text{CH}.$$

Moreover (M_0, E_0) is the minimum inner model of (M, E) such that $(M_0, E_0) \models$ ZF. □

After this seminal theorem of Gödel, a key question remained. Suppose

$$(M, E) \models \text{ZFC}.$$

Can there exist a nontrivial outer model of (M, E)? If not then by Gödel's theorem, ZF must *prove* both the Axiom of Choice and the Continuum Hypothesis.

Suppose $(M, E) \models$ ZF. For each $p \in M$, let X_p be the set of all $q \in M$ such that $(q, p) \in E$. Define the model (M, E) to be *full* if for each $p \in M$ and for each $A \subseteq X_p$ there exists $p_A \in M$ such that

$$A = \{q \in M \mid (q, p_A) \in E\}.$$

If $(M, E) \models$ ZF and (M, E) is full then there are no nontrivial outer models of (M, E).

However, if $(M, E) \models$ ZF and M is countable then (M, E) cannot be full.

THEOREM 1.2 (Cohen). *Suppose that $(M, E) \models$ ZFC and that M is countable. Then there exists an outer model (M_1, E_1) of (M, E) such that*

$$(M_1, E_1) \models \text{ZFC} + \neg\text{CH}. \qquad \square$$

Cohen also proved the analogous theorem for the Axiom of Choice.

THEOREM 1.3 (Cohen). *Suppose that $(M, E) \models$ ZFC and M is countable. Then there exists an outer model (M_1, E_1) of (M, E) such that*

$$(M_1, E_1) \models \text{ZF} + \neg\text{AC}. \qquad \square$$

We will conclude this account with one additional section which contains some additional remarks and for completeness, also contains one key definition which was not previously given.

2. The Universe of Sets

The universe of sets, denoted V, is a fundamental conception in modern mathematics. This generalizes the conception of the integers and the ZFC axioms are almost universally now accepted as holding for V. In fact the ZFC axioms are the natural generalization of the Peano Axioms.

These axioms show that V is generated by the transfinite iteration of the *Power Set* operation which is the operaation that when applied to X yields $\mathcal{P}(X)$, where $\mathcal{P}(X)$ denotes the set of all subsets of X.

Before giving the definition we review some basic notions and this will also lead us directly to the problem of Cantor's Continuum Hypothesis.

A set X is transitive if for all $Y \in X$, $Y \subset X$. Equivalently, if $Z \in Y$ and $Y \in X$, then $Z \in X$. Thus the emptyset, $\emptyset$, is transitive, but $\{\{\emptyset\}\}$ is not. If X is transitive then $\mathcal{P}(X)$ is transitive. Thus all the sets which can be generated from the trivial set, $\emptyset$, by a finite number of iterations of taking powersets, are transitive.

An ordinal is a transitive set which is linearly ordered by $\in$. Thus $\emptyset$ is an ordinal and if α is an ordinal then so is $\alpha \cup \{\alpha\}$. If α and β are ordinals then either $\alpha \in \beta$, $\alpha = \beta$, or $\beta \in \alpha$. We write $\alpha < \beta$ to indicate that $\alpha \in \beta$.

The ordinals themselves are ordered by the membership relation and the first three ordinals in this order are

$$\emptyset, \{\emptyset\}, \{\emptyset, \{\emptyset\}\}.$$

More generally, the finite ordinals correspond to the non-negative integers.

The importance of the ordinals is the following. Suppose that $(L, <)$ is a linear order which is a wellorder. More precisely, suppose that $(L, <)$ is a linear order for which every nonempty set $X \subseteq L$ has a $<$-least member. Then there exists an ordinal α such that

$$(\alpha, \in) \cong (L, <)$$

as linear orders. Both α and the isomorphism are unique, and so ordinals measure unambiguously the length of any wellorder. Thus if α and β are ordinals then $\alpha < \beta$ if and only if $(\alpha, \in)$ has shorter length than $(\beta, \in)$.

Two sets X and Y have the same cardinality if there is a bijection

$$\pi : X \to Y.$$

This relation is denoted, $|X| = |Y|$. Similarly, $|X| \leq |Y|$ indicates that there is a 1-to-1 function

$$\pi : X \to Y.$$

Without assuming the Axiom of Choice, there must exist sets X and Y such that neither $|X| \leq |Y|$ or $|Y| \leq |X|$ holds. But nevertheless, if $|X| \leq |Y|$ and $|Y| \leq |X|$ then $|X| = |Y|$. This is the Schroeder-Bernstein Theorem which simplifies the problem in many cases of showing that $|X| = |Y|$.

For example, first note that if $|X| = |Y|$ then trivially $|\mathcal{P}(X)| = |\mathcal{P}(Y)|$. Thus

$$|\mathbb{R}| \leq |\mathcal{P}(\mathbb{N})|$$

since $|\mathbb{N}| = |\mathbb{Q}|$ and since the function

$$\pi : \mathbb{R} \to \mathcal{P}(\mathbb{Q})$$

defined by $\pi(x) = \{r \in \mathbb{Q} \mid r < x\}$, is clearly 1-to-1. For the reverse direction, define for each $A \subseteq \mathbb{N}$,

$$x_A = \Sigma_{k \in A}\ 10^{-k}$$

if $A \neq \emptyset$, and $x_A = 0$ otherwise. Then the function

$$e : \mathcal{P}(\mathbb{N}) \to \mathbb{R}$$

defined by $\pi(A) = X_A$ is 1-to-1. Therefore $|\mathbb{R}| \leq |\mathcal{P}(\mathbb{N})|$ and $|\mathcal{P}(\mathbb{N}| \leq |\mathbb{R}|$, and so by the Schroeder-Bernstein Theorem, $|\mathbb{R}| = |\mathcal{P}(\mathbb{N})|$.

Thus Cantor's Theorem that $|\mathbb{R}| \neq |\mathbb{N}|$ now easily follows from the (now) more general theorem (of Cantor) that for any set X, $|X| \neq |\mathcal{P}(X)|$. This is almost completely trivial noting that if

$$F : X \to \mathcal{P}(X),$$

then $\{a \in X \mid a \notin F(a)\}$ is not in range of F, and so F cannot be a bijection.

The Axiom of Choice has many equivalent formulations. These range from the mysterious (suspicious) version:

- For every set X there exists a wellordering of the set X,

to (by a theorem of Tarski) the arguably more plausible assertion:

- Suppose that X and Y are infinite sets.Then

$$|X \times Y| = |X \cup Y|.$$

An ordinal α is a *cardinal*, if $|\beta| < |\alpha|$ for all $\beta < \alpha$. The least infinite ordinal, denoted ω, is a cardinal and it is the least infinite cardinal.

As is customary, we will abuse notation and write $|X| = \kappa$, where κ is a cardinal, to indicate that $|X| = |\kappa|$. This $|X| < \omega$ indicates that X is finite, and a set X is *countable* if $|X| \leq \omega$. A set X is uncountable if $\omega < |X|$.

The least uncountable cardinal is vastly more complicated to describe than ω, since the ordinals α such that $|\alpha| = \omega$ have very strong closure properties. Thus ω_1 denotes the first uncountable cardinal.

For every ordinal $\alpha > 0$ there exists a cardinal κ such that α measures the ordertype of the set of infinite cardinals λ such that $\lambda < \kappa$. This cardinal is denoted ω_α. Thus $\omega_0 = \omega$, etc.

By the Axiom of Choice, for every set X there exists an ordinal α such that

$$|X| = \omega_\alpha.$$

Thus $|\mathbb{R}| = \omega_\alpha$ for some ordinal α and a fundamental question is: Which α? The entire subject of Set Theory began with this question together with Cantor's theorem that $\alpha > 0$.

DEFINITION 2.1 (**Cantor's Continuum Hypothesis:** CH). $|\mathbb{R}| = \omega_1$. □

There are many equivalent formulations of CH and a useful one is the following. Suppose $X \subseteq \mathbb{R}$ is infinite and X is not countable. Then $|X| = |\mathbb{R}|$.

Our position is that any coherent conception of the V requires that CH hold or fail. It cannot be a question with no answer. The difficulty, as is now well known, is that by the theorems of Gödel and then Cohen, this question *cannot* be answered on the basis of the ZFC axioms alone.

The fundamental issue then is what axiom(s) are we missing beyond the ZFC axioms which will reveal the answer to the question of whether CH holds or fails in V? The trivial answer is either the axiom, CH holds, or its negation. But that hides the key question.

The key question is whether the conception of the universe of sets, based on the founding intuitions of that conception, leads in a convincing fashion to a sharpening of that conception which reveals the answer to the problem of CH. Of course this is not a precise mathematical question.

But that does not mean one cannot transform the question into a precise mathematical question, or at the very least pose a precise mathematical question, the answer to which if positive would arguably resolve the problem of CH.

3. The cumulative hierarchy

DEFINITION 3.1 (Cumulative Hierarchy). The sets V_α where α is an ordinal are defined by induction on α as follows.

(1) $V_0 = \emptyset$.
(2) $V_{\alpha+1} = \mathcal{P}(V_\alpha)$.
(3) If $\alpha > 0$ and α is a limit ordinal then $V_\alpha = \cup\{V_\beta \mid \alpha < \beta\}$. □

Thus $V_1 = \{\emptyset\}$, $V_2 = \{\emptyset, \{\emptyset\}\}$, etc. For each finite ordinal n, V_n is finite and

$$|V_{n+1}| = 2^{|V_n|}.$$

However, V_ω is infinite and it contains all the finite transitive sets. It is therefore simply the union of all the finite transitive sets.

In general, the V_α are all transitive since $\mathcal{P}(M)$ is transitive if M is transitive and since the union of transitive sets is transitive.

Further the V_α are strictly increasing in the sense that if $\alpha < \beta$ then $V_\alpha \subseteq V_\beta$ and $V_\alpha \neq V_\beta$.

Finally as implied by the ZFC axioms, for every set X there exists an ordinal α such that $X \in V_\alpha$.

The classical structures of mathematics therefore appear at some stage α. In fact, the structure

$$(V_\omega, \in)$$

is bi-interpretable with

$$(\mathbb{N}, +, \cdot)$$

the standard structure for Number Theory. The structure

$$(V_{\omega+1}, \in)$$

is bi-interpretable with

$$(\mathcal{P}(\mathbb{N}), \mathbb{N}, +, \cdot, \in),$$

which is the standard structure of Second Order Number Theory.

The statement of the Continuum Hypothesis is at worst a question of the structure

$$(V_{\omega+2}, \in).$$

We note that there are (consistent) sentences which if true in

$$(V_{\omega+1}, \in)$$

imply the Continuum Hypothesis. However, by an application of Cohen's method, there is no such sentence which can refute the Continuum Hypothesis.

4. Cohen's method

A model of ZFC is just a non-empty set M and a binary relation

$$E \subset M \times M$$

such that all the formal axioms of ZFC hold in

$$(M, E)$$

when E is interpreted as the membership relation. This is written

$$(M, E) \models \text{ZFC},$$

and for $a, b \in M$, we write $a\,E\,b$ to indicate $(a, b) \in E$.

By the Gödel Completness Theorem of mathematical logic, if the ZFC are formally consistent then there exists a model (M, E) such that

$$(M, E) \models \text{ZFC}.$$

If $(M, E) \models \text{ZFC}$ then Ord^M denotes the set of all $a \in M$ such that

$$(M, E) \models \text{“}a \text{ is an ordinal”}.$$

Thus $(\text{Ord}^M, E|\text{Ord}^M)$ is a linear order. The model (M, E) is *wellfounded* if

$$(\text{Ord}^M, E|\text{Ord}^M)$$

is wellorder, and this is equivalent to the assertion that

$$(M, E) \cong (N, \in)$$

for some transitve set N.

An inner model of (M, E) is a structure of the form $(N, E|M)$ where $N \subseteq M$ and such two key additional conditions are satisfied.

(1) For all $a \in N$, $\{b \in M \mid b\, E\, a\} \subseteq N$.
(2) $\text{Ord}^M = \text{Ord}^N$.

Gödel proved the relative consistency of CH with the ZFC axioms by proving the following theorem. This theorem also shows that the Axiom of Choice is relatively consistent with the ZF axioms. Here ZF denotes the ZFC axioms without the Axiom of Choice.

THEOREM 4.1 (Gödel [**4**], [**5**]). *Suppose* $(M, E) \models \text{ZF}$. *Then there is an inner model* $(N, E|N)$ *of* (M, E) *such that*

$$(N, E|N) \models \text{ZFC} + \text{CH}. \qquad \square$$

After this theorem it remained open whether there even can exist a model

$$(M, E) \models \text{ZFC}$$

such that (M, E) has a nontrivial inner model of ZFC.

The missing technique was one for constructing *outer models* of a given countable model $(M, E) \models \text{ZFC}$. Here a model, $(M^*, E^*) \models \text{ZFC}$, is an *outer model* of (M, E) if (M, E) is an inner model of (M^*, E^*).

THEOREM 4.2 (Cohen [**1**], [**2**]). *Suppose that* $(M, E) \models \text{ZFC}$ *and* M *is countable. Then there exists*

$$(M^*, E^*) \models \text{ZFC} + (\neg\text{CH})$$

such that (M, E) *is an inner model of* (M^*, E^*). $\square$

The technique Cohen introduced is the technique of *forcing*. It is an extremely powerful method which has been enormously developed in the 50+ years since Cohen's original work.

The restriction to countable models is necessary since if

$$(M, E) \cong (V_\alpha, \in)$$

for some ordinal α, then for any model

$$(M^*, E^*) \models \text{ZFC}$$

such that (M, E) is an inner model of (M^*, E^*), necessarily $M = M^*$.

What Cohen actually showed is that for any model $(M, E) \models \text{ZFC}$, there is a "blueprint" which can be constructed within (M, E) for an outer model of (M, E) in which CH is false.

If (M, E) is countable then that blueprint can actually be realized and the outer model constructed.

5. Beyond the ZFC axioms

Given the results of Gödel and Cohen, any resolution of CH requires new axioms. A compelling source of new axioms are *Large Cardinal Axioms*. These axioms assert the existence of sets so large that they cannot otherwise be proved to exist.

A very simple example is the large cardinal axiom which asserts the existence of a *strongly inaccessible cardinal*, where a cardinal $\kappa > \omega$ is strongly inaccessible if

(1) $|V_\alpha| < \kappa$ for all $\alpha < \kappa$,
(2) For all functions $\pi : V_\alpha \to \kappa$, where $\alpha < \kappa$, there exists $\beta < \kappa$ such that $\pi(x) < \beta$ for all $x \in V_\alpha$.

Thus if κ is strongly inaccessible then $V_\kappa \models \text{ZFC}$. Note that ω is nearly strongly inaccessible, satisfying all the conditions except of course the condition $\kappa > \omega$.

Much stronger and with a completely different intuition, κ is a *measurable cardinal* if there exists a uniform ultrafilter U on $\mathcal{P}(\kappa)$ such that for all $X \subset U$, if $|X| < \kappa$ then $\cap X \in U$. Measurable cardinals are strongly inaccessible and much more. Here an ultrafilter U on κ is a uniform ultrafilter if for all $X \in U$, $|X| = \kappa$.

The list of current large cardinal axioms is long, having been developed over the last 50 years. We list 6 large cardinal axioms, in increasing order of strength.

For the large cardinal axioms on this list, and for essentially all large cardinal axioms, the ordering of strength is *rank reflection.*

Here one axiom rank reflects another axiom, if the first axiom implies that there exists an ordinal α such that $V_\alpha \models \text{ZFC}$ and such that second axiom holds in V_α.

- There is a Woodin cardinal.
- There is a strong cardinal.
- There is a supercompact cardinal.
- There is an extendible cardinal.
- There is a huge cardinal.
- There is an ω-huge cardinal.
- There is an Axiom I_0 cardinal.

The concept of a Woodin cardinal will be particularly important in the discussion that follows and a detailed definition is given in Section 20 below. In fact all these large cardinal axioms are relevant or have been relevant, to the story of CH as it is currently unfolding, and the last axiom is one of the strongest axioms known.

Gödel speculated that large cardinal axioms might resolve CH. Unfortunately by Cohen's method and by its adaptation to also produce outer models of a given countable model, in which CH holds, this cannot happen. Cohen's outer model construction preserves the existence of large cardinals (for example it preserves any on the previous list).

What about the generalizations of Cohen's construction? The same negative result applies to these constructions provided that in the initial model (M, E), a simple variation of the large cardinal axiom holds. This is the variation which asserts that for all ordinals α there exists $\kappa > \alpha$ with the designated feature (for example that κ is a huge cardinal).

6. Perhaps CH simply has no answer

Perhaps the problem of CH is simply an unusal case. But this is most certainly not accurate. There are now quite a number of statements which have been proved to be unsolvable on the basis of the ZFC axioms using variations and enhancements of Cohen's method. A vast majority of these statements, like CH, originate in Set Theory, but there are notable examples which do not.

These include:

- From group theory: The Whitehead Problem (Shelah)
- From analysis: Kaplansky's Conjecture (Solovay)
- Also from analysis: Suslin's Problem (Solovay-Tennenbaum, Jensen, Jech)
- From measure theory: The Borel Conjecture (Laver)
- From operator algebras: The Brown-Douglas-Filmore Automorphism Problem (Phillips-Weaver, Farah)

Even seemingly innocuous statements can be unsolvable on the basis of ZFC axioms. For example consider the following claim which is a natural generalization of the Fubini Theorem for measurable subsets of the unit square, $[0, 1] \times [0, 1]$.

Claim: *Suppose that*

$$A \subset [0, 1] \times [0, 1]$$

and that for Lebesgue almost all $z \in [0, 1]$, the horizontal and vertical sections of A given by z are Lebesgue measurable. Suppose that for Lebesgue almost all $z \in [0, 1]$, the horizontal section of A given by z is of Lebesgue measure 0. Then for Lebesgue almost all $z \in [0, 1]$, the vertical section of A given by z is of Lebesgue measure 0. □

This claim is refuted by CH but it is relatively consistent with the ZFC axioms. In fact if $(M, E) \models$ ZFC is a countable model, then there is an outer model of (M, E) (satisfying ZFC) in which this claim holds. This is proved by using Cohen's method.

The same is true for the Baire category version of the claim, that claim is also refuted by CH, but by Cohen's method it is relatively consistent with the ZFC axioms. In fact, Cohen's original construction suffices here.

Therefore both of these claims imply that CH is false. Perhaps such claims could provide the resolution to CH since at first glance they each seem to be reasonable candidates for new axioms. We conjecture that these two claims are mutually inconsistent, more precisely we conjecture that each claim implies that the other claim is false. What would be the basis for choosing one over the other?

This example also illustrates a completely different aspect of the power of Cohen's method. It is a recent theorem that if there is a probability measure on $[0, 1]$ which measures *all* subsets of $[0, 1]$ and which is non-atomic, then the Claim above *holds*, even though it must fail if reformulated for that probability measure. But the proof which one would naturally expect to just use basic methods from analysis, uses instead Cohen's method of forcing. There is analogous result for the property of Baire.

This ubiquity of unsolvable problems suggests it is hopeless to resolve the problem of CH and so that there is no answer beyond just a personal preference.

But this "resolution" itself runs into a serious obstruction. Large cardinal axioms are not provable since by Gödel's Second Incompleteness Theorem, ZFC cannot prove there is a model of ZFC, unless ZFC inconsistent. Similarly, for any proposed new axiom ϕ, ZFC + ϕ cannot prove that there is a model of ZFC + ϕ, unless ZFC + ϕ is inconsistent.

But large cardinal axioms are falsifiable. Thus the conception of the universe of sets, V, in which large cardinals exist, makes *number theoretic* claims which are falsifiable. To emphasize this, we make the following claim.

In the next 1000 years, no contradiction in the formal theory

$$\text{ZFC} + \text{"There are infinitely many Woodin cardinals"}$$

will be discovered, not by any means whatsoever.

The real claim of course is that the formal theory,

$$\text{ZFC} + \text{"There are infinitely many Woodin cardinals"},$$

is consistent. Such claims (if true) show that there is mathematical truth which is beyond the reach of formal proof.

Thus the skepticism about there being a resolution to CH based on the skepticism about the entire conception of the universe of sets must surely be wrong, for how else can one justify these claims. But then the tables turn, and the skeptic can now demand that CH have an answer.

It is an incoherent position that claims the existence of very large sets and yet accepts that the most basic of questions about small sets has no answer.

7. Back to the problem of CH

Definability in a structure is a fundamental notion in Mathematical Logic. If (M, E) is a structure with $E \subset M \times M$ then a set $A \subset M$ if logically definable from parameters in M, if there is a formula $\phi(x_0, x_1, \ldots, x_n)$ in the language of set theory, and elements $a_1, \ldots, a_n \in M$ such that

$$A = \{a \in M \mid (M, E) \models \phi[a, a_1, \ldots, a_n]\}.$$

If A can be defined by just a formula $\phi(x_0)$, then A is logically definable in (M, E) without parameters. Thus if $(M, E) \models \text{ZFC}$ then Ord^M is a definable subset of M, and no parameters are required.

For example if $(M, E) = (\mathbb{R}, <)$ then the only sets $A \subseteq \mathbb{R}$ which are logically definable in $(\mathbb{R}, <)$ without parameters are $\mathbb{R}$ and $\emptyset$; and the only sets $A \subseteq \mathbb{R}$ which are logically definable in $(\mathbb{R}, <)$ from parameters, are the finite unions of intervals.

We can similarly define when a binary relation $R \subset M \times M$ is logically definable in (M, E) from parameters, or without parameters

A natural speculation is that one can better understand the problem of CH be examining special cases. But does this even make sense?

In fact it does, the simple idea is to narrow the scope of consideration to the "simplest" uncountable sets.

DEFINITION 7.1. A set $A \subseteq V_{\omega+1}$ is a **projective set** if A can be logically defined in the structure $(V_{\omega+1}, \in)$ from parameters. □

We can easily extend the definition to relations on $V_{\omega+1}$:

DEFINITION 7.2. A set $A \subseteq V_{\omega+1} \times V_{\omega+1}$ is a **projective set** if A can be logically defined as a binary relation in the structure $(V_{\omega+1}, \in)$ from parameters. □

The countable subsets of $V_{\omega+1}$ and $V_{\omega+1} \times V_{\omega+1}$ are projective sets but so are $V_{\omega+1}$ and $V_{\omega+1} \times V_{\omega+1}$ themselves, and these sets are not countable.

The Continuum Hypothesis is equivalent to the statement:

- Suppose $A \subseteq V_{\omega+1}$ is uncountable. Then A and $V_{\omega+1}$ have the same cardinality.

This is a statement about **all** subsets of $V_{\omega+1}$ which is why CH is expressible in $V_{\omega+2}$. But we can naturally narrow the scope of consideration to just the projective sets and this leads to the formulation of Projective CH.

DEFINITION 7.3 (**Projective** CH). Suppose $A \subseteq V_{\omega+1}$ is an uncountable projective set. Then here is a bijection,

$$F : V_{\omega+1} \to A$$

such that F is a projective set. □

There is an important point. One might reasonably argue that Projective CH should just be the statement that if $A \subseteq V_{\omega+1}$ is infinite then $|A| = |V_{\omega+1}|$. But this is apriori a much weaker statement, and it narrows the scope of consideration in a very different (and asymmetric) way.

One can also narrow the scope of consideration to more carefully examine the Axiom of Choice. This will actually highlight the rather fundamental differences in the two possible formulations of Projective CH.

DEFINITION 7.4. Suppose that $A \subseteq X \times Y$. A function

$$F : X \to Y$$

is a **choice function** for A if for all $a \in X$, either

$$\{b \in a \mid (a, b) \in A\} = \emptyset,$$

or $(a, F(a)) \in A$. □

The Axiom of Choice (AC) is the assertion that for every set $A \subset X \times Y$, there is a choice function for A. We can naturally narrow the scope and define Projective AC.

DEFINITION 7.5 (**Projective** AC). Suppose $A \subseteq V_{\omega+1} \times V_{\omega+1}$ is a projective set. Then there is a function

$$F : V_{\omega+1} \to V_{\omega+1}$$

such that F is a choice function for A and such that F is a projective set. □

Notice that even in the context of ZFC, the question of whether Projective AC holds is entirely unclear. This is because of the requirement that the choice function be projective. Without that requirement, Projective AC would be a trivial consequence of the Axiom of Choice.

Thus the problem of Projective CH and Projective AC are taken together, defne the projective analog of Hilbert's first question. Here we are drawing a distinction between the Axiom of Choice versus the Wellordering Principle. The latter also has a projective incarnation but unlike the case of the Axiom of Choice and the Wellordering Principle, the projective versions are *not equivalent.*

There were many attempts in the early 1900s to solve both the problem of projective Continuum Hypothesis and the problem of the projective Axiom of Choice and success was achieved for the simplest cases (where in effect, the logical complexity of the logical definitions allowed in the definition of the projective sets was restricted).

However, by 1925 these problems both looked hopeless and we now know why. Both the problems of Projective CH and Projectice AC are unsolvable on the basis of the ZFC axioms, just like the case of CH. In fact the very constructions of Gödel and Cohen suffice for this.

If $(M, E) \models \text{ZFC}$ and $(N, E{\restriction}N)$ is the inner model of (M, E) defined by Gödel, then

$$(N, E{\restriction}N) \models \text{Projective CH} + \text{Projective AC}.$$

Further if M is countable and (M^*, E^*) is the outer model of $(N, E{\restriction}N)$ defined by Cohen (and so starting not from (M, E) but from the model Gödel defined) then

$$(M^*, E^*) \not\models \text{Projective CH}$$

and

$$(M^*, E^*) \not\models \text{Projective AC}.$$

Thus we seemed to have achieved nothing by narrowing the scope. But something quite unexpected and astonishing happened.

8. An unexpected entanglement

THEOREM 8.1 ([**16**]). *Suppose there are infinitely many Woodin cardinals. Then* Projective CH *holds.* □

One also obtains the following theorem which answers the question of whether the projective version of the Wellordering Principle holds.

THEOREM 8.2 ([**16**]). *Suppose there are infinitely many Woodin cardinals. Then there is no projective wellordering of the reals.* □

This nevertheless leaves open the question of Projective AC. The following theorem answers that as well and shows that in the projective realm, the Wellordering Principle is *not* equivalent to the Axiom of Choice.

THEOREM 8.3 (Martin-Steel). *Suppose there are infinitely many Woodin cardinals. Then* Projective AC *holds.* □

What is surprising and unexpected is that the existence of very large sets can affect the structure of the projective sets.

The explanation involves yet another family of axioms, the determinacy axioms. For every set $A \subset \mathbb{R}$, there is naturally associated to A an infinite game G_A which can be defined in any of number of ways. The two players alternate choosing the digits of a decimal expansion and so produce after infinitely many rounds, a real number $x \in [0, 1]$. Player I wins if $x \in A$ and Player II wins if $x \notin A$.

The game G_A is *determined*, and we say A is determined, if there is a winning strategy for one of the players in the game G_A. Here a strategy τ is a function from the set of all finite decimal expansions (i.e., the set of finite sequences of numbers from the set $\{0, 1, \ldots, 9\}$) to the set $\{0, 1, \ldots, 9\}$. The strategy τ is a winning strategy for Player I, if in using τ to choose their move at every stage, Player I wins the G_A no matter how Player II moves. Similarly define that τ is a winning strategy for Player II.

The Axiom of Determinacy (AD), proposed by Mycielski and Stenhaus [**14**] as an alternative to the Axiom of Choice, asserts that every set $A \subseteq \mathbb{R}$ is determined. It is not difficult to show that the Axiom of Choice implies there is a set $A \subseteq \mathbb{R}$ which is not determined.

Narrowing the scope, one defines Projective Determinacy (PD) as the axiom which asserts that if $A \subseteq \mathbb{R}$ is a projective set then A is determined. One can give a more traditional definition of the projective subsets of $\mathbb{R}$, since $A \subseteq \mathbb{R}$ is projective if and only if A can be generated from the initial set of $\mathbb{R}$ itself, in finitely many steps of taking complements or images by functions

$$f : \mathbb{R} \to \mathbb{R}$$

whose restriction to $\mathbb{R}\backslash\mathbb{Q}$ is continuous. In the same fashion, one can define the projective subsets of the plane, and so define when a function $F : \mathbb{R} \to \mathbb{R}$ is projective.

THEOREM 8.4 (Martin-Steel [**9**]). *Suppose there are infinitely many Woodin cardinals. Then* PD *holds.*

It was previously known that PD answers the questions of Projective CH and Projective AC, and in very strong ways. Note that is $A \subseteq \mathbb{R}$ is a projective set and A contains a perfect subset, then there is a bijection

$$F : A \to \mathbb{R}$$

which is a projective function.

THEOREM 8.5. *Assume* PD. *Then for every projective set, $A \subseteq \mathbb{R}$, the following hold.*

(1) ([**14**]) *A has the property of Baire.*
(2) (Mycielski–Swierczkowski [**13**]) *A is Lebesgue measurable.*
(3) ([**14**]) *Suppose A is uncountable. Then A contains a perfect subset.* □

THEOREM 8.6 (Moschovakis [**12**]). *Assume* PD. *Then* Projective AC *holds.* □

The axiom PD is the axiom for the standard structure of Second Number Theory,

$$(\mathcal{P}(\mathbb{N}), \mathbb{N}, +, \cdot, \in),$$

which when added to the standard axioms of Second Order Number Theory, yields what is arguably the correct analog of the Peano Axioms. In this context the axiom PD is stated as a logical *scheme* of axioms just as the formal axiom of mathematical induction is stated as a logical scheme of axioms within Peano Axioms.

The connection with large cardinal axioms is the remarkable fact that if ϕ is a sentence of Second Order Number Theory, then this expanded set of axioms logically proves ϕ if and only if for some $n \in \mathbb{N}$, the statment

$$(\mathcal{P}(\mathbb{N}), \mathbb{N}, +, \cdot, \in) \models \phi,$$

as statement of set theory, is proved by the axioms,

$$\text{ZFC} + \text{“There are } n \text{ many Woodin cardinals”}.$$

Another version of the connection is given by the following “converse” to the Martin-Steel Theorem, Theorem 8.4.

THEOREM 8.7 (Woodin [**8**]). *Suppose that*

$$(M, E) \models \text{ZFC} + \text{PD}.$$

Then for each $n \in \mathbb{N}$, *there exists an inner model* (M_n, E_n) *of* (M, E) *such that*

$$(M_n, E_n) \models \text{ZFC} + \text{"There are } n \text{ many Woodin cardinals"}. \qquad \square$$

This is in some sense the best possible converse. The axioms ZFC + PD cannot prove the existence of any large cardinal, for then these axioms would prove their own consistency and so by Gödel's Second Incompleteness Theorem, ZFC + PD would be inconsistent.

9. The effective cumulative hierarchy: Gödel's universe L

Perhaps the solution to the projective version of CH can lead us to the solution to CH itself.

DEFINITION 9.1. Suppose X is a set. Then $\mathcal{P}_{\text{Def}}(X)$ is the set of all $Y \subseteq X$ such that Y is logically definable in the structure

$$(X, \in)$$

from parameters. □

Thus $\mathcal{P}_{\text{Def}}(V_{\omega+1})$ is just the collection of all the projective sets. One can now naturally modify the definition of the cumulative hierarchy, by using the definable powerset operation, in place of the powerset operation.

DEFINITION 9.2 (Gödel). Define L_α by induction on the ordinal α as follows.

(1) $L_0 = \emptyset$.
(2) $L_{\alpha+1} = \mathcal{P}_{\text{Def}}(L_\alpha)$.
(3) if α is a limit ordinal then $L_\alpha = \cup\{L_\beta \mid \beta < \alpha\}$. □

L is the class of all sets X such that $X \in L_\alpha$ for some α, and the axiom $V = L$, is the formal axiom which asserts the every set belongs to L. This axiom resolves the problems of CH, Projective CH, and Projective AC, in one theorem.

THEOREM 9.3 (Gödel). *Assume* $V = L$. *Then there is a projective wellordering of* $\mathbb{R}$ *of length* ω_1. □

From the perspective of inner models of a given model $(M, E) \models \text{ZF}$, the following theorem shows explicitly what Gödel's construction of L achieved.

THEOREM 9.4 (Gödel). *Suppose that*

$$(M, E) \models \text{ZF}.$$

Let N *be the set of* $a \in N$ *such that* $(N, E) \models$ *"*$a \in L$*". Then* $(N, E{\restriction}N)$ *is an inner model of* (M, E) *and*

$$(N, E{\restriction}N) \models \text{ZFC} + \text{"}V = L\text{"}.$$

Further if

$$(M, E) \models \text{"}V = L\text{"},$$

then $M = N$. □

As we have already noted, Cohen's method of forcing is the primary tool for showing statements are unsolvable on the basis of the ZFC axioms. But by Gödel's theorem, if $(M, E) \models \text{ZFC}$ and (M^*, E^*) is an outer model of (M, E) such that

$$(M^*, E^*) \models \text{ZFC} + \text{“}V = L\text{”},$$

then $M = M^*$ and so

$$(M, E) \models \text{ZFC} + \text{“}V = L\text{”}.$$

Thus in the context of axioms, ZFC + "$V = L$", Cohen's method of forcing *cannot* be used to establish that sentences are unsolvable.

10. The axiom $V = L$ and large cardinals

By Gödel's theorem that if $V = L$ then there is a projective wellordering of the reals, and since the existence of infinitely many Woodin cardinals implies there is no projective wellordering of the reals, if $V = L$ then there cannot exist infinitely many Woodin cardinals.

But the fundamental conflict between assuming $V = L$ and large cardinal axioms was proved much earlier.

THEOREM 10.1 (Scott [**15**]). *Assume $V = L$. Then there are no measurable cardinals.* □

This theorem of Scott provided the first indication that not only was theory

$$\text{ZFC} + \text{“}V \neq L\text{”}$$

consistent, the axiom $V = L$ is *false*. Scott's theorem predates Cohen results by several years.

Thus began the program to construct "enlargements" of L in which large cardinal axioms can hold. The methodology for this is to use the existence of large cardinals to expand the notion of the definable powerset. Then one obtains the enlargement of L by the transfinite iteration of this enhanced version on $\mathcal{P}_{\text{Def}}(X)$.

This is the Inner Model Program which through the efforts many mathematicians over the last 50 years has been remarkably successful. Enlargements of L were first constructed at the level of a single measurable cardinal and then at the level of a single Woodin cardinal. The current threshold is well beyond the existence of a single Woodin cardinal.

The survey [**7**] is an excellent source of the historical development of the Inner Model Program.

Thus for example one has the following theorem.

THEOREM 10.2 (Martin-Steel [**10**]). *Suppose that $n \in \mathbb{N}$ and*

$$(M, E) \models \text{ZFC} + \text{“There are } n \text{ many Woodin cardinals”}.$$

Then there is an inner model (M_n, E_n) of (M, E) such that

$$(M_n, E_n) \models \text{ZFC} + \text{“There are } n \text{ many Woodin cardinals”}$$

and such that

$$(M, E) \models \text{“There is a projective wellordering of the reals of length } \omega_1\text{”}. \quad \square$$

Enlargements of L at the level of infinitely many Woodin cardinals cannot have projective wellorderings of the reals since PD must hold within these inner models. But that is not a serious obstruction to the construction of these enlargements of L, it simply indicates increasing complexity. The basic template for these constructions was first defined by Mitchell-Steel [**11**].

These constructions do not however yield easily specified generalizations of the axiom $V = L$. The reason is that construction of say the enlargement of L for the existence of infinitely many Woodin cardinals, if repeated within that enlargement does *not* yield the same inner model. This can be managed and the axiom generalizing $V = L$ defined, but the problem gets more and more complicated as one examines the enlargements of L at higher levels of the large cardinal hierarchy.

At the just level of the existence of a Woodin cardinal which is a limit of Woodin cardinals, very serious obstructions start to appear.

But there is another far more serious issue and this concerns all the enlargements which have been constructed at *any* level. Every specific enlargement which has been constructed, is subject to a generalized Scott's theorem. For example for any enlargement of L constructed using the Mitchell–Steel methodology, that enlargement must satisfy the sentence which asserts that there are no supercompact cardinals.

Thus prospects of an ultimate enlargement of L seem from this perspective very remote at best. But there is another approach which is to try to formulate the axiom $V =$ Ultimate-L *before* having to specify the construction of Ultimate-L.

Given the remarks above, this seems beyond hopeless. But it turns that it is not hopeless at all. The key is to generalize the projective sets. The point here is that the projective sets are the sets in $\mathcal{P}_{\mathrm{Def}}(V_{\omega+1})$. If there is an ultimate version of L then likely there is an ultimate enhancement of $\mathcal{P}_{\mathrm{Def}}(X)$. But then applied to $X = V_{\omega+1}$, it would yield a generalization of the projective sets.

Thus if one could define, by other means, that ultimate generalization of the projective sets, then maybe one could use that to identify the axiom $V =$ Ultimate-L.

11. The universally Baire sets

$V_{\omega+1}$ is homeomorphic to the Cantor set, with the topology on $V_{\omega+1}$ given by the sets

$$O_{n,a} = \{X \subseteq V_\omega \mid X \cap V_n = a\}$$

as basic open sets, where $n < \omega$ and $a \in V_{n+1}$.

The projective subsets of $V_{\omega+1}$ are exactly the sets generated from the open sets and closing under the operations:

(1) Taking images by continuous functions $F : V_{\omega+1} \to V_{\omega+1}$.
(2) Taking complements.

This definition generalizes to any topological space and in particular this extends the notion of the projective sets to the Euclidean spaces $\mathbb{R}^n$. Interpreting $\mathbb{R}^n \subseteq V_{\omega+1}$ in any reasonable fashion, so that $\mathbb{R}^n \in \mathcal{P}_{\mathrm{Def}}(V_{\omega+1})$, the two notions of projective subsets of $\mathbb{R}^n$ coincide.

DEFINITION 11.1 (Feng-Magidor-Woodin [**3**]). A set $A \subseteq \mathbb{R}^n$ is **universally Baire** if for all topological spaces Ω and for all continuous functions

$$\pi : \Omega \to \mathbb{R}^n;$$

the preimage of A by π has the property of Baire in the space Ω. □

By taking $\Omega = \mathbb{R}^n$ and π to be the identity, every universally Baire set $A \subset \mathbb{R}^n$ has the property of Baire. By letting Ω range over the comeager subsets of the Stone space of the measure algebra given by Lebesgue measure on the spaces $\mathbb{R}^n$, if $A \subseteq \mathbb{R}^n$ is universally Baire then A is Lebesgue measurable.

However, one cannot hope to prove outright that the universally Baire sets $A \subseteq \mathbb{R}$, are even just closed under images by continuous functions $F : \mathbb{R} \to \mathbb{R}$.

THEOREM 11.2. *Assume $V = L$. Then every set $A \subseteq \mathbb{R}$ is the image of a universally Baire set by a continuous function*

$$F : \mathbb{R} \to \mathbb{R}. \qquad \square$$

But for the universally Baire sets to provide the sought after ultimate generalization of the projective sets, one needs strong closure conditions. A natural closure condition is given by the following generalization of the definition of L.

DEFINITION 11.3. Suppose $A \subseteq \mathbb{R}$. Define $L_\alpha(A, \mathbb{R})$ by induction on α by:

(1) $L_0(A, \mathbb{R}) = V_{\omega+1} \cup \{A\}$,
(2) (Successor case) $L_{\alpha+1}(A, \mathbb{R}) = \mathcal{P}_{\text{Def}}(L_\alpha(A, \mathbb{R}))$,
(3) (Limit case) $L_\alpha(A, \mathbb{R}) = \cup \{L_\beta(A, \mathbb{R}) \mid \beta < \alpha\}$. $\square$

$L(A, \mathbb{R})$ is the class of all sets X such that $X \in L_\alpha(A, \mathbb{R})$ for some ordinal α. The question becomes under what conditions, if any, can one prove that if $A \subseteq \mathbb{R}$ is universally Baire then so every set

$$B \in \mathcal{P}(\mathbb{R}) \cap L(A, \mathbb{R}).$$

12. The universally Baire sets as the ultimate generalization of the projective sets

If every projective set $A \subseteq \mathbb{R}$ is universally Baire then every projective set is Lebesgue measurable and there is no projective wellordering of the reals. We already have a large cardinal axiom which implies this, and in fact the natural and mild strengthening of that axiom suffices to prove every projective set is universally Baire.

There is a proper class of Woodin cardinals if for every ordinal α, there is a Woodin cardinal δ such that $\delta > \alpha$. For technical reasons the existence of infinitely many Woodin cardinals cannot imply that every projective set is universally Baire.

THEOREM 12.1. *Suppose that there is a proper class of Woodin cardinals and suppose $A \subseteq \mathbb{R}$ is universally Baire. Then every set $B \in L(A, \mathbb{R}) \cap \mathcal{P}(\mathbb{R})$ is universally Baire.* $\square$

Thus if there is a proper class of Woodin cardinals then every projective set $A \subseteq \mathbb{R}$ is universally Baire. In fact, as the following theorem shows, again assuming there is a proper class of Woodin cardinals, then the universally Baire sets form a true generalization of the projective sets.

THEOREM 12.2. *Suppose that there is a proper class of Woodin cardinals. Then the following hold.*

(1) *Suppose $A \subseteq \mathbb{R} \times \mathbb{R}$ is universally Baire and uncountable. Then A contains a perfect set.*
(2) (Steel) *Suppose $A \subseteq \mathbb{R} \times \mathbb{R}$ is universally Baire. Then A has a choice function which is universally Baire.*

(3) (Martin-Steel) *Suppose $A \subseteq \mathbb{R}$ is universally Baire. Then A is determined.* □

Thus if A is universally Baire and there is a proper class of Woodin cardinals then $L(A, \mathbb{R}) \models$ AD, where AD is the Axiom of Determinacy.

But is there any evidence that even assuming that there is a proper class of Woodin cardinals that the universally Baire sets have any kind of rich internal structure? This would seem to be a necessary feature if one is gong to really be able to extract the axiom $V =$ Ultimate-L from the universally Baire sets.

Definition 12.3. Suppose A and B are subsets of $\mathbb{R}$.

(1) A is **weakly Wadge reducible** to B, $A \leq_{\text{Wadge}} B$, if there is a function

$$\pi : \mathbb{R} \to \mathbb{R}$$

such that π is continuous on $\mathbb{R} \backslash \mathbb{Q}$, and such that either A is the preimage of B under π, $A = \pi^{-1}[B]$, or A is the complement of that preimage, $A = \mathbb{R} \backslash \pi^{-1}[B]$.

(2) A and B are **weakly Wadge bi-reducible** if $A \leq_{\text{Wadge}} B$ and $B \leq_{\text{Wadge}} A$. □

The **weak Wadge degree** of a set $A \subseteq \mathbb{R}$ is the equivalence class of all sets $B \subseteq \mathbb{R}$ which are weakly Wadge bi-reducible with A.

Note that if A is weakly Wadge reducible to B and B is universally Baire then A is universally Baire.

Theorem 12.4 (Martin-Steel, Martin, Wadge [**12**]). *Assume there is a proper class of Woodin cardinals. Then the weak Wadge degrees of the universally Baire sets are linearly ordered by weak Wadge reducibility, and this order is a wellorder.* □

This shows that assuming the existence of a proper class of Woodin cardinals, the universally Baire sets define a wellordered hierarchy under a very simple notion of *complexity*. This hierarchy is extremely fine, for example there are borel sets of uncountable rank.

This structure really does suggest that the universally Baire sets can lead us to the ultimate generalization of the axiom $V = L$. But how?

The key lies in yet another construction of Gödel.

13. Gödel's transitive class HOD

Definition 13.1. HOD is the class of all sets X such that there exist $\alpha \in \text{Ord}$ and $M \in V_\alpha$ such that

(1) $X \in M$ and M is transitive.

(2) Every element of M is definable in $(V_\alpha, \in)$ from ordinal parameters. □

Suppose $(M, E) \models$ ZF. Recall Ord^M denotes the set of $a \in M$ such that

$$(M, E) \models \text{“}a \text{ is an ordinal”}.$$

Note that if $a \in M$ then

$$(M, E) \models \text{“}a \text{ is transitive”}$$

if and only if for $b, c \in M$ if $c\,E\,b$ and $b\,E\,a$ then $c\,E\,a$.

LEMMA 13.2. *Suppose* $(M, E) \models \mathrm{ZF}$ *and* $a \in M$. *Suppose that*

$$(M, E) \models \text{“}a \text{ is transitive”}.$$

Then the following are equivalent.

(1) $(M, E) \models \text{“}a \in \mathrm{HOD}\text{”}$.
(2) *For all* $b \in M$ *such that* $b\,E\,a$, b *is definable in* (M, E) *with parameters from* Ord^M. □

The previous lemma shows that the definition of HOD captures exactly what Tarski's theorem on the undefinability of truth suggests cannot be formalized. Suppose

$$(M, E) \models \mathrm{ZFC}$$

and let N be the set of all $a \in M$ such that a is definable without parameters in (M, E), then N is *not* definable in (M, E) without parameters. In contrast, if N is HOD as defined in (M, E). Then N *is* definable in (M, E) without parameters. This was Gödel's main point in defining HOD.

The other main point of the definition of HOD is the following theorem which provides a completely different proof that for every model $(M, E) \models \mathrm{ZF}$, there is an inner model of (M, E) in which the Axiom of Choice holds.

THEOREM 13.3 (after Gödel). *Suppose* $(M, E) \models \mathrm{ZFC}$. *Let*

$$N = \{a \in M \mid (M, E) \models \text{“}a \in \mathrm{HOD}\text{”}\}.$$

Then $(N, E{\restriction}M)$ *is an inner model of* (M, E) *and*

$$(N, E{\restriction}N) \models \mathrm{ZFC}.$$

□

14. $\mathrm{HOD}^{L(A,\mathbb{R})}$ and large cardinals

We fix some notation and our focus will be the case where $A \subseteq \mathbb{R}$ is universally Baire. The speculation is that if the universally Baire sets (in the context of a proper class of Woodin cardinals) *are* truly the correct transfinite generalization of the projective sets, and so form fundamental hierarchy, then the inner models given by defining HOD within $L(A, \mathbb{R})$ should also form a fundamental hierarchy. But these are inner models in which the Axiom of Choice holds, as opposed to the inner models $L(A, \mathbb{R})$ in which the Axiom of Choice necessarily fails.

DEFINITION 14.1. Suppose that $A \subseteq \mathbb{R}$. Then $\mathrm{HOD}^{L(A,\mathbb{R})}$ is the class HOD as defined within $L(A, \mathbb{R})$. □

Thus by Theorem 13.3, the Axiom of Choice must hold in $\mathrm{HOD}^{L(A,\mathbb{R})}$ for any set $A \subseteq \mathbb{R}$, even if

$$L(A, \mathbb{R}) \models \mathrm{AD}.$$

The first connection of AD with large cardinal axioms was discovered by Solovay over 50 years ago.

THEOREM 14.2 (Solovay [**12**]). *Suppose that* $A \subseteq \mathbb{R}$ *and* $L(A, \mathbb{R}) \models \mathrm{AD}$. *Then* ω_1^V *is a measurable cardinal in* $\mathrm{HOD}^{L(A,\mathbb{R})}$. □

Thus, suppose that there is a proper class of Woodin cardinals and that $A \subseteq \mathbb{R}$ is universally Baire. Then

$$L(A, \mathbb{R}) \models \mathrm{AD}$$

and so

$$\mathrm{HOD}^{L(A,\mathbb{R})} \models \text{“}\omega_1^V \text{ is a measurable cardinal”}$$

where ω_1^V denote the ordinal ω_1 as defined in V and not as defined in $\mathrm{HOD}^{L(A,\mathbb{R})}$.

The key issue becomes what other large cardinal axioms might hold in $\mathrm{HOD}^{L(A,\mathbb{R})}$ and where in $\mathrm{HOD}^{L(A,\mathbb{R})}$ might they occur.

THEOREM 14.3. *Suppose there is a proper class of Woodin cardinals and that A is universally Baire. Then ω_1^V is the least measurable cardinal in* $\mathrm{HOD}^{L(A,\mathbb{R})}$. □

Thus, if stronger large cardinals exist in $\mathrm{HOD}^{L(A,\mathbb{R})}$, they can only occur above ω_1^V.

DEFINITION 14.4. Suppose that $A \subseteq \mathbb{R}$ is universally Baire. Then $\Theta^{L(A,\mathbb{R})}$ is the supremum of the ordinals α such that there exists a surjection, $\pi : \mathbb{R} \to \alpha$, such that $\pi \in L(A, \mathbb{R})$. □

Note that ω_1^V is the supremum of the ordinals α such that there is a surjection

$$\pi : \omega \to \alpha,$$

it makes no difference whether or not one restricts to $\pi \in L(A, \mathbb{R})$.

Defining Θ to be the supremum of the ordinals α such that there is a surjection

$$\pi : \mathbb{R} \to \alpha,$$

one has a natural generalization of the definition of ω_1. But now restricting where the functions π are chosen affects the value of Θ. Defining Θ in V, $\Theta = c^+$ where $c = |\mathbb{R}|$. But $\Theta^{L(A,\mathbb{R})}$ is Θ as defined in $L(A, \mathbb{R})$ and this can be very different.

For example, assuming there is a proper class of Woodin cardinals and that $A \subseteq \mathbb{R}$ is universally Baire, it is a prominent open problem whether it is even possible that $\Theta^{L(A,\mathbb{R})} \geq \omega_3$. Here ω_3 is as defined in V and by the results of Cohen, c can be vastly larger than ω_3. It is known that $\Theta^{L(A,\mathbb{R})} > \omega_2$ is possible.

The following lemma gives an upper bound on where large cardinals, measurable cardinals and beyond, can possibly occur in $\mathrm{HOD}^{L(A,\mathbb{R})}$.

LEMMA 14.5. *Suppose that there is a proper class of Woodin cardinals, A is universally Baire, and $\kappa \geq \Theta^{L(A,\mathbb{R})}$. Then κ is not a measurable cardinal in* $\mathrm{HOD}^{L(A,\mathbb{R})}$. □

THEOREM 14.6. *Suppose that there is a proper class of Woodin cardinals and that A is universally Baire. Then $\Theta^{L(A,\mathbb{R})}$ is a Woodin cardinal in* $\mathrm{HOD}^{L(A,\mathbb{R})}$. □

The previous lemma and theorem, taken together, highlight a key feature of the large cardinal axiom that δ is a Woodin cardinal. The strength of this axiom lies not at δ but in the consequences for V_δ.

15. The axiom $V =$ Ultimate-L

We now formulate the axiom $V =$ Ultimate-L.

DEFINITION 15.1 (**The axiom $V =$ Ultimate-L**). (1) For every ordinal α there exists a Woodin cardinal δ such that $\delta > \alpha$.

(2) For each sentence ψ, if $V_\alpha \models \psi$, for some ordinal α, then there exists a universally Baire set $A \subseteq \mathbb{R}$ such that

$$\mathrm{HOD}^{L(A,\mathbb{R})} \models \text{“}V_\alpha \models \psi, \text{ for some ordinal } \alpha\text{”}$$ □

Thus the axiom $V =$ Ultimate-L simply asserts a form a *resemblance* between the sets V_α as defined in V, (downward) to the sets V_α as defined in

$$\mathrm{HOD}^{L(A,\mathbb{R})}$$

for some universally Baire set $A \subseteq \mathbb{R}$.

The convention in set theory is that a sentence ϕ is a Σ_2-sentence if for some sentence ψ, it asserts that

$$V_\alpha \models \psi$$

for some ordinal α. Thus the usual formulation of the axiom $V =$ Ultimate-L asserts in the second clause: Suppose ϕ is a Σ_2-sentence which is true in V. Then there exists a universally Baire set $A \subseteq \mathbb{R}$ such that

$$\mathrm{HOD}^{L(A,\mathbb{R})} \models \phi.$$

Both CH and its negation are Σ_2-sentences. A vast majority of large cardinal axioms are expressible as Σ_2-sentences. But several important large cardinal axioms cannot be. These include the axiom that there is a supercompact cardinal and the axiom that there is an extendible cardinal. These particular large cardinal axioms play a central role in the current story of CH.

Closely related to the notion of a Σ_2-sentence is the notion that a class be Σ_2-definable in V from a parameter p.

DEFINITION 15.2. Suppose $X \subset V$ and $p \in V$. Then X is Σ_2-definable from p if there is a formula $\psi(x_0, x_1)$ such that $y \in X$ if and only if there exists an ordinal α such that $y, p \in V_\alpha$ and

$$V_\alpha \models \psi[y, p]. \qquad \square$$

Thus $L(A, \mathbb{R})$ is Σ_2-definable from $(A, V_{\omega+1})$ and HOD is Σ_2-definable from the trivial parameter $p = \emptyset$.

Even though Σ_2-definable classes may not be sets, for example V itself is such a class, developing a theory of all Σ_2-definable classes from parameters makes perfect sense. Just as in Number Theory, one can develop the theory of infinite sets of integers within just Number Theory, if one restricts to the sets which can be defined by formulas of some given fixed logical complexity.

The power of the axiom $V =$ Ultimate-L derives from the expressible power of Σ_2-sentences coupled with the rich structure theory of the universally Baire sets which flows from the fact that if $A \subset \mathbb{R}$ is universally Baire (and there is a proper class of Woodin cardinals) then

$$L(A, \mathbb{R}) \models \mathrm{AD}.$$

In fact the relevant axiom is not AD, but a specific refinement of that axiom which is denoted AD^+. The distinction is not really relevant to this exposition. It is conjectured that for any set $A \subseteq \mathbb{R}$, if $L(A, \mathbb{R}) \models \mathrm{AD}$ then $L(A, \mathbb{R}) \models \mathrm{AD}^+$, and this conjecture is verified in many cases, including for example the case $A = \emptyset$.

THEOREM 15.3. *Assume $V =$ Ultimate-L. Then the following hold.*

(1) CH.
(2) *For every set X, $X \in \mathrm{HOD}$.*
(3) *Suppose $x \in \mathbb{R}$. Then there exists a universally Baire set $A \subset \mathbb{R}$ such that*

$$x \in \mathrm{HOD}^{L(A,\mathbb{R})}. \qquad \square$$

The conclusion (3) is the generalization for the axiom V = Ultimate-L, of Gödel's theorem that if $V = L$ then there is a projective wellordering of the reals.

Cohen's method of forcing cannot be used to establish that sentences are unsolvable on the basis the axioms, ZFC + "V = Ultimate-L".

THEOREM 15.4. *Suppose that*

$$(M, E) \models \text{ZFC}$$

and that (M^*, E^*) *is an outer model of* (M, E) *such that*

$$(M^*, E^*) \models \text{ZFC} + \text{"}V = \text{Ultimate-}L\text{"}$$

and such that (M^*, E^*) *is a forcing extension of* (M, E). *Then* $M = M^*$. □

Thus the key issue is whether there is a generalization of Scott's theorem to the case of V = Ultimate-L. If not and it is possible to somehow prove this, then the axiom V = Ultimate-L becomes a rather serious candidate for an axiom which must be true in V and which renders Cohen's method powerless for establishing unsolvability. This would fulfill Gödel's hope that all questions of Set Theory can be reduced to axioms of infinity.

16. The language of large cardinals: elementary embeddings

DEFINITION 16.1. Suppose X and Y are transitive sets. A function

$$j : X \to Y$$

is an **elementary embedding** if for all logical formulas $\phi[x_0, \ldots, x_n]$ and all $a_0, \ldots, a_n \in X$,

$$(X, \in) \models \phi[a_0, \ldots, a_n]$$

if and only if

$$(Y, \in) \models \phi[j(a_0), \ldots, j(a_n)].$$ □

Isomorphisms are elementary embeddings but the only isomorphisms of $(X, \in)$ and $(Y, \in)$ are trivial. Equivalently, if X is a transitive set then there are no non-trivial automorphisms of the the structure $(X, \in)$.

LEMMA 16.2. *Suppose that* $j : V_\alpha \to V_\beta$ *is an elementary embedding. Then the following are equivalent.*

(1) j *is not the identity.*
(2) *There is an ordinal* $\eta < \alpha$ *such that* $j(\eta) \neq \eta$. □

If

$$j : V_\alpha \to V_\beta$$

is an elementary embedding and j is not the identity, then $\text{CRT}(j)$ denotes the least ordinal κ such that $j(\kappa) \neq \kappa$.

One can show, [**7**], that $\text{CRT}(j)$ is strongly inaccessible and moreover, $\text{CRT}(j)$ is a measurable cardinal which is a limit of measurable cardinals and much more. For example, $\text{CRT}(j)$ must be a Woodin cardinal which is a limit of Woodin cardinals.

DEFINITION 16.3 (Reinhardt). Suppose that δ is a cardinal. Then δ is an **extendible cardinal** if for each $\lambda > \delta$ there exists an elementary embedding

$$j : V_{\lambda+1} \to V_{j(\lambda)+1}$$

such that $\text{CRT}(j) = \delta$ and $j(\delta) > \lambda$. □

We need to introduce some more notation.

DEFINITION 16.4. For each (infinite) cardinal γ, $H(\gamma)$ denotes the union of all transitive sets M such that $|M| < \gamma$. □

Thus $H(\omega) = V_\omega$ and further, $H(\omega_1)$ and $V_{\omega+1}$ are logically equivalent structures. However, $H(\omega_2)$ and $V_{\omega+2}$ are logically equivalent if and only if CH holds.

More generally, for every ordinal $\alpha > \omega$, $V_{\alpha+1}$ and $H(\lambda^+)$ are logically equivalent where

$$\lambda = |V_\alpha|$$

and where λ^+ is the least cardinal γ such that $\gamma > \lambda$.

Assuming the Axiom of Choice

$$H(\omega_1) \models \text{ZFC}\backslash\text{Powerset}$$

and more generally, for any cardinal $\lambda \geq \omega$,

$$H(\lambda^+) \models \text{ZFC}\backslash\text{Powerset}.$$

Here ZFC\Powerset denotes the axioms ZFC without the axiom that for all X, $\mathcal{P}(X)$ exists.

A class $N \subseteq V$ is *transitive* if for all $M \in N$, $M \subseteq N$. Thus the class of ordinals is a transitive class but the class of all cardinals is not.

The class N is a proper class if N is not a set.

DEFINITION 16.5. Suppose that N is a transitive class containing the ordinals and that N is Σ_2-definable from some parameter p. Then N is an **inner model** if for each infinite cardinal γ,

$$M \cap H(\gamma^+) \models \text{ZFC}\backslash\text{Powerset}. \qquad \square$$

Both L and HOD are inner models.

The following lemma shows that the two notions of inner model that we have defined are compatible.

LEMMA 16.6. *Suppose that*

$$(M, E) \models \text{ZFC}.$$

Suppose $\phi(x_0, x_1)$ is a Σ_2-formula, $p \in M$ and that

$$(M, E) \models \text{“ The class of all } A \text{ such that } \phi[A, p] \text{ holds is an inner model”}.$$

Let N be the set of all $A \in M$ such that $(M, E) \models \phi[A, p]$. Then $(N, E{\restriction}N)$ is an inner model of (M, E) and

$$(N, E{\restriction}N) \models \text{ZFC}. \qquad \square$$

17. The δ-cover and δ-approximation properties

We begin with Hamkins' conditions. It is now convenient to introduce the notion that a cardinal κ is a *regular* cardinal. This is the property of κ that for all $X \subset \kappa$, if $|X| < \kappa$ then $X \subset \beta$ for some $\beta < \kappa$.

Thus ω is a regular cardinal and assuming the Axiom of Choice, for each cardinal $\delta \geq \omega$, δ^+ is a regular cardinal.

If $\gamma > \omega$ is a regular cardinal then

$$H(\gamma) \models \text{ZFC}\backslash\text{Powerset}.$$

DEFINITION 17.1 (Hamkins, [**6**]). Suppose that $\delta > \omega$ is a regular cardinal and that N is a transitive inner model of ZFC containing the ordinals. Then:

(1) N has the *δ-cover property* if for all $\sigma \subset N$ with $|\sigma| < \delta$, there exists $\tau \in N$ such that $\sigma \subset \tau$ and such that $|\tau| < \delta$.
(2) N has the *δ-approximation property* if for all $X \subset N$ the following are equivalent.
 (a) $X \in N$.
 (b) $X \cap \tau \in N$ for all $\tau \in N$ with $|\tau| < \delta$. □

Suppose that N is an inner model of V and that V is a forcing extension of N. Then for all sufficiently large regular cardinals δ, N has the δ-approximation property and N has the δ-cover property.

In general the converse is not true.

THEOREM 17.2 (Hamkins Uniqueness Theorem). *Suppose $\delta > \omega$ is a regular cardinal and that N_0, N_1 are transitive inner models with the δ-cover property and with the δ-approximation property. Suppose*

$$N_0 \cap H(\delta^+) = N_1 \cap H(\delta^+).$$

Then $N_0 = N_1$. □

A key consequence of the proof of the Hamkins Uniqueness Theorem is that if N is an inner model such that N has δ-cover property and the δ-approximation property. then N is Σ_2 -definable from the parameter p where

$$p = N \cap H(\delta^+).$$

Thus, and this is a key point, the theory of theory of inner models N of V which have the δ-approximation property and the δ-cover property, for some δ, is part of the theory of V.

THEOREM 17.3 (Hamkins Universality Theorem). *Suppose that N is an inner model with the δ-cover and δ-approximation properties, $\kappa > \delta$, and that κ is an extendible cardinal. Then*

$$N \models \text{“}\kappa \text{ is an extendible cardinal”}. \quad \square$$

Hamkins actually proved the generalization of this for almost all large cardinal axioms and this is the Hamkins Universality Theorem. However, there is a counterexample in the uppermost reaches of the large cardinal hierarchy.

The assertion that λ is an Axiom I_0 cardinal is one of the strongest large cardinal axioms known, and the Hamkins Universality Theorem is false for this large cardinal notion. A third property is needed to obtain full (strong) universality. But even so the precise statement of universality must be altered.

18. The δ-genericity property and strong universality

Suppose N is an inner model and σ is a set of ordinals. Then $N[\sigma]$ denotes the smallest inner model N^* such that $N \subseteq N$ and $\sigma \in N^*$. If N is Σ_2-definable from some parameter p, then N^* exists and N^* is Σ_2-definable from the parameter (p, σ).

With this notation, we state the δ-genericity property and we restrict the definition to just the special case that δ is a strongly inaccessible cardinal since that is all we require here.

DEFINITION 18.1. Suppose that N is an inner model and δ is strongly inaccessible. Then N has the **δ-genericity property** if for all $\sigma \subset \delta$, if $|\sigma| < \delta$ then $N[\sigma] \cap V_\delta$ is a Cohen extension of $N \cap V_\delta$.

The following remarkable theorem of Vopěnka shows that if $V \neq \mathrm{HOD}$ then in some weak sense V is a forcing extension of HOD.

THEOREM 18.2 (Vopěnka). *Suppose that σ is a set of ordinals and $\sigma \notin \mathrm{HOD}$. Then $\mathrm{HOD}[\sigma]$ is a forcing extension of* HOD. □

The proof of Vopěnka's theorem easily yields the following variation of that theorem.

THEOREM 18.3 (after Vopěnka). *Suppose that δ is strongly inaccessible. Then* HOD *has the δ-genericity property.* □

By adding the third property of δ-genericity, one obtains the Strong Universality Theorem which strengthen the conclusion of the Hamkins Universality Theorem to cover the case of the existence of Axiom I_0 cardinals, and much more.

THEOREM 18.4. *Suppose that N has the δ-approximation property, the δ-cover property, and the δ-genericity property. Suppose that $\lambda > \delta$ and that λ is a limit of* Axiom I_0 *cardinals. Then*

$$N \models \text{“ } \lambda \text{ is a limit of Axiom } \mathrm{I}_0 \text{ cardinals”}. \qquad \square$$

But why focus on any of this? Suppose that one can prove from some large cardinal hypothesis that there must exist an inner model N such that

$$N \models \text{“}V = \text{Ultimate-}L\text{”}$$

and such that for some strongly inaccessible cardinal δ, N has the δ-approximation property, the δ-cover property, and the δ-genericity property.

Then by the universality theorems, one will have proved that there is no generalization of Scott's theorem for the axiom, $V =$ Ultimate-L.

19. The Ultimate-L Conjecture and the two futures of Set Theory

The key conjecture now is the following conjecture and by the definability theorem which is a corollary of the (the proof of) Hamkins Uniqueness Theorem, this conjecture is expressible as a conjecture of Set Theory.

DEFINITION 19.1 (**The** Ultimate-L **Conjecture**). Suppose that there is an extendible cardinal. Then **provably** there exists an inner model N such that for some δ:

(1) N has the δ-cover and δ-approximation properties.
(2) N has the δ-genericity property.
(3) $N \models$ “$V =$ Ultimate-L”. □

The choice of the large cardinal hypothesis on which to base the formulation of the Ultimate-L Conjecture is not really the issue here. The significance would be the same if one proved the conjecture with the hypothesis that there is an extendible cardinal, replaced by the hypothesis that there is an extendible cardinal and a proper class of Axiom I_0 cardinals.

The Ultimate-L Conjecture is an existential number theoretic statement. Therefore, if it is unsolvable then it is *false*. In particular, the Ultimate-L Conjecture **must** be either true or false. It cannot be meaningless.

The Ultimate-L Conjecture reduces the **entire** post-Cohen debate on Set Theoretic truth to a single question which **must** have an answer. Thus Set Theory faces one of two futures.

Future 1: The Ultimate-L Conjecture is true.

In this future, the axiom $V =$ Ultimate-L is very likely the key missing axiom for V since there is **no** generalization of Scott's Theorem for the axiom $V =$ Ultimate-L.

We emphasize that the axioms ZFC + "$V =$ Ultimate-L" resolve **all** the questions which have been shown to be unsolvable by Cohen's method modulo large cardinal axioms.

Even without asserting $V =$ Ultimate-L, just the large cardinal hypothesis that there is an extendible cardinal will be verified to have far reaching implications. For example, it will imply that for some strongly inaccessible cardinal δ, HOD has the δ-approximation property, the δ-cover property, and the δ-genericity property.

If one assumes there is a proper class of (a mild strengthening of) extendible cardinals then one obtains the same conclusion *without* assuming the Axiom of Choice. That any large axiom could tame the wild chaos of the universe without choice, seems almost beyond belief.

Future 2: The Ultimate-L Conjecture is false.

In this future, the program to understand V by generalizing the success in understanding $V_{\omega+1}$ and the projective sets, utterly fails. Further in this future, and one can make this precise, the Inner Model Program, which is the program to construct enlargements of L for specified large cardinal axioms, ultimately also utterly fails.

So which is it?

20. Concluding remarks

It is important that we note what we did not cover. One central point that we have not addressed is why there is any evidence whatsoever that the Ultimate-L Conjecture might be true. For example, is it even reasonable to expect that the generalizations of the currently known constructions would (or even could) yield inner models satisfying $V =$ Ultimate-L with the key properties of approximation, cover, and genericity? In fact we claim it is very reasonable speculation, but that is another lecture.

We have also not discussed the reasons why if the Ultimate-L Conjecture is true that (we believe) the axiom $V =$ Ultimate-L is inevitable. Here we simply take note and review what has happened with determinacy axioms.

The axiom of Projective Determinacy quickly emerged as a natural special case of the Axiom of Determinacy. Within the first decade of the study of Projective Determinacy, the richness of the structure that PD yields for the projective sets became apparent. But it took another two decades for the *truth* of the axiom to become clear, and even now PD is not universally accepted as true. But that will change.

There is within Set Theory a duality program, it is (here named), the AD^+ Duality Program. This program deals with the correspondence of models of $\mathrm{ZF}+\mathrm{AD}^+$

with models of ZFC in which large cardinals exist. Both the axiom $V = \text{Ultimate-}L$ and the Ultimate-L Conjecture are products of the AD^+ Duality Program.

The axiomatic power (and promise) of $V = \text{Ultimate-}L$ lies in the fact that it directly connects the structure of V with the rich structure of models of AD^+, and in fact the connection is to the structure of a canonical family of models of AD^+. These are the inner models of AD^+ given by the universally Baire sets and here the context is the existence of a proper class of Woodin cardinals. The latter is the modern incarnation of the rich structure of the projective sets in the context of PD.

We have given complete definitions for both the axiom $V = \text{Ultimate-}L$ and the Ultimate-L Conjecture, except we have not defined what a Woodin cardinal is. We do that now.

The definition of a Woodin cardinal is rather unintuitive in part because its origin is not the usual one for large cardinal axioms. Large cardinal axioms are usually formulated as principles of infinity in tandem with arguments for why they should be considered. Here the bigger the better but one seeks to avoid the specter of inconsistency.

The original goal in defining Woodin cardinals was completely different. The goal was to isolate an optimal hypothesis for a specific series of theorems, a notion of infinity which is both large enough and yet not too large. But as the survey given in the previous sections suggests, the notion has turned out to be relevant to quite a number of other theorems.

So what exactly is a Woodin cardinal? Suppose δ is a strongly inaccessible cardinal. This (as we have already defined) is the assertion that δ is an uncountable cardinal such that the following hold.

(1) If $|X| < \delta$ then $|\mathcal{P}(X)| < \delta$.
(2) If $|X| < \delta$ and if for each $Y \in X$, $|Y| < \delta$, then $|\cup X| < \delta$.

Thus ω fails to be a strongly inaccessible cardinal only because ω is countable and there are quite a number of arguments that the existence of strongly inaccessible cardinals is inevitable, however one sharpens the conception of V through refinements of the ZFC axioms.

Suppose δ is strongly inaccessible. Then if follows that

$$V_\delta \models \text{ZFC},$$

and so existence of a strongly inaccessible cardinal is a large cardinal notion.

The definition of Woodin cardinals is formulated in terms of elementary embeddings

$$j : V_\delta \to M$$

where M is transitive and j is nontrivial (so $j \restriction \delta$ is not the identity).

Suppose that δ is strongly inaccessible. Then δ is a Woodin cardinal if for each function

$$f : \delta \to \delta$$

there exists a transitive set M and an elementary embedding

$$j : V_\delta \to M$$

such that j is not the identity and such that the following hold where κ is the least ordinal such that $j(\kappa) \neq \kappa$.

(1) For all $\alpha < \kappa$, $f(\alpha) < \kappa$.
(2) Let $g = f \restriction \kappa$. Then $V_{j(g)(\kappa)} \subset M$.

Thus if δ is a Woodin cardinal then δ is a limit of measurable cardinals, but the least Woodin cardinal is not a measurable cardinal. The strength of δ being a Woodin cardinal lies within V_δ and not at δ itself.

References

[1] Paul Cohen, *The independence of the continuum hypothesis*, Proc. Nat. Acad. Sci. U.S.A. **50** (1963), 1143–1148, DOI 10.1073/pnas.50.6.1143. MR157890

[2] Paul J. Cohen, *The independence of the continuum hypothesis. II*, Proc. Nat. Acad. Sci. U.S.A. **51** (1964), 105–110, DOI 10.1073/pnas.51.1.105. MR159745

[3] Qi Feng, Menachem Magidor, and Hugh Woodin, *Universally Baire sets of reals*, Set theory of the continuum (Berkeley, CA, 1989), Math. Sci. Res. Inst. Publ., vol. 26, Springer, New York, 1992, pp. 203–242, DOI 10.1007/978-1-4613-9754-0_15. MR1233821

[4] Kurt F. Gödel. The consistency of the Axiom of Choice and of the Generalized Continuum Hypothesis. *Proc. Nat. Acad. Sci. U.S.A.*, 24:556–557, 1938.

[5] Kurt F. Gödel. Consistency-proof for the Generalized Continuum Hypothesis. *Proc. Nat. Acad. Sci. U.S.A.*, 25:220–224, 1939.

[6] Joel David Hamkins, *Extensions with the approximation and cover properties have no new large cardinals*, Fund. Math. **180** (2003), no. 3, 257–277, DOI 10.4064/fm180-3-4. MR2063629

[7] Akihiro Kanamori, *The higher infinite*, Perspectives in Mathematical Logic, Springer-Verlag, Berlin, 1994. Large cardinals in set theory from their beginnings. MR1321144

[8] Peter Koellner and W. Hugh Woodin, *Large cardinals from determinacy*, Handbook of set theory. Vols. 1, 2, 3, Springer, Dordrecht, 2010, pp. 1951–2119, DOI 10.1007/978-1-4020-5764-9_24. MR2768702

[9] Donald A. Martin and John R. Steel, *A proof of projective determinacy*, J. Amer. Math. Soc. **2** (1989), no. 1, 71–125, DOI 10.2307/1990913. MR955605

[10] D. A. Martin and J. R. Steel, *Iteration trees*, J. Amer. Math. Soc. **7** (1994), no. 1, 1–73, DOI 10.2307/2152720. MR1224594

[11] William J. Mitchell and John R. Steel, *Fine structure and iteration trees*, Lecture Notes in Logic, vol. 3, Springer-Verlag, Berlin, 1994, DOI 10.1007/978-3-662-21903-4. MR1300637

[12] Yiannis N. Moschovakis, *Descriptive set theory*, 2nd ed., Mathematical Surveys and Monographs, vol. 155, American Mathematical Society, Providence, RI, 2009, DOI 10.1090/surv/155. MR2526093

[13] Jan Mycielski and S. Świerczkowski, *On the Lebesgue measurability and the axiom of determinateness*, Fund. Math. **54** (1964), 67–71, DOI 10.4064/fm-54-1-67-71. MR161788

[14] Jan Mycielski and H. Steinhaus, *A mathematical axiom contradicting the axiom of choice*, Bull. Acad. Polon. Sci. Sér. Sci. Math. Astronom. Phys. **10** (1962), 1–3. MR140430

[15] Dana Scott, *Measurable cardinals and constructible sets*, Bull. Acad. Polon. Sci. Sér. Sci. Math. Astronom. Phys. **9** (1961), 521–524. MR143710

[16] W. Hugh Woodin, *Supercompact cardinals, sets of reals, and weakly homogeneous trees*, Proc. Nat. Acad. Sci. U.S.A. **85** (1988), no. 18, 6587–6591, DOI 10.1073/pnas.85.18.6587. MR959110

Departments of Philosophy and Mathematics, Harvard University, Cambridge MA 02138

Email address: woodin@math.harvard.edu

104 **A. Kechris, N. Makarov, D. Ramakrishnan, and X. Zhu, Editors,** Nine Mathematical Challenges, 2021

103 **Sergey Novikov, Igor Krichever, Oleg Ogievetsky, and Senya Shlosman, Editors,** Integrability, Quantization, and Geometry, 2021

102 **David T. Gay and Weiwei Wu, Editors,** Breadth in Contemporary Topology, 2019

101 **Avraham Aizenbud, Dmitry Gourevitch, David Kazhdan, and Erez M. Lapid, Editors,** Representations of Reductive Groups, 2019

100 **Chiu-Chu Melissa Liu and Motohico Mulase, Editors,** Topological Recursion and its Influence in Analysis, Geometry, and Topology, 2018

99 **Vicente Muñoz, Ivan Smith, and Richard P. Thomas, Editors,** Modern Geometry, 2018

98 **Amir-Kian Kashani-Poor, Ruben Minasian, Nikita Nekrasov, and Boris Pioline, Editors,** String-Math 2016, 2018

97 **Tommaso de Fernex, Brendan Hassett, Mircea Mustaţă, Martin Olsson, Mihnea Popa, and Richard Thomas, Editors,** Algebraic Geometry: Salt Lake City 2015 (Parts 1 and 2), 2018

96 **Si Li, Bong H. Lian, Wei Song, and Shing-Tung Yau, Editors,** String-Math 2015, 2017

95 **Izzet Coskun, Tommaso de Fernex, and Angela Gibney, Editors,** Surveys on Recent Developments in Algebraic Geometry, 2017

94 **Mahir Bilen Can, Editor,** Algebraic Groups: Structure and Actions, 2017

93 **Vincent Bouchard, Charles Doran, Stefan Méndez-Diez, and Callum Quigley, Editors,** String-Math 2014, 2016

92 **Kailash C. Misra, Daniel K. Nakano, and Brian J. Parshall, Editors,** Lie Algebras, Lie Superalgebras, Vertex Algebras and Related Topics, 2016

91 **V. Sidoravicius and S. Smirnov, Editors,** Probability and Statistical Physics in St. Petersburg, 2016

90 **Ron Donagi, Sheldon Katz, Albrecht Klemm, and David R. Morrison, Editors,** String-Math 2012, 2015

89 **D. Dolgopyat, Y. Pesin, M. Pollicott, and L. Stoyanov, Editors,** Hyperbolic Dynamics, Fluctuations and Large Deviations, 2015

88 **Ron Donagi, Michael R. Douglas, Ljudmila Kamenova, and Martin Rocek, Editors,** String-Math 2013, 2014

87 **Helge Holden, Barry Simon, and Gerald Teschl, Editors,** Spectral Analysis, Differential Equations and Mathematical Physics: A Festschrift in Honor of Fritz Gesztesy's 60th Birthday, 2013

86 **Kailash C. Misra, Daniel K. Nakano, and Brian J. Parshall, Editors,** Recent Developments in Lie Algebras, Groups and Representation Theory, 2012

85 **Jonathan Block, Jacques Distler, Ron Donagi, and Eric Sharpe, Editors,** String-Math 2011, 2012

84 **Alex H. Barnett, Carolyn S. Gordon, Peter A. Perry, and Alejandro Uribe, Editors,** Spectral Geometry, 2012

83 **Hisham Sati and Urs Schreiber, Editors,** Mathematical Foundations of Quantum Field Theory and Perturbative String Theory, 2011

82 **Michael Usher, Editor,** Low-dimensional and Symplectic Topology, 2011

81 **Robert S. Doran, Greg Friedman, and Jonathan Rosenberg, Editors,** Superstrings, Geometry, Topology, and C^*-algebras, 2010